Graduate Texts in Mathematics 206

Graduate Texts in Mathematics

1 TAKEUTI/ZARING. Introduction to Axiomatic Set Theory. 2nd ed.
2 OXTOBY. Measure and Category. 2nd ed.
3 SCHAEFER. Topological Vector Spaces. 2nd ed.
4 HILTON/STAMMBACH. A Course in Homological Algebra. 2nd ed.
5 MAC LANE. Categories for the Working Mathematician. 2nd ed.
6 HUGHES/PIPER. Projective Planes.
7 J.-P. Serre. A Course in Arithmetic.
8 TAKEUTI/ZARING. Axiomatic Set Theory.
9 HUMPHREYS. Introduction to Lie Algebras and Representation Theory.
10 COHEN. A Course in Simple Homotopy Theory.
11 CONWAY. Functions of One Complex Variable I. 2nd ed.
12 BEALS. Advanced Mathematical Analysis.
13 ANDERSON/FULLER. Rings and Categories of Modules. 2nd ed.
14 GOLUBITSKY/GUILLEMIN. Stable Mappings and Their Singularities.
15 BERBERIAN. Lectures in Functional Analysis and Operator Theory.
16 WINTER. The Structure of Fields.
17 ROSENBLATT. Random Processes. 2nd ed.
18 HALMOS. Measure Theory.
19 HALMOS. A Hilbert Space Problem Book. 2nd ed.
20 HUSEMOLLER. Fibre Bundles. 3rd ed.
21 HUMPHREYS. Linear Algebraic Groups.
22 BARNES/MACK. An Algebraic Introduction to Mathematical Logic.
23 GREUB. Linear Algebra. 4th ed.
24 HOLMES. Geometric Functional Analysis and Its Applications.
25 HEWITT/STROMBERG. Real and Abstract Analysis.
26 MANES. Algebraic Theories.
27 KELLEY. General Topology.
28 ZARISKI/SAMUEL. Commutative Algebra. Vol. I.
29 ZARISKI/SAMUEL. Commutative Algebra. Vol. II.
30 JACOBSON. Lectures in Abstract Algebra I. Basic Concepts.
31 JACOBSON. Lectures in Abstract Algebra II. Linear Algebra.
32 JACOBSON. Lectures in Abstract Algebra III. Theory of Fields and Galois Theory.
33 HIRSCH. Differential Topology.
34 SPITZER. Principles of Random Walk. 2nd ed.
35 ALEXANDER/WERMER. Several Complex Variables and Banach Algebras. 3rd ed.
36 KELLEY/NAMIOKA et al. Linear Topological Spaces.
37 MONK. Mathematical Logic.
38 GRAUERT/FRITZSCHE. Several Complex Variables.
39 ARVESON. An Invitation to C-Algebras.
40 KEMENY/SNELL/KNAPP. Denumerable Markov Chains. 2nd ed.
41 APOSTOL. Modular Functions and Dirichlet Series in Number Theory. 2nd ed.
42 J.-P. SERRE. Linear Representations of Finite Groups.
43 GILLMAN/JERISON. Rings of Continuous Functions.
44 KENDIG. Elementary Algebraic Geometry.
45 LOÈVE. Probability Theory I. 4th ed.
46 LOÈVE. Probability Theory II. 4th ed.
47 MOISE. Geometric Topology in Dimensions 2 and 3.
48 SACHS/WU. General Relativity for Mathematicians.
49 GRUENBERG/WEIR. Linear Geometry. 2nd ed.
50 EDWARDS. Fermat's Last Theorem.
51 KLINGENBERG. A Course in Differential Geometry.
52 HARTSHORNE. Algebraic Geometry.
53 MANIN. A Course in Mathematical Logic.
54 GRAVER/WATKINS. Combinatorics with Emphasis on the Theory of Graphs.
55 BROWN/PEARCY. Introduction to Operator Theory I: Elements of Functional Analysis.
56 MASSEY. Algebraic Topology: An Introduction.
57 CROWELL/FOX. Introduction to Knot Theory.
58 KOBLITZ. p-adic Numbers, p-adic Analysis, and Zeta-Functions. 2nd ed.
59 LANG. Cyclotomic Fields.
60 ARNOLD. Mathematical Methods in Classical Mechanics. 2nd ed.
61 WHITEHEAD. Elements of Homotopy Theory.
62 KARGAPOLOV/MERIZJAKOV. Fundamentals of the Theory of Groups.
63 BOLLOBAS. Graph Theory.
64 EDWARDS. Fourier Series. Vol. I. 2nd ed.
65 WELLS. Differential Analysis on Complex Manifolds. 2nd ed.
66 WATERHOUSE. Introduction to Affine Group Schemes.
67 SERRE. Local Fields.
68 WEIDMANN. Linear Operators in Hilbert Spaces.
69 LANG. Cyclotomic Fields II.

(continued after index)

M. Ram Murty

Problems in Analytic Number Theory

Second Edition

M. Ram Murty
Department of Mathematics & Statistics
Queen's University
99 University Avenue
Kingston ON K7L 3N6
Canada
murty@mast.queensu.ca

ISBN 978-1-4419-2477-3 e-ISBN 978-0-387-72350-1
DOI: 10.1007/978-0-387-72350-1

Printed on acid-free paper

9 8 7 6 5 4 3 2 1

springer.com

Like fire in a piece of flint, knowledge exists in the mind.
Suggestion is the friction which brings it out.

Vivekananda

Preface to the Second Edition

This expanded and corrected second edition has a new chapter on the important topic of equidistribution. Undoubtedly, one cannot give an exhaustive treatment of the subject in a short chapter. However, we hope that the problems presented here are enticing that the student will pursue further and learn from other sources.

A problem style presentation of the fundamental topics of analytic number theory has its virtues, as I have heard from those who benefited from the first edition. Mere theoretical knowledge in any field is insufficient for a full appreciation of the subject and one often needs to grapple with concrete questions in which these ideas are used in a vital way. Knowledge and the various layers of "knowing" are difficult to define or describe. However, one learns much and gains insight only through practice. Making mistakes is an integral part of learning. Indeed, "it is practice first and knowledge afterwards."

Kingston, Ontario, Canada, September 2007 M. Ram Murty

Acknowledgments for the Second Edition

I would like to thank several people who have assisted me in correcting and expanding the first edition. They are Amir Akbary, Robin Chapman, Keith Conrad, Chantal David, Brandon Fodden, Sanoli Gun, Wentang Kuo, Yu-Ru Liu, Kumar Murty, Purusottam Rath and Michael Rubinstein.

Kingston, Ontario, Canada, September 2007 M. Ram Murty

Preface to the First Edition

"In order to become proficient in mathematics, or in any subject," writes André Weil, "the student must realize that most topics involve only a small number of basic ideas." After learning these basic concepts and theorems, the student should "drill in routine exercises, by which the necessary reflexes in handling such concepts may be acquired.... There can be no real understanding of the basic concepts of a mathematical theory without an ability to use them intelligently and apply them to specific problems." Weil's insightful observation becomes especially important at the graduate and research level. It is the viewpoint of this book. Our goal is to acquaint the student with the methods of analytic number theory as rapidly as possible through examples and exercises.

Any landmark theorem opens up a method of attacking other problems. Unless the student is able to sift out from the mass of theory the underlying techniques, his or her understanding will only be academic and not that of a participant in research. The prime number theorem has given rise to the rich Tauberian theory and a general method of Dirichlet series with which one can study the asymptotics of sequences. It has also motivated the development of sieve methods. We focus on this theme in the book. We also touch upon the emerging Selberg theory (in Chapter 8) and p-adic analytic number theory (in Chapter 10).

This book is a collection of about five hundred problems in analytic number theory with the singular purpose of training the beginning graduate student in some of its significant techniques. As such, it is expected that the student has had at least a semester course in each of real and complex analysis. The problems have been organized with the purpose of self-instruction. Those who exercise their mental muscles by grappling with these problems on a daily basis will develop not only a knowledge of analytic number theory but also the discipline needed for self-instruction, which is indispensable at the research level.

The book is ideal for a first course in analytic number theory either at the senior undergraduate level or the graduate level. There are several ways to give such a course. An introductory course at the senior undergraduate level can focus on chapters 1, 2, 3, 9, and 10. A beginning graduate course can in addition cover chapters 4, 5, and 8. An intense graduate course can easily cover the entire text in one semester, relegating some of the routine chapters such as chapters 6, 7, and 10 to student presentations. Or one can take up a chapter a week during a semester course with the instructor focusing on the main theorems and illustrating them with a few worked examples.

In the course of training students for graduate research, I found it tedious to keep repeating the cyclic pattern of courses in analytic and algebraic number theory. This book, along with my other book "Problems in Algebraic Number Theory" (written jointly with J. Esmonde), which appears as *Graduate Texts in Mathematics, Vol.* **190**, are intended to enable the student gain a quick initiation into the beautiful subject of number theory. No doubt, many important topics have been left out. Nevertheless, the material included here is a "basic tool kit" for the number theorist and some of the harder exercises reveal the subtle "tricks of the trade."

Unless the mind is challenged, it does not perform. The student is therefore advised to work through the questions with some attention to the time factor. "Work expands to fill the time allotted to it" and so if no upper limit is assigned, the mind does not get focused. There is no universal rule on how long one should work on a problem. However, it is a well-known fact that self-discipline, whatever shape it may take, opens the door for inspiration. If the mental muscles are exercised in this fashion, the nuances of the solution become clearer and significant. In this way, it is hoped that many, who do not have access to an "external teacher" will benefit by the approach of this text and awaken their "internal teacher."

Princeton, November 1999 M. Ram Murty

Acknowledgments for the First Edition

I would like to thank Roman Smirnov for his excellent job of typesetting this book into LaTeX. I also thank Amir Akbary, Kalyan Chakraborty, Alina Cojocaru, Wentang Kuo, Yu-Ru Liu, Kumar Murty, and Yiannis Petridis for their comments on an earlier version of the manuscript. The text matured from courses given at Queen's University, Brown University, and the Mehta Research Institute. I thank the students who participated in these courses. Since it was completed while the author was at the Institute for Advanced Study in the fall of 1999, I thank IAS for providing a congenial atmosphere for the work. I am grateful to the Canada Council for their award of a Killam Research Fellowship, which enabled me to devote time to complete this project.

Princeton, November 1999 M. Ram Murty

Contents

Part I

Problems

1
Arithmetic Functions

$\mathbb{N}$ will denote the natural numbers. An **arithmetic function** f is a complex-valued function defined on the natural numbers $\mathbb{N}$. Such an f is called an **additive function** if

$$f(mn) = f(m) + f(n) \tag{1.1}$$

whenever m and n are coprime. If (1.1) holds for all m, n, then f is called **completely additive**. A **multiplicative** function is an arithmetic function f satisfying $f(1) = 1$ and

$$f(mn) = f(m)f(n) \tag{1.2}$$

whenever m and n are coprime. If (1.2) holds for all m, n, then f is called **completely multiplicative**. The notation (m, n) will be frequently used to denote the greatest common divisor of m and n. Thus, $(m, n) = 1$ if and only if m and n are coprime.

Let $\nu(n)$ denote the number of distinct prime divisors of n. Let $\Omega(n)$ be the number of prime divisors of n counted with multiplicity. Then ν and Ω are examples of additive functions. Moreover, Ω is completely additive, whereas ν is not.

Let $s \in \mathbb{C}$ and consider the divisor functions

$$\sigma_s(n) = \sum_{d|n} d^s,$$

where the summation is over the sth powers of the positive divisors of n. The special case $s = 0$ gives the number of divisors of n, usually denoted by $d(n)$. It is not difficult to see that for each $s \in \mathbb{C}$, $\sigma_s(n)$ is a multiplicative function that is not completely multiplicative. We also have a tendency to use the letter p to denote a prime number.

An important multiplicative function is the **Möbius function**, defined by

$$\mu(n) = \begin{cases} (-1)^{\nu(n)} & \textit{if } n \textit{ is square-free,} \\ 0 & \textit{otherwise.} \end{cases}$$

We set $\mu(1) = 1$.
The **Euler totient function** given by

$$\varphi(n) = n \cdot \prod_{p|n} \left(1 - \frac{1}{p}\right)$$

is another well-known multiplicative function which enumerates the number of coprime residue classes (mod n).

The **von Mangoldt function**, defined by $\Lambda(1) = 0$ and

$$\Lambda(n) = \begin{cases} \log p & \textit{if } n = p^\alpha \textit{ for some } \alpha \geq 1, \textit{and } p \textit{ prime} \\ 0 & \textit{otherwise,} \end{cases}$$

is neither additive nor multiplicative. Still, it plays a central role in the study of the distribution of prime numbers.

1.1 The Möbius Inversion Formula and Applications

Exercise 1.1.1 *Prove that*

$$\sum_{d|n} \mu(d) = \begin{cases} 1 & \textit{if} \quad n = 1, \\ \\ 0 & \textit{otherwise.} \end{cases}$$

Exercise 1.1.2 (The Möbius inversion formula) *Show that*

$$f(n) = \sum_{d|n} g(d) \qquad \forall n \in \mathbb{N}$$

if and only if

$$g(n) = \sum_{d|n} \mu(d) f(n/d) \qquad \forall n \in \mathbb{N}.$$

Exercise 1.1.3 *Show that*

$$\sum_{d|n} \varphi(d) = n.$$

Exercise 1.1.4 *Show that*

$$\frac{\varphi(n)}{n} = \sum_{d|n} \frac{\mu(d)}{d}.$$

Exercise 1.1.5 *Let f be multiplicative. Suppose that*

$$n = \prod_{p^\alpha \| n} p^\alpha$$

is the unique factorization of n into powers of distinct primes. Show that

$$\sum_{d|n} f(d) = \prod_{p^\alpha \| n} (1 + f(p) + f(p^2) + \cdots + f(p^\alpha)).$$

Deduce that the function $g(n) = \sum_{d|n} f(d)$ is also multiplicative. The notation $p^\alpha \| n$ means that p^α is the exact power of p dividing n.

Exercise 1.1.6 *Show that*

$$\sum_{d|n} \Lambda(d) = \log n.$$

Deduce that

$$\Lambda(n) = -\sum_{d|n} \mu(d) \log d.$$

Exercise 1.1.7 *Show that*

$$\sum_{d^2|n} \mu(d) = \begin{cases} 1 & \text{if } n \text{ is square-free,} \\ 0 & \text{otherwise.} \end{cases}$$

Exercise 1.1.8 *Show that for any natural number k,*

$$\sum_{d^k|n} \mu(d) = \begin{cases} 1 & \text{if } n \text{ is } k\text{th power-free,} \\ 0 & \text{otherwise.} \end{cases}$$

Exercise 1.1.9 *If for all positive x,*

$$G(x) = \sum_{n \leq x} F\left(\frac{x}{n}\right),$$

show that

$$F(x) = \sum_{n \leq x} \mu(n) G\left(\frac{x}{n}\right)$$

and conversely.

Exercise 1.1.10 *Suppose that*

$$\sum_{k=1}^{\infty} d_3(k)|f(kx)| < \infty,$$

where $d_3(k)$ denotes the number of factorizations of k as a product three numbers. Show that if

$$g(x) = \sum_{m=1}^{\infty} f(mx),$$

then

$$f(x) = \sum_{n=1}^{\infty} \mu(n) g(nx)$$

and conversely.

Exercise 1.1.11 *Let $\lambda(n)$ denote Liouville's function given by $\lambda(n) = (-1)^{\Omega(n)}$, where $\Omega(n)$ is the total number (counting multiplicity) of prime factors of n. Show that*

$$\sum_{d|n} \lambda(d) = \begin{cases} 1 & \text{if } n \text{ is a square,} \\ 0 & \text{otherwise.} \end{cases}$$

Exercise 1.1.12 (Ramanujan sums) *The Ramanujan sum $c_n(m)$ is defined as*

$$c_n(m) = \sum_{\substack{1 \leq h \leq n \\ (h,n)=1}} e\left(\frac{hm}{n}\right),$$

where $e(t) = e^{2\pi i t}$. Show that

$$c_n(m) = \sum_{d|(m,n)} d\mu(n/d).$$

Exercise 1.1.13 *Show that*

$$\mu(n) = \sum_{\substack{1 \leq h \leq n \\ (h,n)=1}} e\left(\frac{h}{n}\right).$$

Exercise 1.1.14 *Let $\delta = (n, m)$. Show that*

$$c_n(m) = \mu(n/\delta)\varphi(n)/\varphi(n/\delta).$$

1.2 Formal Dirichlet Series

If f is an arithmetic function, the formal Dirichlet series attached to f is given by

$$D(f, s) = \sum_{n=1}^{\infty} f(n) n^{-s}.$$

We define the sum and product of two such series in the obvious way:

$$D(f, s) + D(g, s) = \sum_{n=1}^{\infty} (f(n) + g(n)) n^{-s}$$

and

$$D(f, s) D(g, s) = \sum_{n=1}^{\infty} h(n) n^{-s},$$

where

$$h(n) = \sum_{de=n} f(d)g(e).$$

We sometimes write $h = f * g$ to denote this equality. It is also useful to introduce $\delta(n) = 1$ if $n = 1$, $\delta(n) = 0$ for $n \neq 1$. Thus $D(\delta, s) = 1$.

Exercise 1.2.1 *Let f be a multiplicative function. Show that*

$$D(f, s) = \prod_p \left(\sum_{\nu=0}^{\infty} f(p^\nu) p^{-\nu s} \right).$$

Exercise 1.2.2 *If*

$$\zeta(s) = D(1, s) = \sum_{n=1}^{\infty} \frac{1}{n^s},$$

show that

$$D(\mu, s) = 1/\zeta(s).$$

Exercise 1.2.3 *Show that*

$$D(\Lambda, s) = \sum_{n=1}^{\infty} \frac{\Lambda(n)}{n^s} = -\frac{\zeta'}{\zeta}(s),$$

where $-\zeta'(s) = \sum_{n=1}^{\infty} (\log n) n^{-s}$.

Exercise 1.2.4 *Suppose that*

$$f(n) = \sum_{d|n} g(d).$$

Show that $D(f, s) = D(g, s)\zeta(s)$.

Exercise 1.2.5 *Let $\lambda(n)$ be the Liouville function defined by $\lambda(n) = (-1)^{\Omega(n)}$, where $\Omega(n)$ is the total number of prime factors of n. Show that*

$$D(\lambda, s) = \frac{\zeta(2s)}{\zeta(s)}.$$

Exercise 1.2.6 *Prove that*

$$\sum_{n=1}^{\infty} \frac{2^{\nu(n)}}{n^s} = \frac{\zeta^2(s)}{\zeta(2s)}.$$

Exercise 1.2.7 *Show that*

$$\sum_{n=1}^{\infty} \frac{|\mu(n)|}{n^s} = \frac{\zeta(s)}{\zeta(2s)}.$$

Exercise 1.2.8 *Let $d(n)$ denote the number of divisors of n. Prove that*

$$\sum_{n=1}^{\infty} \frac{d^2(n)}{n^s} = \frac{\zeta^4(s)}{\zeta(2s)},$$

(This example is due to Ramanujan.)

Exercise 1.2.9 *For any complex numbers a, b, show that*

$$\sum_{n=1}^{\infty} \frac{\sigma_a(n)\sigma_b(n)}{n^s} = \frac{\zeta(s)\zeta(s-a)\zeta(s-b)\zeta(s-a-b)}{\zeta(2s-a-b)}.$$

Exercise 1.2.10 *Let $q_k(n)$ be 1 if n is kth power-free and 0 otherwise. Show that*

$$\sum_{n=1}^{\infty} \frac{q_k(n)}{n^s} = \frac{\zeta(s)}{\zeta(ks)}.$$

1.3 Orders of Some Arithmetical Functions

The order of an arithmetic function refers to its rate of growth. There are various ways of measuring this rate of growth. The most common way is to find some nice continuous function that serves as a universal upper bound. For example, $d(n) \leq n$, but this is not the best possible bound, as the exercises below illustrate.

We will also use freely the "big O" notation. We will write $f(n) = O(g(n))$ if there is a constant K such that $|f(n)| \leq Kg(n)$ for all values of n. Sometimes we use the notation $\gg$ and write $g(n) \gg f(n)$ to indicate the same thing. We may also indicate this by $f(n) \ll g(n)$. This is just for notational convenience. Thus $d(n) = O(n)$. However, $d(n) = O(\sqrt{n})$, and in fact is $O(n^\epsilon)$ for any $\epsilon > 0$ as the exercises below show. We also have $\varphi(n) = O(n)$.

It is also useful to introduce the "little o" notation. We will write $f(x) = o(g(x))$ to mean

$$f(x)/g(x) \to 0$$

as $x \to \infty$. Thus $d(n) = o(n^2)$, and in fact, $d(n) = o(n^\epsilon)$ for any $\epsilon > 0$ by Exercise 1.3.3 below. We also write $p^\alpha \| n$ to mean $p^\alpha | n$ and $p^{\alpha+1} \nmid n$.

Exercise 1.3.1 *Show that $d(n) \leq 2\sqrt{n}$, where $d(n)$ is the number of divisors of n.*

Exercise 1.3.2 *For any $\epsilon > 0$, there is a constant $C(\epsilon)$ such that $d(n) \leq C(\epsilon)n^\epsilon$.*

Exercise 1.3.3 *For any $\eta > 0$, show that*

$$d(n) < 2^{(1+\eta)\log n/\log\log n}$$

for all n sufficiently large.

Exercise 1.3.4 *Prove that $\sigma_1(n) \leq n(\log n + 1)$.*

Exercise 1.3.5 *Prove that*

$$c_1 n^2 \leq \phi(n)\sigma_1(n) \leq c_2 n^2$$

for certain positive constants c_1 and c_2.

Exercise 1.3.6 *Let $\nu(n)$ denote the number of distinct prime factors of n. Show that*

$$\nu(n) \leq \frac{\log n}{\log 2}.$$

1.4 Average Orders of Arithmetical Functions

Let $f(n)$ be an arithmetical function and $g(x)$ a monotonic increasing function of x. Suppose

$$\sum_{n \leq x} f(n) = xg(x) + o(xg(x))$$

as $x \to \infty$. We say that $g(n)$ is the average order of $f(n)$.

Exercise 1.4.1 *Show that the average order of $d(n)$ is $\log n$.*

Exercise 1.4.2 *Show that the average order of $\phi(n)$ is cn for some constant c.*

Exercise 1.4.3 *Show that the average order of $\sigma_1(n)$ is $c_1 n$ for some constant c_1.*

Exercise 1.4.4 *Let $q_k(n) = 1$ if n is kth power-free and zero otherwise. Show that*

$$\sum_{n \le x} q_k(n) = c_k x + O\left(x^{1/k}\right),$$

where

$$c_k = \sum_{n=1}^{\infty} \frac{\mu(n)}{n^k}.$$

1.5 Supplementary Problems

Exercise 1.5.1 *Prove that*

$$\sum_{\substack{n \le x \\ (n,k)=1}} \frac{1}{n} \sim \frac{\phi(k)}{k} \log x$$

as $x \to \infty$.

Exercise 1.5.2 *Let $J_r(n)$ be the number of r-tuples of integers $(a_1, a_2, \ldots, a_r)$ satisfying $1 \le a_i \le n$ and $\gcd(a_1, \ldots, a_r, n) = 1$. Show that*

$$J_r(n) = n^r \prod_{p|n} \left(1 - \frac{1}{p^r}\right)$$

($J_r(n)$ is called Jordan's totient function. For $r = 1$, this is, of course, Euler's ϕ-function.)

Exercise 1.5.3 *For $r \ge 2$, show that there are positive constants c_1 and c_2 such that*

$$c_1 n^r \le J_r(n) \le c_2 n^r.$$

Exercise 1.5.4 *Show that the average order of $J_r(n)$ is cn^r for some constant $c > 0$.*

Exercise 1.5.5 *Let $d_k(n)$ be the number of ways of writing n as a product of k positive numbers. Show that*

$$\sum_{n=1}^{\infty} \frac{d_k(n)}{n^s} = \zeta^k(s).$$

Exercise 1.5.6 *If $d_k^*(n)$ denotes the number of factorizations of n as a product of k positive numbers each greater than 1, show that*

$$\sum_{n=1}^{\infty} \frac{d_k^*(n)}{n^s} = (\zeta(s) - 1)^k.$$

Exercise 1.5.7 *Let $\Delta(n)$ be the number of nontrivial factorizations of n. Show that*

$$\sum_{n=1}^{\infty} \frac{\Delta(n)}{n^s} = (2 - \zeta(s))^{-1}$$

as a formal Dirichlet series.

Exercise 1.5.8 *Show that*

$$\sum_{\substack{n \leq x \\ (n,k)=1}} n = \frac{\phi(k)}{2k} x^2 + O(d(k)x),$$

where $d(k)$ denotes the number of divisors of k.

Exercise 1.5.9 *Prove that*

$$\sum_{\substack{d|n \\ \nu(d) \leq r}} \mu(d) = (-1)^r \binom{\nu(n) - 1}{r},$$

where $\nu(n)$ denotes the number of distinct prime factors of n.

Exercise 1.5.10 *Let $\pi(x, z)$ denote the number of $n \leq x$ coprime to all the prime numbers $p \leq z$. Show that*

$$\pi(x, z) = x \prod_{p \leq z} \left(1 - \frac{1}{p}\right) + O(2^z).$$

Exercise 1.5.11 *Prove that*

$$\sum_{p \leq x} \frac{1}{p} \geq \log \log x + c$$

for some constant c.

Exercise 1.5.12 *Let $\pi(x)$ be the number of primes less than or equal to x. Choosing $z = \log x$ in Exercise 1.5.10, deduce that*

$$\pi(x) = O\left(\frac{x}{\log \log x}\right).$$

Exercise 1.5.13 *Let* $M(x) = \sum_{n \le x} \mu(n)$. *Show that*

$$\sum_{n \le x} M\left(\frac{x}{n}\right) = 1.$$

Exercise 1.5.14 *Let* $\mathbb{F}_p[x]$ *denote the polynomial ring over the finite field of* p *elements. Let* N_d *be the number of monic irreducible polynomials of degree* d *in* $\mathbb{F}_p[x]$. *Using the fact that every monic polynomial in* $\mathbb{F}_p[x]$ *can be factored uniquely as a product of monic irreducible polynomials, show that*

$$p^n = \sum_{d|n} dN_d.$$

Exercise 1.5.15 *With the notation as in the previous exercise, show that*

$$N_n = \frac{1}{n} \sum_{d|n} \mu(d) p^{n/d}$$

and that $N_n \ge 1$. *Deduce that there is always an irreducible polynomial of degree* n *in* $\mathbb{F}_p[x]$.

Exercise 1.5.16 (Dual Möbius inversion formula) *Suppose* $f(d) = \sum_{d|n} g(n)$, *where the summation is over all multiples of* d. *Show that*

$$g(d) = \sum_{d|n} \mu\left(\frac{n}{d}\right) f(n)$$

and conversely (assuming that all the series are absolutely convergent).

Exercise 1.5.17 *Prove that*

$$\sum_{n \le x} \frac{\varphi(n)}{n} = cx + O(\log x)$$

for some constant $c > 0$.

Exercise 1.5.18 *For* $\mathrm{Re}(s) > 2$, *prove that*

$$\sum_{n=1}^{\infty} \frac{\varphi(n)}{n^s} = \frac{\zeta(s-1)}{\zeta(s)}.$$

Exercise 1.5.19 *Let k be a fixed natural number. Show that if*

$$f(n) = \sum_{d^k|n} g(n/d^k),$$

then

$$g(n) = \sum_{d^k|n} \mu(d) f(n/d^k),$$

and conversely.

Exercise 1.5.20 *The mth cyclotomic polynomial is defined as*

$$\phi_m(x) = \prod_{\substack{1 \le i \le m \\ (i,m)=1}} (x - \zeta_m^i),$$

where ζ_m denotes a primitive mth root of unity. Show that

$$x^m - 1 = \prod_{d|m} \phi_d(x).$$

Exercise 1.5.21 *With the notation as in the previous exercise, show that the coefficient of*

$$x^{\varphi(m)-1}$$

in $\phi_m(x)$ is $-\mu(m)$.

Exercise 1.5.22 *Prove that*

$$\phi_m(x) = \prod_{d|m} (x^d - 1)^{\mu(m/d)}.$$

Exercise 1.5.23 *If $\phi_m(x)$ is the mth cyclotomic polynomial, prove that*

$$\phi_m(1) = \begin{cases} p & \text{if } m = p^\alpha \\ 1 & \text{otherwise,} \end{cases}$$

where p is a prime number.

Exercise 1.5.24 *Prove that $\phi_m(x)$ has integer coefficients.*

Exercise 1.5.25 *Let q be a prime number. Show that any prime divisor p of $a^q - 1$ satisfies $p \equiv 1 \pmod{q}$ or $p|(a-1)$.*

Exercise 1.5.26 *Let q be a prime number. Show that any prime divisor p of $1 + a + a^2 + \cdots + a^{q-1}$ satisfies $p \equiv 1 \pmod{q}$ or $p = q$. Deduce that there are infinitely many primes $p \equiv 1 \pmod{q}$.*

Exercise 1.5.27 *Let q be a prime number. Show that any prime divisor p of*

$$1 + b + b^2 + \cdots + b^{q-1}$$

with $b = a^{q^{k-1}}$ satisfies $p \equiv 1 \pmod{q^k}$ or $p = q$.

Exercise 1.5.28 *Using the previous exercise, deduce that there are infinitely many primes $p \equiv 1 \pmod{q^k}$, for any positive integer k.*

Exercise 1.5.29 *Let p be a prime not dividing m. Show that $p|\phi_m(a)$ if and only if the order of $a \pmod{p}$ is m. (Here $\phi_m(x)$ is the mth cyclotomic polynomial.)*

Exercise 1.5.30 *Using the previous exercise, deduce the infinitude of primes $p \equiv 1 \pmod{m}$.*

2
Primes in Arithmetic Progressions

In 1837 Dirichlet proved by an ingenious analytic method that there are infinitely many primes in the arithmetic progression

$$a, \quad a+q, \quad a+2q, \quad a+3q, \quad \ldots$$

in which a and q have no common factor and q is prime. The general case, for arbitrary q, was completed only later by him, in 1840, when he had finished proving his celebrated class number formula. In fact, many are of the view that the subject of analytic number theory begins with these two papers. It is also accurate to say that character theory of finite abelian groups begins here.

In this chapter we will derive Dirichlet's theorem, not exactly following his approach, but at least initially tracing his inspiration.

2.1 Summation Techniques

A very useful result is the following.

Theorem 2.1.1 *Suppose* $\{a_n\}_{n=1}^{\infty}$ *is a sequence of complex numbers and* $f(t)$ *is a continuously differentiable function on* $[1, x]$. *Set*

$$A(t) = \sum_{n \leq t} a_n.$$

Then

$$\sum_{n \le x} a_n f(n) = A(x)f(x) - \int_1^x A(t)f'(t)dt.$$

Proof. First, suppose x is a natural number. We write the left-hand side as

$$\begin{aligned}
\sum_{n \le x} a_n f(n) &= \sum_{n \le x} \{A(n) - A(n-1)\} f(n) \\
&= \sum_{n \le x} A(n)f(n) - \sum_{n \le x-1} A(n)f(n+1) \\
&= A(x)f(x) - \sum_{n \le x-1} A(n) \int_n^{n+1} f'(t)dt \\
&= A(x)f(x) - \sum_{n \le x-1} \int_n^{n+1} A(t)f'(t)dt,
\end{aligned}$$

since $A(t)$ is a step function. Also,

$$\sum_{n \le x-1} \int_n^{n+1} A(t)f'(t)dt = \int_1^x A(t)f'(t)dt,$$

and we have proved the result if x is an integer. If x is not an integer, write $[x]$ for the greatest integer less than or equal to x, and observe that

$$A(x)\{f(x) - f([x])\} - \int_{[x]}^x A(t)f'(t)dt = 0,$$

which completes the proof.

Remark. Theorem 2.1.1 is often referred to as "partial summation."

Exercise 2.1.2 *Show that*

$$\sum_{n \le x} \log n = x \log x - x + O(\log x).$$

Exercise 2.1.3 *Show that*

$$\sum_{n \le x} \frac{1}{n} = \log x + O(1).$$

In fact, show that

$$\lim_{x\to\infty}\left(\sum_{n\le x}\frac{1}{n}-\log x\right)$$

exists. (The limit is denoted by γ and called **Euler's constant**.)

Exercise 2.1.4 *Let $d(n)$ denote the number of divisors of a natural number n. Show that*

$$\sum_{n\le x} d(n) = x\log x + O(x).$$

Exercise 2.1.5 *Suppose $A(x) = O(x^\delta)$. Show that for $s > \delta$,*

$$\sum_{n=1}^{\infty}\frac{a_n}{n^s} = s\int_1^\infty \frac{A(t)}{t^{s+1}}dt.$$

Hence the Dirichlet series converges for $s > \delta$.

Exercise 2.1.6 *Show that for $s > 1$,*

$$\zeta(s) = \frac{s}{s-1} - s\int_1^\infty \frac{\{x\}}{x^{s+1}}dx,$$

where $\{x\} = x - [x]$. Deduce that $\lim_{s\to 1^+}(s-1)\zeta(s) = 1$.

Consider the sequence $\{b_r(x)\}_{r=0}^\infty$ of polynomials defined recursively as follows:

$$\begin{aligned} b_0(x) &= 1, \\ b_r'(x) &= rb_{r-1}(x) \quad (r \ge 1), \\ \int_0^1 b_r(x)dx &= 0 \quad (r \ge 1). \end{aligned}$$

Thus, from the penultimate equation, $b_r(x)$ is obtained by integrating $rb_{r-1}(x)$, and the constant of integration is determined from the last condition.

Exercise 2.1.7 *Prove that*

$$F(x,t) = \sum_{r=0}^{\infty} b_r(x)\frac{t^r}{r!} = \frac{te^{xt}}{e^t - 1}.$$

It is easy to see that

$$b_0(x) = 1,$$

$$b_1(x) = x - \tfrac{1}{2},$$

$$b_2(x) = x^2 - x - \tfrac{1}{6},$$

$$b_3(x) = x^3 - \tfrac{3}{2}x^2 + \tfrac{1}{2}x,$$

$$b_4(x) = x^4 - 2x^3 + x^2 - \tfrac{1}{30},$$

$$b_5(x) = x^5 - \tfrac{5}{2}x^4 + \tfrac{5}{3}x^3 - \tfrac{1}{6}x.$$

These are called the **Bernoulli polynomials**. One defines the rth **Bernoulli function** $B_r(x)$ as the periodic function that coincides with $b_r(x)$ on $[0, 1)$. The number $B_r := B_r(0)$ is called the rth **Bernoulli number**. Note that if we denote by $\{x\}$ the quantity $x - [x]$, $B_r(x) = b_r(\{x\})$.

Exercise 2.1.8 *Show that* $B_{2r+1} = 0$ *for* $r \geq 1$.

The Bernoulli polynomials are useful in deriving the **Euler - Maclaurin summation formula** (Theorem 2.1.9 below).

Let $a, b \in \mathbb{Z}$. We will use the Stieltjes integral with respect to the measure $d[t]$. Then

$$\sum_{a<n\leq b} f(n) = \int_a^b f(t)d[t].$$

Notice that the interval of summation is $a < n \leq b$, so that

$$\sum_{a<n\leq b} f(n) = \int_a^b f(t)dt - \int_a^b f(t)dB_1(t)$$

because $d[t] = dt - d\{t\}$ and $B_1(t) = \{t\} - \frac{1}{2}$, by the theory of the Stieltjes integral. We can evaluate the last integral by parts:

$$\int_a^b f(t)dB_1(t) = (f(b) - f(a))B_1 - \int_a^b B_1(t)f'(t)dt,$$

since $B_1(b) = B_1(a) = B_1(0)$. From $B_2^{'}(t) = 2B_1(t)$, we can write

$$\int_a^b f(t)dB_1(t) = (f(b) - f(a))B_1 - \frac{1}{2!}\int_a^b f^{'}(t)dB_2(t),$$

provided that f is differentiable on $[a, b]$. We can iterate this procedure to deduce he following theorem:

Theorem 2.1.9 (Euler-Maclaurin summation formula) *Let k be a nonnegative integer and f be $(k+1)$ times differentiable on $[a, b]$ with $a, b \in \mathbb{Z}$. Then*

$$\begin{aligned}\sum_{a<n\leq b} f(n) &= \int_a^b f(t)dt + \sum_{r=0}^{k} \frac{(-1)^{r+1}}{(r+1)!}(f^{(r)}(b) - f^{(r)}(a))B_{r+1} \\ &+ \frac{(-1)^k}{(k+1)!}\int_a^b B_{k+1}(t)f^{(k+1)}(t)dt.\end{aligned}$$

Example 2.1.10 *For integers $x \geq 1$,*

$$\sum_{n\leq x}\frac{1}{n} = \log x + \gamma + \frac{1}{2x} + \frac{1}{12x^2} + O\left(\frac{1}{x^3}\right).$$

Solution. Put $f(t) = 1/t$ in Theorem 2.1.9, $a = 1$, $b = x$, and $k = 2$. Then

$$\sum_{2\leq n\leq x}\frac{1}{n} = \log x + \frac{1}{2}\Big(\frac{1}{x} - 1\Big) + \frac{1}{12}\Big(\frac{1}{x^2} - 1\Big) - \int_1^x \frac{B_3(t)}{t^4}dt,$$

so that

$$\sum_{n\leq x}\frac{1}{n} = \log x + \frac{1}{2} - \frac{1}{12} - \int_1^x \frac{B_3(t)}{t^4}dt + \frac{1}{2x} - \frac{1}{12x^2}.$$

Since

$$-\gamma = \lim_{x\to\infty}\left(\log x - \sum_{n\leq x}\frac{1}{n}\right),$$

we must have

$$\gamma = \frac{1}{2} - \frac{1}{12} - \int_1^\infty \frac{B_3(t)dt}{t^4}.$$

Also,

$$\int_x^\infty \frac{B_3(t)dt}{t^4} = O\Big(\frac{1}{x^3}\Big),$$

so that the result is now immediate.

Exercise 2.1.11 *Show that for some constant B,*

$$\sum_{n\le x}\frac{1}{\sqrt{n}}=2\sqrt{x}+B+O\Big(\frac{1}{\sqrt{x}}\Big).$$

Exercise 2.1.12 *For $z\in\mathbb{C}$, and $|\arg z|\le\pi-\delta$, where $\delta>0$, show that*

$$\sum_{j=0}^{n}\log(z+j) = \Big(z+n+\frac{1}{2}\Big)\log(z+n) -n-\Big(z-\frac{1}{2}\Big)\log z+\int_0^n\frac{B_1(x)dx}{z+x}.$$

2.2 Characters mod q

Consider the group $(\mathbb{Z}/q\mathbb{Z})^*$ of coprime residue classes mod q. A homomorphism

$$\chi:(\mathbb{Z}/q\mathbb{Z})^*\to\mathbb{C}^*$$

into the multiplicative group of complex numbers is called a **character** (mod q). Since $(\mathbb{Z}/q\mathbb{Z})^*$ has order $\varphi(q)$, then by Euler's theorem we have

$$a^{\varphi(q)}\equiv 1\ (\mathrm{mod}\ q),$$

and so we must have $\chi^{\varphi(q)}(a)=1$ for all $a\in(\mathbb{Z}/q\mathbb{Z})^*$. Thus $\chi(a)$ must be a $\varphi(q)$th root of unity.

We extend the definition of χ to all natural numbers by setting

$$\chi(n)=\begin{cases}\chi(n\,(\mathrm{mod}\,q)) & \text{if } (n,q)=1,\\ 0 & \text{otherwise.}\end{cases}$$

Exercise 2.2.1 *Prove that χ is a completely multiplicative function.*

We now define the L-series,

$$L(s,\chi)=\sum_{n=1}^{\infty}\frac{\chi(n)}{n^s}.$$

Since $|\chi(n)|\le 1$, the series is absolutely convergent for $\mathrm{Re}(s)>1$.

Exercise 2.2.2 *Prove that for* $\mathrm{Re}(s) > 1$,

$$L(s,\chi) = \prod_p \left(1 - \frac{\chi(p)}{p^s}\right)^{-1},$$

where the product is over prime numbers p.

The character

$$\chi_0 : (\mathbb{Z}/q\mathbb{Z})^* \to \mathbb{C}^*$$

satisfying $\chi_0(a) = 1$ for all $(a,q) = 1$ is called the **trivial** character. Moreover, if χ and ψ are characters, so is $\chi\psi$, as well as $\overline{\chi}$ defined by

$$\overline{\chi}(a) = \overline{\chi(a)},$$

which is clearly a homomorphism of $(\mathbb{Z}/q\mathbb{Z})^*$. Thus, the set of characters forms a group. This is a finite group, as the value of $\chi(a)$ is a $\varphi(q)$th root of unity for $(a,q) = 1$.

But more can be said. If we write

$$q = p_1^{\alpha_1} \cdots p_k^{\alpha_k}$$

as the unique factorization of q as a product of prime powers, then by the Chinese remainder theorem,

$$\mathbb{Z}/q\mathbb{Z} \simeq \oplus_i \mathbb{Z}/p_i^{\alpha_i}\mathbb{Z}$$

is an isomorphism of rings. Thus,

$$(\mathbb{Z}/q\mathbb{Z})^* \simeq \oplus_i (\mathbb{Z}/p_i^{\alpha_i}\mathbb{Z})^*.$$

Exercise 2.2.3 *Show that* $(\mathbb{Z}/p\mathbb{Z})^*$ *is cyclic if* p *is a prime.*

An element g that generates $(\mathbb{Z}/p\mathbb{Z})^*$ is called a **primitive root** (mod p).

Exercise 2.2.4 *Let* p *be an odd prime. Show that* $(\mathbb{Z}/p^a\mathbb{Z})^*$ *is cyclic for any* $a \geq 1$.

In the previous exercise it is crucial that p is odd. For instance, $(\mathbb{Z}/8\mathbb{Z})^*$ is not cyclic but rather isomorphic to the Klein four-group $\mathbb{Z}/2\mathbb{Z} \times \mathbb{Z}/2\mathbb{Z}$. However, one can show that $(\mathbb{Z}/2^\alpha\mathbb{Z})^*$ is isomorphic to a direct product of a cyclic group and a group of order 2 for $\alpha \geq 3$.

Exercise 2.2.5 *Let $a \geq 3$. Show that $5 \pmod{2^a}$ has order 2^{a-2}.*

Exercise 2.2.6 *Show that $(\mathbb{Z}/2^a\mathbb{Z})^*$ is isomorphic to $(\mathbb{Z}/2\mathbb{Z}) \times (\mathbb{Z}/2^{a-2}\mathbb{Z})$, for $a \geq 3$.*

Exercise 2.2.7 *Show that the group of characters $\pmod q$ has order $\varphi(q)$.*

Exercise 2.2.8 *If $\chi \neq \chi_0$, show that*

$$\sum_{a \pmod q} \chi(a) = 0.$$

Exercise 2.2.9 *Show that*

$$\sum_{\chi \pmod q} \chi(n) = \begin{cases} \varphi(q) & \text{if } n \equiv 1 \pmod q, \\ 0 & \text{otherwise.} \end{cases}$$

2.3 Dirichlet's Theorem

The central idea of Dirichlet's argument is to show that

$$\lim_{s \to 1^+} \sum_{p \equiv a \pmod q} \frac{1}{p^s} = +\infty,$$

where the summation is over primes $p \equiv a \pmod q$.

If $q = 1$, this is clear, because

$$\zeta(s) = \prod_p \left(1 - \frac{1}{p^s}\right)^{-1}$$

and

$$\begin{aligned} \log \zeta(s) &= -\sum_p \log\left(1 - \frac{1}{p^s}\right) \\ &= \sum_p \left(\sum_{n=1}^{\infty} \frac{1}{np^{ns}}\right) \end{aligned}$$

upon using the expression

$$-\log(1-x) = \sum_{n=1}^{\infty} \frac{x^n}{n}.$$

Observing that

$$\lim_{s\to 1^+} \zeta(s) = +\infty$$

by virtue of the divergence of the harmonic series, we get

$$\lim_{s\to 1^+} \log \zeta(s) = +\infty.$$

Consequently,

$$\lim_{s\to 1^+} \left(\sum_p \frac{1}{p^s} + \sum_p \sum_{n\geq 2} \frac{1}{np^{ns}} \right) = +\infty.$$

In view of the fact for $s \geq 1$,

$$\sum_p \sum_{n\geq 2} \frac{1}{np^{ns}} \leq \sum_p \sum_{n\geq 2} \frac{1}{np^n} \leq \sum_p \frac{1}{p(p-1)} < \infty,$$

we deduce

$$\lim_{s\to 1^+} \sum_p \frac{1}{p^s} = +\infty.$$

Exercise 2.3.1 *Let* $\chi = \chi_0$ *be the trivial character* $(\bmod\, q)$*. Show that*

$$\lim_{s\to 1^+} \log L(s, \chi_0) = +\infty.$$

Exercise 2.3.2 *Show that for* $s > 1$*,*

$$\sum_{\chi (\bmod\, q)} \log L(s, \chi) = \varphi(q) \sum_{n\geq 1} \sum_{p^n \equiv 1 (\bmod\, q)} \frac{1}{np^{ns}}.$$

Exercise 2.3.3 *Show that for* $s > 1$ *the Dirichlet series*

$$\sum_{n=1}^{\infty} \frac{a_n}{n^s} := \prod_{\chi (\bmod\, q)} L(s, \chi)$$

has the property that $a_1 = 1$ *and* $a_n \geq 0$ *for* $n \geq 2$.

Exercise 2.3.4 *For* $\chi \neq \chi_0$*, a Dirichlet character* $(\bmod\, q)$*, show that* $|\sum_{n\leq x} \chi(n)| \leq q$*. Deduce that*

$$L(s, \chi) = \sum_{n=1}^{\infty} \frac{\chi(n)}{n^s}$$

converges for $s > 0$.

Exercise 2.3.5 *If $L(1,\chi) \neq 0$, show that $L(1,\overline{\chi}) \neq 0$, for any character $\chi \neq \chi_0 \bmod q$.*

Exercise 2.3.6 *Show that*

$$\lim_{s \to 1^+} (s-1)L(s,\chi_0) = \varphi(q)/q.$$

Exercise 2.3.7 *If $L(1,\chi) \neq 0$ for every $\chi \neq \chi_0$, deduce that*

$$\lim_{s \to 1^+} (s-1) \prod_{\chi (\mathrm{mod}\, q)} L(s,\chi) \neq 0$$

and hence

$$\sum_{p \equiv 1 (\mathrm{mod}\, q)} \frac{1}{p} = +\infty.$$

Conclude that there are infinitely many primes $p \equiv 1 (\mathrm{mod}\, q)$.

This exercise shows that the essential step in establishing the infinitude of primes congruent to 1 $(\mathrm{mod}\, q)$ is the nonvanishing of $L(1,\chi)$. The exercise below establishes the same for other progressions (mod q).

Exercise 2.3.8 *Fix $(a,q) = 1$. Show that*

$$\sum_{\chi (\mathrm{mod}\, q)} \overline{\chi(a)}\chi(n) = \begin{cases} \varphi(q) & \text{if } n \equiv a \,(\mathrm{mod}\, q), \\ 0 & \text{otherwise}. \end{cases}$$

Exercise 2.3.9 *Fix $(a,q) = 1$. If $L(1,\chi) \neq 0$, show that*

$$\lim_{s \to 1^+} (s-1) \prod_{\chi (\mathrm{mod}\, q)} L(s,\chi)^{\overline{\chi(a)}} \neq 0.$$

Deduce that

$$\sum_{p \equiv a (\mathrm{mod}\, q)} \frac{1}{p} = +\infty.$$

The essential thing now is to show that $L(1,\chi) \neq 0$ for $\chi \neq \chi_0$. Historically, this was a difficult step to surmount. Now, there are many ways to establish this. We will take the most expedient route. We will exploit the fact that

$$F(s) := \prod_{\chi (\mathrm{mod}\, q)} L(s,\chi)$$

is a Dirichlet series $\sum_{n=1}^{\infty} a_n n^{-s}$ with $a_1 = 1$ and $a_n \geq 0$. If for some χ_1, $L(1, \chi_1) = 0$, we want to establish a contradiction.

Exercise 2.3.10 *Suppose $\chi_1 \neq \overline{\chi_1}$ (that is, χ_1 is not real-valued). Show that $L(1, \chi_1) \neq 0$ by considering $F(s)$.*

It remains to show that $L(1, \chi) \neq 0$ when χ is real and not equal to χ_0.

We will establish this in the next section by developing an interesting technique discovered by Dirichlet that was first developed by him not to tackle this question, but rather another problem, namely the Dirichlet divisor problem.

2.4 Dirichlet's Hyperbola Method

Suppose we have an arithmetical function $f = g * h$. That is,

$$f(n) = \sum_{d|n} g(d)h(n/d)$$

for two arithmetical functions g and h. Define

$$G(x) = \sum_{n \leq x} g(n),$$

$$H(x) = \sum_{n \leq x} h(n).$$

Theorem 2.4.1 *For any $y > 0$,*

$$\sum_{n \leq x} f(n) = \sum_{d \leq y} g(d) H\left(\frac{x}{d}\right) + \sum_{d \leq \frac{x}{y}} h(d) G\left(\frac{x}{d}\right) - G(y) H\left(\frac{x}{y}\right).$$

Proof. We have

$$\begin{aligned}
\sum_{n \leq x} f(n) &= \sum_{de \leq x} g(d)h(e) \\
&= \sum_{\substack{de \leq x \\ d \leq y}} g(d)h(e) + \sum_{\substack{de \leq x \\ d > y}} g(d)h(e) \\
&= \sum_{d \leq y} g(d) H\left(\frac{x}{d}\right) + \sum_{e \leq \frac{x}{y}} h(e) \left\{ G\left(\frac{x}{e}\right) - G(y) \right\} \\
&= \sum_{d \leq y} g(d) H\left(\frac{x}{d}\right) + \sum_{e \leq \frac{x}{y}} h(e) G\left(\frac{x}{e}\right) - G(y) H\left(\frac{x}{y}\right). \qquad \square
\end{aligned}$$

The method derives its name from the fact that the inequality $de \leq x$ is the area underneath a hyperbola. Historically, this method was first applied to the problem of estimating the error term $E(x)$ defined as

$$E(x) = \sum_{n \leq x} \sigma_0(n) - \{x(\log x) + (2\gamma - 1)x\},$$

where $\sigma_0(n)$ is the number of divisors of n and γ is Euler's constant.

Exercise 2.4.2 *Prove that*

$$\sum_{n \leq x} \sigma_0(n) = x \log x + (2\gamma - 1)x + O(\sqrt{x}).$$

Exercise 2.4.3 *Let χ be a real character* (mod q). *Define*

$$f(n) = \sum_{d|n} \chi(d).$$

Show that $f(1) = 1$ and $f(n) \geq 0$. In addition, show that $f(n) \geq 1$ whenever n is a perfect square.

Exercise 2.4.4 *Using Dirichlet's hyperbola method, show that*

$$\sum_{n \leq x} \frac{f(n)}{\sqrt{n}} = 2L(1, \chi)\sqrt{x} + O(1),$$

where $f(n) = \sum_{d|n} \chi(d)$ and $\chi \neq \chi_0$.

Exercise 2.4.5 *If $\chi \neq \chi_0$ is a real character, deduce from the previous exercise that $L(1, \chi) \neq 0$.*

Exercise 2.4.6 *Prove that*

$$\sum_{n > x} \frac{\chi(n)}{n} = O\left(\frac{1}{x}\right)$$

whenever χ is a nontrivial character (mod q).

Exercise 2.4.7 *Let*

$$a_n = \sum_{d|n} \chi(d)$$

where χ is a nonprincipal character $(\bmod q)$. *Show that*

$$\sum_{n \le x} a_n = xL(1, \chi) + O(\sqrt{x}).$$

Exercise 2.4.8 *Deduce from the previous exercise that $L(1, \chi) \neq 0$ for χ real.*

Thus, we have proved the following Theorem:

Theorem 2.4.9 (Dirichlet) *For any natural number q, and a coprime residue class a* $(\bmod q)$, *there are infinitely many primes $p \equiv a \pmod q$.*

2.5 Supplementary Problems

Exercise 2.5.1 *Let $d_k(n)$ be the number of ways of writing n as a product of k numbers. Show that*

$$\sum_{n \le x} d_k(n) = \frac{x(\log x)^{k-1}}{(k-1)!} + O(x(\log x)^{k-2})$$

for every natural number $k \ge 2$.

Exercise 2.5.2 *Show that*

$$\sum_{n \le x} \log \frac{x}{n} = x + O(\log x).$$

Exercise 2.5.3 *Let $A(x) = \sum_{n \le x} a_n$. Show that for x a positive integer,*

$$\sum_{n \le x} a_n \log \frac{x}{n} = \int_1^x \frac{A(t)dt}{t}.$$

Exercise 2.5.4 *Let $\{x\}$ denote the fractional part of x. Show that*

$$\sum_{n \le x} \left\{\frac{x}{n}\right\} = (1 - \gamma)x + O(x^{1/2}),$$

where γ is Euler's constant.

Exercise 2.5.5 *Prove that*

$$\sum_{n\le x} \log^k \frac{x}{n} = O(x)$$

for any $k > 0$.

Exercise 2.5.6 *Show that for* $x \geq 3$,

$$\sum_{3\le n\le x} \frac{1}{n\log n} = \log\log x + B + O\Big(\frac{1}{x\log x}\Big).$$

Exercise 2.5.7 *Let* χ *be a nonprincipal character* $(\bmod q)$. *Show that*

$$\sum_{n\ge x} \frac{\chi(n)}{\sqrt{n}} = O\Big(\frac{1}{\sqrt{x}}\Big).$$

Exercise 2.5.8 *For any integer* $k \geq 0$, *show that*

$$\sum_{n\le x} \frac{\log^k n}{n} = \frac{\log^{k+1} x}{k+1} + O(1).$$

Exercise 2.5.9 *Let* $d(n)$ *be the number of divisors of* n. *Show that for some constant* c,

$$\sum_{n\le x} \frac{d(n)}{n} = \frac{1}{2}\log^2 x + 2\gamma \log x + c + O\Big(\frac{1}{\sqrt{x}}\Big)$$

for $x \geq 1$.

Exercise 2.5.10 *Let* $\alpha \geq 0$ *and suppose* $a_n = O(n^\alpha)$ *and*

$$A(x) := \sum_{n\le x} a_n = O(x^\delta)$$

for some fixed $\delta < 1$. *Define*

$$b_n = \sum_{d|n} a_d.$$

Prove that

$$\sum_{n\le x} b_n = cx + O\left(x^{(1-\delta)(1+\alpha)/(2-\delta)}\right),$$

for some constant c.

Exercise 2.5.11 *Let χ be a nontrivial character $(\bmod q)$ and set*

$$f(n) = \sum_{d|n} \chi(d).$$

Show that

$$\sum_{n \leq x} f(n) = xL(1, \chi) + O(q\sqrt{x}),$$

where the constant implied is independent of q.

Exercise 2.5.12 *Suppose that $a_n \geq 0$ and that for some $\delta > 0$, we have*

$$\sum_{n \leq x} a_n \ll \frac{x}{(\log x)^\delta}.$$

Let b_n be defined by the formal Dirichlet series

$$\sum_{n=1}^{\infty} \frac{b_n}{n^s} = \left(\sum_{n=1}^{\infty} \frac{a_n}{n^s} \right)^2.$$

Show that

$$\sum_{n \leq x} b_n \ll x(\log x)^{1-2\delta}.$$

Exercise 2.5.13 *Let $\{a_n\}$ be a sequence of nonnegative numbers. Show that there exists $\sigma_0 \in \mathbb{R}$ (possibly infinite) such that*

$$f(s) = \sum_{n=1}^{\infty} \frac{a_n}{n^s}$$

converges for $\mathrm{Re}(s) > \sigma_0$ and diverges for $\mathrm{Re}(s) < \sigma_0$. Moreover, show that the series converges uniformly in $\mathrm{Re}(s) \geq \sigma_0 + \delta$ for any $\delta > 0$ and that

$$f^{(k)}(s) = (-1)^k \sum_{n=1}^{\infty} \frac{a_n (\log n)^k}{n^s}$$

for $\mathrm{Re}(s) > \sigma_0$ (σ_0 is called the **abscissa of convergence of the Dirichlet series** *$\sum_{n=1}^{\infty} a_n/n^s$).*

Exercise 2.5.14 (Landau's theorem) *Let $a_n \geq 0$ be a sequence of non-negative numbers. Let σ_0 be the abscissa of convergence of*

$$f(s) = \sum_{n=1}^{\infty} \frac{a_n}{n^s}.$$

Show that $s = \sigma_0$ is a singular point of $f(s)$ (that is, $f(s)$ cannot be extended to define an analytic function at $s = s_0$).

Exercise 2.5.15 *Let χ be a nontrivial character* $\pmod q$ *and define*

$$\sigma_{a,\chi} = \sum_{d|n} \chi(d) d^a.$$

If χ_1, χ_2 are two characters $\pmod q$, *prove that for $a, b \in \mathbb{C}$,*

$$\sum_{n=1}^{\infty} \sigma_{a,\chi_1}(n)\sigma_{b,\chi_2}(n)n^{-s}$$

$$= \frac{\zeta(s)L(s-a,\chi_1)L(s-b,\chi_2)L(s-a-b,\chi_1\chi_2)}{L(2s-a-b,\chi_1\chi_2)}.$$

as formal Dirichlet series.

Exercise 2.5.16 *Let χ be a nontrivial character* $\pmod q$. *Set $a = \overline{b}$, $\chi_1 = \chi$ and $\chi_2 = \overline{\chi}$ in the previous exercise to deduce that*

$$\sum_{n=1}^{\infty} |\sigma_{a,\chi}(n)|^2 n^{-s} = \frac{\zeta(s)L(s-a,\chi)L(s-\overline{a},\overline{\chi})L(s-a-\overline{a},\chi_0)}{L(2s-a-\overline{a},\chi_0)}$$

Exercise 2.5.17 *Using Landau's theorem and the previous exercise, show that $L(1,\chi) \neq 0$ for any non-trivial real character* $\pmod q$.

Exercise 2.5.18 *Show that $\zeta(s) \neq 0$ for* $\mathrm{Re}(s) > 1$.

Exercise 2.5.19 (Landau's theorem for integrals) *Let $A(x)$ be right continuous for $x \geq 1$ and of bounded finite variation on each finite interval. Suppose that*

$$f(s) = \int_1^{\infty} \frac{A(x)}{x^{s+1}} dx,$$

with $A(x) \geq 0$. Let σ_0 be the infimum of all real s for which the integral converges. Show that $f(s)$ has a singularity at $s = \sigma_0$.

Exercise 2.5.20 *Let λ denote Liouville's function and set*

$$S(x) = \sum_{n \leq x} \lambda(n).$$

Show that if $S(x)$ is of constant sign for all x sufficiently large, then $\zeta(s) \neq 0$ for $\mathrm{Re}(s) > \frac{1}{2}$. (The hypothesis is an old conjecture of Pólya. It was shown by Haselgrove in 1958 that $S(x)$ changes sign infinitely often.)

Exercise 2.5.21 *Prove that*

$$b_n(x) = \sum_{k=0}^{n} \binom{n}{k} B_{n-k} x^k,$$

where $b_n(x)$ is the nth Bernoulli polynomial and B_n denotes the nth Bernoulli number.

Exercise 2.5.22 *Prove that*

$$b_n(1-x) = (-1)^n b_n(x),$$

where $b_n(x)$ denotes the nth Bernoulli polynomial.

Exercise 2.5.23 *Let*

$$s_k(n) = 1^k + 2^k + 3^k + \cdots + (n-1)^k.$$

Prove that for $k \geq 1$,

$$(k+1)s_k(n) = \sum_{i=0}^{k} \binom{k+1}{i} B_i n^{k+i-i}.$$

3
The Prime Number Theorem

Let $\pi(x)$ denote the number of primes $p \leq x$. The prime number theorem is the assertion that

$$\lim_{x \to \infty} \frac{\pi(x)}{x/\log x} = 1.$$

It was proved independently by Hadamard and de la Vallée Poussin in 1896. It is the goal of this chapter to prove this theorem following a method evolved by Wiener and Ikehara in the early twentieth century.

As far as we know, it was Legendre who first conjectured that for large x, $\pi(x)$ is approximately

$$\frac{x}{\log x - 1.08}.$$

This suggests the truth of the prime number theorem. In a letter of 1849, Gauss related that as a boy he had thought about this question and felt that a good approximation to $\pi(x)$ is given by the **logarithmic integral**

$$\mathrm{li}\, x := \int_2^x \frac{dt}{\log t}.$$

This is closer to the truth. Indeed, one can prove

$$\pi(x) = \operatorname{li} x + O\left(xe^{-c\sqrt{\log x}}\right)$$

for some constant c. Integrating the logarithmic integral by parts, we see that

$$\operatorname{li} x = \frac{x}{\log x} + \frac{x}{(\log x)^2} + \cdots + \frac{n!x}{(\log x)^{n+1}} + (n+1)! \int_2^x \frac{dt}{(\log t)^{n+1}},$$

from which it is easily deduced that if we interpret Legendre's statement as

$$\pi(x) = \frac{x}{\log x - A(x)},$$

where $A(x) \to 1.08$, then the above analysis shows that it is false, since $A(x) \to 1$.

Chebyshev in 1851 obtained by very elementary methods upper and lower bounds for $\pi(x)$. He proved that

$$\liminf \frac{\pi(x)}{x/\log x} \leq 1 \leq \limsup \frac{\pi(x)}{x/\log x},$$

so that if the limit exists, then it must be 1.

3.1 Chebyshev's Theorem

The elementary method of Chebyshev begins by observing that the binomial coefficient

$$\binom{2n}{n}$$

is divisible by every prime between n and $2n$.

Exercise 3.1.1 *Let*

$$\theta(n) = \sum_{p \leq n} \log p,$$

where the summation is over primes. Prove that

$$\theta(n) \leq 4n \log 2.$$

Exercise 3.1.2 *Prove that* $\theta(2m+1) - \theta(m) \leq 2m \log 2$. *Deduce that*

$$\theta(n) \leq 2n \log 2.$$

Exercise 3.1.3 *Let*

$$\psi(x) = \sum_{p^\alpha \le x} \log p = \sum_{n \le x} \Lambda(n),$$

where Λ *is the von Mangoldt function. Show that*

$$\mathrm{lcm}[1, 2, \cdots, n] = e^{\psi(n)}.$$

Exercise 3.1.4 *Show that*

$$e^{\psi(2n+1)} \int_0^1 x^n(1-x)^n dx$$

is a positive integer. Deduce that $\psi(2n+1) \ge 2n \log 2$. (The method of deriving this is due to M. Nair.)

Exercise 3.1.5 *Prove that there are positive constants A and B such that*

$$\frac{Ax}{\log x} \le \pi(x) \le \frac{Bx}{\log x}$$

for all x *sufficiently large.* This result was first proved by Chebyshev.

Exercise 3.1.6 *Prove that*

$$T(x) := \sum_{n \le x} \log n = x \log x - x + \frac{1}{2} \log x + c + O(1/x)$$

for some constant c *(this improves Exercise* 2.1.2*).*

Exercise 3.1.7 *Using the fact*

$$\log n = \sum_{d|n} \Lambda(d),$$

prove that

$$\sum_{n \le x} \frac{\Lambda(n)}{n} = \log x + O(1).$$

Exercise 3.1.8 *Prove that*

$$\sum_{p \le x} \frac{1}{p} = \log \log x + O(1).$$

Theorem 3.1.9 (Bertrand's postulate) *For n sufficiently large, there is a prime between n and $2n$.*

Proof: (S. Ramanujan) Observe that if

$$a_0 \geq a_1 \geq a_2 \geq \cdots$$

is a decreasing sequence of real numbers tending to zero, then

$$a_0 - a_1 \leq \sum_{n=0}^{\infty} (-1)^n a_n \leq a_0 - a_1 + a_2.$$

This is the starting point of Ramanujan's proof. We can write

$$T(x) = \sum_{n \leq x} \log n = \sum_{de \leq x} \Lambda(d) = \sum_{e \leq x} \psi\left(\frac{x}{e}\right).$$

We know that $T(x) = x \log x - x + O(\log x)$ by Exercise 2.1.2. On the other hand,

$$T(x) - 2T\left(\frac{x}{2}\right) = \sum_{n \leq x} (-1)^{n-1} \psi\left(\frac{x}{n}\right) \leq \psi(x) - \psi\left(\frac{x}{2}\right) + \psi\left(\frac{x}{3}\right)$$

by the observation above. Hence

$$\psi(x) - \psi\left(\frac{x}{2}\right) + \psi\left(\frac{x}{3}\right) \geq (\log 2)x + O(\log x).$$

On the other hand,

$$\psi(x) - \psi\left(\frac{x}{2}\right) \leq (\log 2)x + O(\log x),$$

from which we deduce inductively

$$\psi(x) \leq 2(\log 2)x + O\left(\log^2 x\right).$$

Thus, $\psi(x) - \psi\left(\frac{x}{2}\right) \geq \frac{1}{3}(\log 2)x + O(\log^2 x)$. Now, $\psi(x) = \theta(x) + O\left(\sqrt{x} \log^2 x\right)$. Hence

$$\theta(x) - \theta\left(\frac{x}{2}\right) \geq \frac{1}{3}(\log 2)x + O\left(\sqrt{x} \log^2 x\right).$$

Therefore, for x sufficiently large, there is a prime between $x/2$ and x.

Remark. This theorem was first proved by Chebyshev by a similar, but more elaborate, method.

Exercise 3.1.10 *Suppose that* $\{a_n\}_{n=1}^{\infty}$ *is a sequence of complex numbers and set*

$$S(x) = \sum_{n \le x} a_n.$$

If

$$\lim_{x \to \infty} \frac{S(x)}{x} = \alpha,$$

show that

$$\sum_{n \le x} \frac{a_n}{n} = \alpha \log x + o(\log x)$$

as $x \to \infty$.

Exercise 3.1.11 *Show that*

$$\lim_{x \to \infty} \frac{\psi(x)}{x} = 1$$

if and only if

$$\lim_{x \to \infty} \frac{\pi(x)}{x/\log x} = 1.$$

Exercise 3.1.12 *If*

$$\lim_{x \to \infty} \frac{\pi(x)}{x/\log x} = \alpha,$$

then show that

$$\sum_{p \le x} \frac{1}{p} = \alpha \log \log x + o(\log \log x).$$

Deduce that if the limit exists, it must be 1.

3.2 Nonvanishing of Dirichlet Series on $\mathrm{Re}(s) = 1$

The proof of the prime number theorem, as given by Hadamard and de la Vallée Poussin, has two ingredients: (a) the analytic continuation of $\zeta(s)$ to $\mathrm{Re}(s) = 1$ and (b) the nonvanishing of $\zeta(s)$ on $\mathrm{Re}(s) = 1$.

It was believed that any proof of the prime number theorem must use the theory of complex variables until Erdös and Selberg independently discovered an "elementary proof" in 1949.

In this section we will discuss nonvanishing results of various Dirichlet series.

Exercise 3.2.1 *Show that*

$$\zeta(s) = \frac{s}{s-1} - s\int_1^\infty \frac{\{x\}}{x^{s+1}}dx$$

for $\mathrm{Re}(s) > 1$. *Since the right-hand side of the equation is analytic for* $\mathrm{Re}(s) > 0$, $s \neq 1$, *we obtain an analytic continuation of* $(s-1)\zeta(s)$.

Exercise 3.2.2 *Show that* $\zeta(s) \neq 0$ *for* $\mathrm{Re}(s) > 1$.

Exercise 3.2.3 *Prove that for* $\sigma > 1$, $t \in \mathbb{R}$,

$$\mathrm{Re}\log\zeta(\sigma + it) = \sum_{n=2}^{\infty} \frac{\Lambda(n)}{n^\sigma \log n}\cos(t\log n).$$

Exercise 3.2.4 *Prove that*

$$\mathrm{Re}(3\log\zeta(\sigma) + 4\log\zeta(\sigma + it) + \log\zeta(\sigma + 2it)) \geq 0,$$

for $\sigma > 1$, $t \in \mathbb{R}$.

Exercise 3.2.5 *Prove that for* $\sigma > 1$, $t \in \mathbb{R}$,

$$|\zeta(\sigma)^3\zeta(\sigma + it)^4\zeta(\sigma + 2it)| \geq 1.$$

Deduce that $\zeta(1 + it) \neq 0$ *for any* $t \in \mathbb{R}$, $t \neq 0$. *Deduce in a similar way, by considering*

$$\zeta(\sigma)^3 L(\sigma, \chi)^4 L(\sigma, \chi^2),$$

that $L(1, \chi) \neq 0$ *for* χ *not real.*

Exercise 3.2.6 *Show that* $-\frac{\zeta'}{\zeta}(s)$ *has an analytic continuation to* $\mathrm{Re}(s) = 1$, *with only a simple pole at* $s = 1$, *with residue* 1.

In the exercises below we will attempt to unravel the essential trigonometric idea underlying the proof of the nonvanishing of $\zeta(s)$ on $\mathrm{Re}(s) = 1$. We begin with a few trigonometric identities.

Exercise 3.2.7 *Prove that*

$$\frac{1}{2} + \cos\theta + \cos 2\theta + \cdots + \cos n\theta = \frac{\sin(n + \frac{1}{2})\theta}{2\sin\frac{\theta}{2}}.$$

Exercise 3.2.8 *Prove that*

$$\cos\theta + \cos 3\theta + \cdots + \cos(2n-1)\theta = \frac{\sin 2n\theta}{2\sin\theta}.$$

Exercise 3.2.9 *Prove that*

$$1 + \frac{\sin 3\theta}{\sin\theta} + \frac{\sin 5\theta}{\sin\theta} + \cdots + \frac{\sin(2n-1)\theta}{\sin\theta} = \left(\frac{\sin n\theta}{\sin\theta}\right)^2.$$

Exercise 3.2.10 *Prove that*

$$(2m+1) + 2\sum_{j=0}^{2m-1} (j+1)\cos(2m-j)\theta = \left(\frac{\sin(m+\frac{1}{2})\theta}{\sin\frac{\theta}{2}}\right)^2,$$

for all integers $m \geq 0$.

Remark. Notice that the case $m = 1$ gives

$$3 + 4\cos\theta + 2\cos 2\theta \geq 0,$$

which would have worked equally well in Exercises 3.2.4 and 3.2.5.

The following exercise gives us a general theorem of nonvanishing of Dirichlet series on $\mathrm{Re}(s) = 1$.

Exercise 3.2.11 *Let* $f(s)$ *be a complex-valued function satisfying:*

1. f *is holomorphic in* $\mathrm{Re}(s) > 1$ *and non-zero there;*

2. $\log f(s)$ *can be written as a Dirichlet series*

$$\sum_{n=1}^{\infty} \frac{b_n}{n^s}$$

with $b_n \geq 0$ *for* $\mathrm{Re}(s) > 1$;

3. *on the line* $\mathrm{Re}(s) = 1$, f *is holomorphic except for a pole of order* $e \geq 0$ *at* $s = 1$.

If f *has a zero on the line* $\mathrm{Re}(s) = 1$, *then prove that the order of the zero is bounded by* $e/2$. (This result is due to Kumar Murty [MM, p.10].)

Exercise 3.2.12 *Let* $f(s) = \prod_\chi L(s,\chi)$, *where the product is over Dirichlet characters* $(\mathrm{mod}\, q)$. *Show that* $f(s)$ *is a Dirichlet series with nonnegative coefficients. Deduce that* $L(s,\chi) \neq 0$ *for* $\mathrm{Re}(s) = 1$.

3.3 The Ikehara - Wiener Theorem

We begin by reviewing certain facts from Fourier analysis. Let

$$S = \left\{ f \in C^{\infty}(\mathbb{R}) : \lim_{|x|\to\infty} x^n \frac{d^m f}{dx^m} = 0 \quad \text{for all} \quad n, m \in \mathbb{Z}^+ \right\}.$$

This space is called the Schwartz space of rapidly decreasing functions. For $f \in S$, we have the Fourier transform

$$\hat{f}(x) = \frac{1}{\sqrt{2\pi}} \int_{-\infty}^{\infty} f(t) e^{-itx} dt.$$

The Fourier inversion formula gives

$$f(x) = \frac{1}{\sqrt{2\pi}} \int_{-\infty}^{\infty} \hat{f}(t) e^{itx} dt.$$

Hence

$$\hat{f}(x-y) = \frac{1}{\sqrt{2\pi}} \int_{-\infty}^{\infty} f(t) e^{ity} e^{-itx} dt,$$

so that $\hat{f}(x-y)$ and $f(t)e^{ity}$ are Fourier transforms of each other. Parseval's formula is

$$\int_{-\infty}^{\infty} f(x) g(x) dx = \int_{-\infty}^{\infty} \hat{f}(t) \hat{g}(t) dt.$$

Though these formulas are first established for $f, g \in S$, they are easily extended to all $f, g \in L^2(\mathbb{R})$. We will employ these facts for such functions.

The Riemann - Lebesgue lemma states that

$$\lim_{\lambda\to\infty} \int_{-\infty}^{\infty} f(t) e^{i\lambda t} dt = 0$$

for absolutely integrable functions. The Fejér kernel

$$K_\lambda(x) = \frac{\sin^2 \lambda x}{\lambda x^2}$$

has Fourier transform

$$\widehat{K}_\lambda(x) = \begin{cases} \sqrt{\pi/2}(1 - \frac{|x|}{2\lambda}) & \text{if} \quad |x| \le 2\lambda, \\ 0 & \text{otherwise.} \end{cases}$$

We begin with the following theorem due to Ikehara and Wiener (see for example, [MM, p.7]).

Theorem 3.3.1 *Let $F(s) = \sum_{n=1}^{\infty} b_n/n^s$ be a Dirichlet series with non-negative coefficients and absolutely convergent for* $\mathrm{Re}(s) > 1$. *Suppose that $F(s)$ can be extended to a meromorphic function in the region* $\mathrm{Re}(s) \geq 1$ *having no poles except for a simple pole at $s = 1$ with residue $R \geq 0$. Then*

$$B(x) := \sum_{n \leq x} b_n = Rx + o(x)$$

as $x \longrightarrow \infty$.

Remark. Without loss of generality, we may suppose $R > 0$, for if $R = 0$, we can consider $F(s) + \zeta(s)$. If $F(s)$ is analytic at $s = 1$, we obtain $\sum_{n \leq x} b_n = o(x)$ as $x \to \infty$.

Proof. Replacing b_n by b_n/R, we may suppose without loss of generality that $R = 1$. Then

$$F(s) = s \int_1^{\infty} \frac{B(x)}{x^{s+1}} dx.$$

Set $x = e^u$. Then

$$\frac{F(s)}{s} = \int_0^{\infty} B(e^u) e^{-us} du.$$

Note that

$$\int_0^{\infty} e^{-u(s-1)} du = \frac{1}{s-1}.$$

Setting $s = 1 + \delta + it$, $\delta > 0$, we get

$$\frac{F(1+\delta+it)}{1+\delta+it} - \frac{1}{s-1} = \int_0^{\infty} (B(e^u)e^{-u} - 1) e^{-u\delta} e^{-iut} du. \quad (3.1)$$

Set

$$\begin{aligned} g(u) &= B(e^u)e^{-u}, \\ h_\delta(t) &= \frac{F(1+\delta+it)}{1+\delta+it} - \frac{1}{s-1}, \end{aligned}$$

and

$$h(t) = \frac{F(1+it)}{1+it} - \frac{1}{s-1} \quad (s = 1+it),$$

which extends to a continuous function for all $t \in \mathbb{R}$. We also put

$$\phi_\delta(t) = \begin{cases} (g(u) - 1)e^{-u\delta} & \text{if } u \geq 0 \\ 0 & \text{if } u < 0. \end{cases}$$

Our goal is to prove $g(u) \to 1$ as $u \to \infty$. The formula (3.1) says that the Fourier transform of $\sqrt{2\pi}\phi_\delta$ is $h_\delta(t)$. Observe that for $\delta > 0$, both of these functions are square-integrable, since

$$B(x) = \sum_{n \leq x} b_n \ll \sum_{n=1}^{\infty} |b_n|(x/n)^c \ll x^c,$$

for every $c > 1$. Applying Parseval's formula gives, for each real v and real $\lambda > 0$,

$$\begin{aligned} \int_0^\infty (g(u)-1)e^{-u\delta} K_\lambda(u-v)du &= \int_{-\infty}^\infty \phi_\delta(u) K_\lambda(u-v)du \\ &= \frac{1}{\sqrt{2\pi}} \int_{-\infty}^\infty h_\delta(t)\widehat{K}_\lambda(t)e^{itv}dt. \end{aligned}$$

Since $\hat{K}_\lambda$ has compact support, the limit as $\delta \to 0$ of the right-hand side exists. The limit of the left side as $\delta \to 0$ is

$$\int_0^\infty (g(u)-1)K_\lambda(u-v)du$$

by the monotone convergence theorem. Hence,

$$\int_0^\infty (g(u)-1)K_\lambda(u-v)du = \frac{1}{\sqrt{2\pi}} \int_{-\infty}^\infty h(t)\widehat{K}_\lambda(t)e^{itv}dt.$$

By the Riemann - Lebesgue lemma, the limit of the integral on the right-hand side as $v \to \infty$ is 0. Thus,

$$\lim_{v \to \infty} \int_0^\infty (g(u)-1)K_\lambda(u-v)du = 0.$$

Since (by Exercise 3.4.13)

$$\int_{-\infty}^\infty \frac{\sin^2 \lambda x}{\lambda x^2} dx = \pi,$$

we obtain

$$\lim_{v \to \infty} \int_0^\infty g(u)K_\lambda(u-v)du = \pi.$$

Making the substitution $u = v + \alpha/\lambda$ and noting that $g(u) = 0$ for $u < 0$ gives

$$\int_0^\infty g(u)K_\lambda(u-v)du = \int_{-\infty}^\infty g(u)K_\lambda(u-v)du = \int_{-\infty}^\infty g(v+\alpha/\lambda)\frac{\sin^2 \alpha}{\alpha^2}d\alpha.$$

Thus,

$$\lim_{v\to\infty}\int_{-\infty}^{\infty} g(v+\alpha/\lambda)\frac{\sin^2\alpha}{\alpha^2}d\alpha = \pi,$$

for each $\lambda > 0$. Since $B(x)$ is monotone increasing, we see that

$$g(u_2) \geq g(u_1)e^{u_1-u_2}, \quad u_1 \leq u_2.$$

Thus, for $|\alpha| \leq \sqrt{\lambda}$, we have

$$g\left(v+\frac{\alpha}{\lambda}\right) \geq g\left(v-\frac{1}{\sqrt{\lambda}}\right)e^{-\frac{1}{\sqrt{\lambda}}-\frac{\alpha}{\lambda}} \geq g\left(v-\frac{1}{\sqrt{\lambda}}\right)e^{-\frac{2}{\sqrt{\lambda}}}.$$

Hence

$$g(v-1/\sqrt{\lambda})e^{-2/\sqrt{\lambda}}\int_{-\sqrt{\lambda}}^{\sqrt{\lambda}}\frac{\sin^2\alpha}{\alpha^2}d\alpha \leq \int_{-\sqrt{\lambda}}^{\sqrt{\lambda}} g(v+\alpha/\lambda)\frac{\sin^2\alpha}{\alpha^2}d\alpha$$

$$\leq \int_{-\infty}^{\infty} g(v+\alpha/\lambda)\frac{\sin^2\alpha}{\alpha^2}d\alpha.$$

Consequently

$$\limsup_{v\to\infty} g(v) = \limsup_{v\to\infty} g(v-1/\sqrt{\lambda}) \leq \frac{\pi e^{2/\sqrt{\lambda}}}{\int_{-\sqrt{\lambda}}^{\sqrt{\lambda}}\frac{\sin^2\alpha}{\alpha^2}d\alpha}.$$

Letting $\lambda \to \infty$, we get

$$\limsup_{v\to\infty} g(v) \leq 1.$$

In particular, we conclude that g is bounded. Let $A = \sup_v g(v)$. Then,

$$\int_{-\infty}^{\infty} g(v+\alpha/\lambda)\frac{\sin^2\alpha}{\alpha^2}d\alpha - \int_{-\sqrt{\lambda}}^{\sqrt{\lambda}} g(v+\alpha/\lambda)\frac{\sin^2\alpha}{\alpha^2}d\alpha$$

$$= \int_{\sqrt{\lambda}}^{\infty}[g(v+\alpha/\lambda) - g(v-\alpha/\lambda)]\frac{\sin^2\alpha}{\alpha^2}d\alpha \leq 2A\int_{\sqrt{\lambda}}^{\infty}\frac{d\alpha}{\alpha^2} \ll \frac{1}{\sqrt{\lambda}}.$$

Hence

$$\liminf_{v\to\infty}\int_{-\sqrt{\lambda}}^{\sqrt{\lambda}} g(v+\alpha/\lambda)\frac{\sin^2\alpha}{\alpha^2}d\alpha = \pi + O(1/\sqrt{\lambda}).$$

For $|\alpha| \leq \sqrt{\lambda}$, we have

$$g(v + \alpha/\lambda) \leq g(v + 1/\sqrt{\lambda})e^{1/\sqrt{\lambda} - \alpha/\sqrt{\lambda}} \leq g(v + 1/\sqrt{\lambda})e^{2/\sqrt{\lambda}}$$

so that

$$g(v + 1/\sqrt{\lambda})e^{2/\sqrt{\lambda}} \int_{-\sqrt{\lambda}}^{\sqrt{\lambda}} \frac{\sin^2 \alpha}{\alpha^2} d\alpha \geq \int_{-\sqrt{\lambda}}^{\sqrt{\lambda}} g(v + \alpha/\lambda) \frac{\sin^2 \alpha}{\alpha^2} d\alpha,$$

and so

$$\liminf_{v \to \infty} g(v) = \liminf_{v \to \infty} g(v + 1/\sqrt{\lambda}) \geq \frac{\pi + O(1/\sqrt{\lambda})}{e^{2/\sqrt{\lambda}} \int_{-\sqrt{\lambda}}^{\sqrt{\lambda}} \frac{\sin^2 \alpha}{\alpha^2} d\alpha}.$$

Letting $\lambda \to \infty$, it follows that

$$\liminf_{v \to \infty} g(v) \geq 1.$$

We therefore conclude that $\lim_{v \to \infty} g(v) = 1$. This implies that

$$\lim_{x \to \infty} \frac{B(x)}{x} = 1,$$

or equivalently

$$B(x) = x + o(x).$$

□

We apply this theorem to the Dirichlet series

$$-\frac{\zeta'}{\zeta}(s) = \sum_{n=1}^{\infty} \frac{\Lambda(n)}{n^s},$$

which has nonnegative coefficients and is absolutely convergent for $\mathrm{Re}(s) > 1$. By virtue of $\zeta(s) \neq 0$ on $\mathrm{Re}(s) = 1$, we see that $-\frac{\zeta'}{\zeta}(s)$ extends to a meromorphic function that has a simple pole at $s = 1$ with residue 1. Indeed, we know that

$$h(s) := (s-1)\zeta(s)$$

is analytic at $s = 1$ and $h(1) = 1$. Moreover, as $\zeta(s) \neq 0$ on $\mathrm{Re}(s) \geq 1$, we get by logarithmic differentiation:

$$\frac{h'(s)}{h(s)} = \frac{1}{s-1} + \frac{\zeta'}{\zeta}(s),$$

for which our assertion is obvious. Applying the Ikehara - Wiener theorem, we obtain the prime number theorem:

Theorem 3.3.2 (The Prime Number Theorem) *Let*

$$\psi(x) = \sum_{n \leq x} \Lambda(n).$$

Then

$$\lim_{n \to \infty} \frac{\psi(x)}{x} = 1.$$

Exercise 3.3.3 *Suppose*

$$f(s) = \sum_{n=1}^{\infty} a_n / n^s$$

is a Dirichlet series with real coefficients that is absolutely convergent for $\mathrm{Re}(s) > 1$. *If* $f(s)$ *extends to a meromorphic function in the region* $\mathrm{Re}(s) \geq 1$, *with only a simple pole at* $s = 1$ *with residue* r, *and* $|a_n| \leq b_n$ *where* $F(s) = \sum_{n=1}^{\infty} b_n / n^s$ *satisfies the hypotheses of Theorem 3.3.1, show that*

$$\sum_{n \leq x} a_n = rx + o(x)$$

as $x \to \infty$.

Exercise 3.3.4 *Show that the conclusion of the previous exercise is still valid if* $a_n \in \mathbb{C}$.

Exercise 3.3.5 *Let* q *be a natural number. Suppose* $(a, q) = 1$. *Show that*

$$\psi(x; q, a) := \sum_{\substack{n \leq x \\ n \equiv a \pmod{q}}} \Lambda(n)$$

satisfies

$$\lim_{x \to \infty} \frac{\psi(x)}{x / \varphi(q)} = 1.$$

Exercise 3.3.6 *Suppose* $F(s) = \sum_{n=1}^{\infty} b_n / n^s$ *is a Dirichlet series with non-negative coefficients and is convergent for* $\mathrm{Re}(s) > c > 0$. *If* $F(s)$ *extends to a meromorphic function in the region* $\mathrm{Re}(s) \geq c$ *with only a simple pole at* $s = c$ *with residue* R, *show that*

$$\sum_{n \leq x} b_n = \frac{Rx^c}{c} + o(x^c)$$

as $x \to \infty$.

Exercise 3.3.7 *Suppose* $f(s) = \sum_{n=1}^{\infty} a_n/n^s$ *is a Dirichlet series with complex coefficients that is absolutely convergent for* $\mathrm{Re}(s) > c$. *If* $f(s)$ *extends to a meromorphic function in the region* $\mathrm{Re}(s) \geq c$ *with only a simple pole at* $s = c$ *and residue* r, *and* $|a_n| \leq b_n$ *where* $f(s) = \sum_{n=1}^{\infty} b_n/n^s$ *satisfies the hypothesis of Exercise 3.3.6, show that*

$$\sum_{n \leq x} a_n = \frac{rx^c}{c} + o(x^c)$$

as $x \to \infty$.

Exercise 3.3.8 *Let* $a(n)$ *be a multiplicative function defined by* $a(1) = 1$ *and*

$$a(p^\alpha) = \begin{cases} p + c_p & \text{if } \alpha = 1, \\ 0 & \text{otherwise}, \end{cases}$$

where $|c_p| \leq p^\theta$ *with* $\theta < 1$. *Show that as* $x \to \infty$,

$$\sum_{n \leq x} a(n) = \frac{rx^2}{2} + o(x^2)$$

for some non-zero constant r.

Exercise 3.3.9 *Suppose* $c_n \geq 0$ *and that*

$$\sum_{n \leq x} c_n = Ax + o(x).$$

Show that

$$\sum_{n \leq x} \frac{c_n}{n} = A \log x + o(\log x)$$

as $x \to \infty$.

3.4 Supplementary Problems

Exercise 3.4.1 *Show that*

$$\sum_{n \leq x} \Lambda(n) \log n = \psi(x) \log x + O(x).$$

Exercise 3.4.2 *Show that*

$$\sum_{d|n} \Lambda(d)\Lambda\Big(\frac{n}{d}\Big) = \Lambda(n)\log n + \sum_{d|n} \mu(d)\log^2 d.$$

Exercise 3.4.3 *Show that*

$$\sum_{d|n} \mu(d)\log^2 \frac{x}{d} =$$

$$\begin{cases} \log^2 x & \text{if} \quad n = 1, \\ 2\Lambda(n)\log x - \Lambda(n)\log n + \sum_{hk=n} \Lambda(h)h(k) & \text{if} \quad n > 1. \end{cases}$$

Exercise 3.4.4 *Let*

$$S(x) = \sum_{n \le x} \Big(\sum_{d|n} \mu(d)\log^2 \frac{x}{d} \Big).$$

Show that

$$S(x) = \psi(x)\log x + \sum_{n \le x} \Lambda(n)\psi\Big(\frac{x}{n}\Big) + O(x).$$

Exercise 3.4.5 *Show that*

$$S(x) - \gamma^2 = \sum_{d \le x} \mu(d)\Big[\frac{x}{d}\Big]\Big\{\log^2 \frac{x}{d} - \gamma^2\Big\},$$

where γ is Euler's constant.

Exercise 3.4.6 *Show that*

$$S(x) = x\sum_{d \le x} \frac{\mu(d)}{d}\Big\{\log^2 \frac{x}{d} - \gamma^2\Big\} + O(x).$$

Exercise 3.4.7 *Using the fact*

$$\sum_{n \le x} \frac{1}{n} = \log x + \gamma + O\Big(\frac{1}{x}\Big),$$

deduce that

$$\frac{S(x)}{x} = \sum_{de \le x} \frac{\mu(d)}{de}\Big(\log \frac{x}{d} - \gamma\Big) + O(1).$$

Exercise 3.4.8 *Prove that*

$$\frac{S(x)}{x} = 2\log x + O(1).$$

Exercise 3.4.9 (Selberg's identity) *Prove that*

$$\psi(x)\log x + \sum_{n \le x} \Lambda(n)\psi\Big(\frac{x}{n}\Big) = 2x\log x + O(x).$$

Exercise 3.4.10 *Show that*

$$\nu(n) = O\Big(\frac{\log n}{\log\log n}\Big),$$

where $\nu(n)$ denotes the number of distinct prime factors of n.

Exercise 3.4.11 *Let $\nu(n)$ be as in the previous exercise. Show that*

$$\sum_{n \le x} \nu(n) = x\log\log x + O(x).$$

Exercise 3.4.12 *Let $\nu(n)$ be as in the previous exercise. Show that*

$$\sum_{n \le x} \nu^2(n) = x(\log\log x)^2 + O(x\log\log x).$$

Exercise 3.4.13 *Prove that*

$$\int_{-\infty}^{\infty} \frac{\sin^2 \lambda x}{\lambda x^2}\,dx = \pi.$$

Exercise 3.4.14 *Let*

$$T(x) := \sum_{n \le x} \log n.$$

Show that for $x > 1$,

$$|T(x) - (x\log x - x)| \le 4 + \log(x+1).$$

Exercise 3.4.15 *Show that*

$$\psi(x) - \psi\Big(\frac{x}{2}\Big) \le (\log 2)x + 12 + 3\log(x+1).$$

Deduce that

$$\psi(x) \le 2(\log 2)x + \frac{12\log x}{\log 2} + \frac{3\log(x+1)\log x}{\log 2}.$$

Exercise 3.4.16 *Show that*

$$\psi(x) - \psi\left(\frac{x}{2}\right) + \psi\left(\frac{x}{3}\right) \geq (\log 2)x - 2\log(x+1) - 7.$$

Exercise 3.4.17 *Prove that for $x \geq e^{12}$,*

$$\psi(x) - \psi\left(\frac{x}{2}\right) \geq \frac{1}{3}(\log 2)x - \frac{5(\log x)\log(x+1)}{\log 2} - 7.$$

Exercise 3.4.18 *Find an explicit constant c_0 such that for $x \geq c_0$,*

$$\psi(x) - \psi\left(\frac{x}{2}\right) > \frac{(\log 2)x}{6} - 7.$$

Exercise 3.4.19 *With c_0 as in the previous exercise, show that for $x \geq c_0$,*

$$\theta(x) - \theta\left(\frac{x}{2}\right) > \frac{(\log 2)x}{6} - \frac{\sqrt{x}(\log x)^2}{\log 2} - 7.$$

Exercise 3.4.20 *Find an explicit constant c_1 such that for $x \geq c_1$,*

$$\theta(x) - \theta\left(\frac{x}{2}\right) > \frac{(\log 2)x}{12} - 7.$$

Exercise 3.4.21 *Find an explicit constant c_3 such that for $x \geq c_3$, $\theta(x) - \theta(x/2) \geq 1$. Deduce that for $x \geq c_3$, there is always a prime between $x/2$ and x.*

Exercise 3.4.22 *Let*

$$F(x) = \sum_{n \leq x} f\left(\frac{x}{n}\right)$$

be a function of bounded variation in every finite interval $[1, x]$. Suppose that as $x \to \infty$,

$$F(x) = x\log x + Cx + O(x^\beta)$$

with C, β constant and $0 \leq \beta < 1$. Show that if $M(x) := \sum_{n \leq x} \mu(n) = o(x)$ as $x \to \infty$, then

$$f(x) = x + o(x).$$

Exercise 3.4.23 *Assuming $M(x) = o(x)$ as in the previous exercise, deduce that*

$$\lim_{x \to \infty} \frac{\psi(x)}{x} = 1.$$

4
The Method of Contour Integration

Given a sequence of complex numbers $\{a_n\}_{n=1}^{\infty}$, one would like to study the asymptotic behavior of

$$A(x) := \sum_{n \le x} a_n$$

as $x \to \infty$. A standard method of analytic number theory is to study instead the associated Dirichlet series

$$f(s) := \sum_{n=1}^{\infty} \frac{a_n}{n^s},$$

derive an analytic continuation to a region containing the line $\text{Re}(s) = 1$, and then apply methods of contour integration to deduce an asymptotic formula for $A(x)$.

4.1 Some Basic Integrals

We shall adopt the following notation:

$$\frac{1}{2\pi i} \int_{c-i\infty}^{c+i\infty} f(s) ds$$

will be abbreviated to

$$\frac{1}{2\pi i}\int_{(c)} f(s)ds.$$

This integral must be interpreted in the principal value sense. That is, we first integrate from $c - iR$ to $c + iR$ and take the limit as R goes to infinity.

Exercise 4.1.1 *If $x > 1$, show that*

$$\frac{1}{2\pi i}\int_{(c)} \frac{x^s}{s} ds = 1$$

for any $c > 0$.

Exercise 4.1.2 *If $0 < x < 1$, show that*

$$\frac{1}{2\pi i}\int_{(c)} \frac{x^s}{s} ds = 0, \qquad c > 0.$$

Exercise 4.1.3 *Show that*

$$\frac{1}{2\pi i}\int_{(c)} \frac{ds}{s} = \frac{1}{2}, \qquad c > 0.$$

We summarize the previous examples and exercises in the following. If

$$\delta(x) = \begin{cases} 0 & \text{if} \quad 0 < x < 1, \\ \frac{1}{2} & \text{if} \quad x = 1, \\ 1 & \text{if} \quad x > 1, \end{cases}$$

then

$$\delta(x) = \frac{1}{2\pi i}\int_{c-i\infty}^{c+i\infty} \frac{x^s}{s} ds.$$

Theorem 4.1.4 *Let $\delta(x)$ be defined as above. Let*

$$I(x, R) = \frac{1}{2\pi i}\int_{c-iR}^{c+iR} \frac{x^s}{s} ds.$$

Then, for $x > 0, c > 0, R > 0$, we have

$$|I(x,R) - \delta(x)| < \begin{cases} x^c \min(1, R^{-1}|\log x|^{-1}) & \textit{if} \quad x \neq 1, \\ \dfrac{c}{R} & \textit{if} \quad x = 1. \end{cases}$$

Proof. Suppose first $0 < x < 1$. Consider the rectangular contour K_U oriented counterclockwise with vertices $c - iR, c + iR, U + iR, U - iR, U > 0$. By Cauchy's theorem

$$\frac{1}{2\pi i}\int_{K_U} \frac{x^s}{s} ds = 0 = \delta(x).$$

To prove the theorem, we must estimate the three integrals

$$\frac{1}{2\pi i}\int_{c+iR}^{U+iR} \frac{x^s}{s} ds, \quad \frac{1}{2\pi i}\int_{c-iR}^{U-iR} \frac{x^s}{s} ds, \quad \frac{1}{2\pi i}\int_{U-iR}^{U+iR} \frac{x^s}{s} ds.$$

Now,

$$\left|\int_{c+iR}^{U+iR} \frac{x^s}{s} ds\right| \leq \frac{1}{R}\int_c^U x^\delta d\delta.$$

As $U \to \infty$, this integral is bounded by

$$\frac{x^c}{R|\log x|}.$$

A similar estimate holds for the other integral. Now,

$$\left|\frac{1}{2\pi i}\int_{U-iR}^{U+iR} \frac{x^s ds}{s}\right| \leq \frac{x^U R}{U},$$

which goes to zero as $U \to \infty$, since $0 < x < 1$. This proves one of the two stated inequalities in the case $0 < x < 1$. For the other inequality, consider the circle of radius $(c^2 + R^2)^{1/2}$ centered at the origin. This circle passes through $c - iR$ and $c + iR$. We can therefore replace the vertical line integral under consideration by a circular path on the right side of the line segment joining $c - iR$ to $c + iR$. The integral is easily estimated:

$$|I(x,R)| \leq \frac{1}{2\pi}\pi R \cdot \frac{x^c}{R} < x^c,$$

since $|x^s| \leq x^c$ on the circular path.

The proof when $x > 1$ is similar but uses a rectangle or a circular arc to the left. The contour then includes the pole at $s = 0$, where the residue is $1 = \delta(x)$. We leave the details as an exercise to the reader.

Finally, the case $x = 1$ is handled directly as in Exercise 4.1.3. We have

$$\frac{1}{2\pi i}\int_{c-iR}^{c+iR}\frac{ds}{s} = \frac{c}{\pi}\int_0^R \frac{dt}{c^2+t^2},$$

which equals

$$\frac{1}{\pi}\int_0^{R/c}\frac{du}{1+u^2} = \frac{1}{2} - \frac{1}{\pi}\int_{R/c}^{\infty}\frac{du}{1+u^2}.$$

The last integral is less than c/R, and this proves the theorem. □

Exercise 4.1.5 *Let*

$$f(s) = \sum_{n=1}^{\infty}\frac{a_n}{n^s}$$

be a Dirichlet series absolutely convergent in $\mathrm{Re}(s) > c - \epsilon$. *Show that if* x *is not an integer, then*

$$\sum_{n<x} a_n = \frac{1}{2\pi i}\int_{(c)} f(s)\frac{x^s}{s}ds.$$

(The integral is taken in the sense of Cauchy's principal value.)

Exercise 4.1.6 *Prove that for* $c > 0$,

$$\frac{1}{2\pi i}\int_{(c)}\frac{x^s}{s^{k+1}}ds = \begin{cases} \frac{1}{k!}(\log x)^k & \text{if } x \geq 1, \\ 0 & \text{if } x \leq 1, \end{cases}$$

for every integer $k \geq 1$.

Exercise 4.1.7 *Let*

$$f(s) = \sum_{n=1}^{\infty}\frac{a_n}{n^s}$$

be a Dirichlet series absolutely convergent in $\mathrm{Re}(s) > c - \epsilon$. *For* $k \geq 1$, *show that*

$$\frac{1}{k!}\sum_{n\le x} a_n\left(\log\frac{x}{n}\right)^k = \frac{1}{2\pi i}\int_{(c)} f(s)\frac{x^s}{s^{k+1}}ds.$$

Exercise 4.1.8 *If k is any positive integer, $c > 0$, show that*

$$\frac{1}{2\pi i}\int_{(c)}\frac{x^s ds}{s(s+1)\cdots(s+k)} = \begin{cases} \frac{1}{k!}\left(1-\frac{1}{x}\right)^k & \text{if } x \ge 1, \\ 0 & \text{if } 0 \le x \le 1. \end{cases}$$

Exercise 4.1.9 *Let*

$$f(s) = \sum_{n=1}^{\infty}\frac{a_n}{n^s}$$

be a Dirichlet series absolutely convergent in $\mathrm{Re}(s) > c - \epsilon$. *Show that*

$$\frac{1}{x^k}\sum_{n\le x} a_n(x-n)^k = \frac{k!}{2\pi i}\int_{c-i\infty}^{c+i\infty}\frac{f(s)x^s ds}{s(s+1)\cdots(s+k)}$$

for any $k \ge 1$.

4.2 The Prime Number Theorem

We will use the ideas of the previous section to give another proof of the prime number theorem. Our derivation is illustrative of a general method of contour integration to derive such formulas. Thus, it can be applied in other contexts. The method also has the advantage of giving an explicit error term.

Our strategy is to begin with the formula

$$\psi(x) := \sum_{n\le x}\Lambda(n) = \frac{1}{2\pi i}\int_{(2)} -\frac{\zeta'}{\zeta}(s)\frac{x^s}{s}ds,$$

which is valid when x is not an integer. We will then move the line of integration to the left and pick up the residue at $s = 1$ coming from the simple pole of $-\zeta'(s)/\zeta(s)$. This residue is x, which is the main term in the formula for $\psi(x)$. Our contour will not include $s = 0$ nor any of the zeros of $-\zeta'(s)/\zeta(s)$, and so the error term comes from estimating the horizontal and vertical integrals of the contour.

Exercise 4.2.1 *Using the Euler - Maclaurin summation formula (Theorem 2.1.9), prove that for $\sigma = \mathrm{Re}(s) > 0$,*

$$\zeta(s) = \sum_{m=1}^{n-1} \frac{1}{m^s} + \frac{n^{-s}}{2} + \frac{n^{1-s}}{s-1} - s\int_n^\infty \frac{x - [x] - \frac{1}{2}}{x^{s+1}} dx,$$

where $[x]$ denotes the greatest integer function.

We will now study $\zeta(s)$ in the region R_T described by the rectangle joining $2 - iT$, $2 + iT$, $\sigma_0 + iT$, $\sigma_0 - iT$, where $\sigma_0 = 1 - 1/\log T$, $T \geq e^2$.

Exercise 4.2.2 *Using the previous exercise, show that*

$$\zeta(s) - \frac{1}{s-1} = O(\log T)$$

for $s \in R_T$.

Exercise 4.2.3 *Show that*

$$\zeta(s) = O(\log T)$$

for s on the boundary of R_T.

Exercise 4.2.4 *Show that for $\sigma \geq \frac{1}{2}$, $\zeta(s) = O(T^{1/2})$, where $T = |\mathrm{Im}(s)| \to \infty$.*

Exercise 4.2.5 *For $s \in R_T$, show that*

$$\zeta'(s) + \frac{1}{(s-1)^2} = O(\log^2 T).$$

Exercise 4.2.6 *Show that*

$$\zeta'(s) = O(\log^2 T),$$

where $T = |\mathrm{Im}(s)|$ and s is on the boundary of R_T.

The method used to show that $\zeta(s) \neq 0$ for $\mathrm{Re}(s) = 1$ can be sharpened to yield a region in which $\zeta(s) \neq 0$.

Theorem 4.2.7 *Let $s = \sigma + it$. There are positive constants c_1 and c_2 such that*

$$1 - \frac{c_1}{(\log T)^9} \leq \sigma \leq 2$$

$$|\zeta(s)| > \frac{c_2}{(\log T)^7}$$

where $1 \leq |\mathrm{Im}(s)| \leq T$.

Proof. In Exercise 3.2.5, we proved

$$|\zeta(\sigma)^3\zeta(\sigma+it)^4\zeta(\sigma+2it)| \geq 1$$

for $\sigma > 1$. Thus,

$$|\zeta(\sigma+it)|^4 \geq |\zeta(\sigma+2it)|^{-1}|\zeta(\sigma)|^{-3}.$$

Now, $\zeta(\sigma)(\sigma-1)$ remains bounded as $\sigma \to 1^+$ and being continuous for $1 \leq \sigma \leq 2$ has an upper bound in that region. By Exercise 4.2.3, for some constant K,

$$|\zeta(\sigma+2it)|^{-1} \geq K(\log T)^{-1}.$$

Thus we get

$$|\zeta(\sigma+it)|^4 \geq K_1(\log T)^{-1}(\sigma-1)^3.$$

If

$$1+\frac{c_1}{(\log T)^9} \leq \sigma \leq 2,$$

then we obtain

$$|\zeta(s)| \gg \frac{1}{(\log T)^7}$$

in this region. We can extend this result to the region

$$1-\frac{c_1}{(\log T)^9} \leq \sigma \leq 1+\frac{c_1}{(\log T)^9}$$

and $1 \leq |\operatorname{Im}(s)| \leq T$, by using the mean value theorem. Indeed, choose s' such that $s' = \sigma' + it$, with

$$\sigma' = 1+\frac{c_1}{(\log T)^9}.$$

Then

$$\zeta(\sigma'+it)-\zeta(\sigma+it) = O((\sigma-\sigma')\log^2 T)$$

by an application of the mean value theorem and Exercise 4.2.5. Thus, if c_1 is chosen sufficiently small, we obtain

$$|\zeta(s)| \gg (\log T)^{-7}.$$

□

Exercise 4.2.8 *Let $s = \sigma + it$, with $1 \leq |t| \leq T$. There is a constant $c > 0$ such that*

$$\frac{\zeta'(s)}{\zeta(s)} = O(\log^9 T)$$

for

$$1 - \frac{c}{(\log T)^9} \leq \sigma \leq 2.$$

We can now prove the prime number theorem in the following form:

Theorem 4.2.9 *Let*

$$\psi(x) = \sum_{n \leq x} \Lambda(n).$$

Then

$$\psi(x) = x + O\left(x \exp\left(-c(\log x)^{1/10}\right)\right)$$

for some positive constant c.

Proof. We have for x which is $1/2$ more than a natural number,

$$\psi(x) = \frac{1}{2\pi i} \int_{(a)} -\frac{\zeta'}{\zeta}(s) \frac{x^s}{s} ds$$

for any $a > 1$. We choose $a = 1 + c/\log^9 T$, with $T \geq 1$ to be determined later. By Theorem 4.1.4, we can replace the infinite line integral by the finite line integral:

$$\begin{aligned} \psi(x) &= \frac{1}{2\pi i} \int_{a-iT}^{a+iT} -\frac{\zeta'}{\zeta}(s) \frac{x^s}{s} ds \\ &\quad + O\Big(\sum_{n=1}^{\infty} \left(\frac{x}{n}\right)^a \Lambda(n) \min\left(1, T^{-1} \left|\log \frac{x}{n}\right|^{-1}\right)\Big). \end{aligned}$$

The O-term is estimated as follows:

if $n < \frac{x}{2}$ or $n > \frac{3x}{2}$, then $|\log \frac{x}{n}| > \log \frac{3}{2}$, and the summation corresponding to such n is bounded by

$$O\Big(\frac{x^a}{T}(a-1)^{-1}\Big) = O\Big(\frac{x^a \log^9 T}{T}\Big),$$

since

$$\sum_{n=1}^{\infty} \frac{\Lambda(n)}{n^a} \ll (a-1)^{-1}$$

for any fixed $a > 1$. For $\frac{x}{2} < n < \frac{3x}{2}$, we put $z = 1 - \frac{n}{x}$ and observe that $|z| \leq 1/2$. Also,

$$\log \frac{x}{n} = -\log(1-z) = z + \frac{z^2}{2} + \cdots,$$

so that for $|z| \leq 1/2$,

$$\left|\log \frac{x}{n}\right| \geq \frac{3}{4}|z|.$$

Thus, for the summation corresponding to this range, we get the estimate

$$(\log x) \sum_{x/2 \leq n \leq 3x/2} 2^a \frac{x}{T|x-n|} \ll \frac{x}{T}(\log x)^2$$

since $|x-n|$ ranges over $\frac{1}{2}, \frac{3}{2}, \cdots, \frac{1}{2} + \frac{[x]}{2}$. Therefore, the O-term is

$$O\Big(x^a \frac{\log^9 T}{T} + \frac{x \log^2 x}{T}\Big).$$

Now, $-\zeta'(s)/\zeta(s)$ has a simple pole at $s = 1$ with residue 1. By Cauchy's theorem,

$$\begin{aligned} &\frac{1}{2\pi i}\int_{a-iT}^{a+iT} -\frac{\zeta'}{\zeta}(s)\frac{x^s}{s}ds \\ &\qquad = x - \frac{1}{2\pi i}\left\{\left(\int_{a+iT}^{b+iT} + \int_{b+iT}^{b-iT} + \int_{b-iT}^{a-iT}\right) - \frac{\zeta'}{\zeta}(s)\frac{x^s}{s}ds\right\}, \end{aligned}$$

where $b = 1 - \frac{c}{\log^9 T}$. The integrals in the above formula are easily estimated using Exercise 4.2.8. Indeed,

$$\left|\frac{1}{2\pi i}\int_{a+iT}^{b+iT} -\frac{\zeta'}{\zeta}(s)\frac{x^s}{s}ds\right| \ll \frac{x^a \log^9 T}{T}$$

with a similar estimate for

$$\frac{1}{2\pi i}\int_{b-iT}^{a+iT} -\frac{\zeta'}{\zeta}(s)\frac{x^s}{s}ds.$$

Also,

$$\left|\int_{b+iT}^{b-iT} -\frac{\zeta'}{\zeta}(s)\frac{x^s}{s}ds\right| \ll x^b \log^{10} T.$$

Therefore,

$$\psi(x) = x + O\left(\frac{x^a \log^9 T}{T} + x^b \log^{10} T + \frac{x(\log x)^2}{T}\right).$$

We choose T such that

$$2c \log x = \log^{10} T.$$

The error term becomes

$$O(x \exp(-c_1 (\log x)^{1/10}))$$

for some constant $c_1 > 0$. This completes the proof. □

In a later chapter we will see that this error term can be improved to

$$O\left(x \exp\left(-c_2 (\log x)^{1/2}\right)\right)$$

for some constant $c_2 > 0$. This can be further improved but not substantially. The Riemann hypothesis would give an estimate of

$$O(x^{1/2} \log^2 x).$$

4.3 Further Examples

The technique introduced in the last two sections can be used to treat other questions. We illustrate this through some examples.

Example 4.3.1 *Prove that*

$$\sum_{n \leq x} d^2(n) \log^3 \frac{x}{n} = x P_3(\log x) + O\left(x^{1/2}\right),$$

where $P_3(t)$ is a polynomial of degree 3 and $d(n)$ denotes the number of divisors of n.

Solution. By Exercise 1.2.8, we have

$$\frac{\zeta^4(s)}{\zeta(2s)} = \sum_{n=1}^{\infty} \frac{d^2(n)}{n^s}.$$

Thus, by Exercise 4.1.6 (with $k = 3$), we have

$$\frac{1}{3!}\sum_{n\leq x} d^2(n)\log^3\frac{x}{n} = \frac{1}{2\pi i}\int_{(a)}\frac{\zeta^4(s)}{\zeta(2s)}\frac{x^s}{s^4}ds,$$

where $a > 1$. We first truncate the infinite line integral at R and estimate the portion of the integral from $-\infty$ to $-R$ and from R to ∞. By Exercise 4.2.4, we have $\zeta^4(s) = O(|t|^2)$, so that

$$\frac{1}{2\pi i}\int_{(a)}\frac{\zeta^4(s)}{\zeta(2s)}\frac{x^s}{s^4}ds = \frac{1}{2\pi i}\int_{a-iR}^{a+iR}\frac{\zeta^4(s)}{\zeta(2s)}\frac{x^s}{s^4}ds + O\Big(\frac{x^a}{R}\Big).$$

Now let C be the rectangular contour joining $a - iR$, $a + iR$, $\frac{1}{2} + iR$, and $\frac{1}{2} - iR$. By Cauchy's theorem,

$$\frac{1}{2\pi i}\int_{C}\frac{\zeta^4(s)}{\zeta(2s)}\frac{x^s}{s^4}ds = \mathrm{Res}_{s=1}\frac{\zeta^4(s)x^s}{\zeta(2s)s^4}.$$

Since

$$\zeta(s) = \frac{1}{s-1} + c_0 + c_1(s-1) + \cdots$$

it is easily seen that

$$\mathrm{Res}_{s=1}\frac{\zeta^4(s)x^s}{\zeta(2s)s^4} = xP_3(\log x)$$

for some polynomial $P_3(t)$ of degree 3. Now we can write

$$\frac{1}{2\pi i}\int_{C}\frac{\zeta^4(s)}{\zeta(2s)}\frac{x^s}{s^4}ds = V_a + H_+ - H_- - V_{1/2},$$

where

$$H_\pm = \frac{1}{2\pi i}\int_{a\pm iR}^{1/2\pm iR}\frac{\zeta^4(s)}{\zeta(2s)}\frac{x^s}{s^4}ds$$

and

$$V_\sigma = \frac{1}{2\pi i}\int_{\sigma-iR}^{\sigma+iR}\frac{\zeta^4(s)}{\zeta(2s)}\frac{x^s}{s^4}ds.$$

The horizontal integrals $H_\pm$ can be bounded using Exercise 4.2.4 and Theorem 4.2.7 to give

$$O\left(\frac{x^a\log^7 R}{R^2\log x}\right).$$

For the vertical integral $V_{1/2}$, we have

$$V_{1/2} \ll x^{1/2}.$$

Choosing $a = 1 + 1/\log x$, we obtain that the sum in question is

$$xP_3(\log x) + O\left(\frac{x \log^7 R}{R^2 \log x}\right) + O\left(x^{1/2}\right).$$

Choosing $R = x$ gives an error term of $O\left(x^{1/2}\right)$ as stated. □

Exercise 4.3.2 *Suppose that for any $\epsilon \geq 0$, we have $a_n = O(n^\epsilon)$. Prove that for any $c > 1$ and x not an integer,*

$$\sum_{n \leq x} a_n = \frac{1}{2\pi i} \int_{c-iR}^{c+iR} \frac{f(s)x^s}{s} ds + O\Big(\frac{x^{c+\epsilon}}{R}\Big) + O\Big(\frac{x^\epsilon \log x}{R}\Big),$$

where

$$f(s) = \sum_{n=1}^{\infty} \frac{a_n}{n^s}.$$

The Lindelöf hypothesis is the assertion that for every $\epsilon > 0$, $\zeta(s) = O(t^\epsilon)$ for $\text{Re}(s) \geq \frac{1}{2}$. One can show that it follows from the Riemann hypothesis. It is, however, a substantially weaker conjecture, which still remains unproved.

Exercise 4.3.3 *Assuming the Lindelöf hypothesis, prove that for any $\epsilon>0$,*

$$\sum_{n \leq x} d_k(n) = xP_{k-1}(\log x) + O(x^{1/2+\epsilon}),$$

where $d_k(n)$ denotes the number of ways of writing n as a product of k natural numbers.

Exercise 4.3.4 *Show that*

$$M(x) := \sum_{n \leq x} \mu(n) = O\left(x \exp\left(-c(\log x)^{1/10}\right)\right)$$

for some positive constant c.

Exercise 4.3.5 *Let $E(x)$ be the number of square-free $n \leq x$ with an even number of prime factors. Prove that*

$$E(x) = \frac{3}{\pi^2} x + O\left(x \exp\left(-c(\log x)^{1/10}\right)\right)$$

for some constant $c > 0$.

4.4 Supplementary Problems

Exercise 4.4.1 *Let $\lambda(n)$ be the Liouville function defined by $\lambda(n) = (-1)^{\Omega(n)}$ where $\Omega(n)$ is the total number of prime factors of n, counted with multiplicity. Show that*

$$\sum_{n \leq x} \lambda(n) = O\left(x \exp\left(-c(\log x)^{1/10}\right)\right)$$

for some constant $c > 0$.

Exercise 4.4.2 *Show that*

$$\sum_{n=1}^{\infty} \frac{\mu(n)}{n^s}$$

converges for every s with $\mathrm{Re}(s) = 1$.

Exercise 4.4.3 *Show that*

$$\sum_{n \leq x} \frac{\Lambda(n)}{n} = \log x + B + O\left(\exp\left(-c(\log x)^{1/10}\right)\right)$$

for some constants B and c, with $c > 0$. [This improves upon Exercise 3.1.7.]

Exercise 4.4.4 *Let $f(s) = \sum_{n=1}^{\infty} A_n/n^s$ be a Dirichlet series absolutely convergent for* $\mathrm{Re}(s) > 1$. *Show that for any $c > 1$,*

$$\sum_{n \leq x} A_n = O(x^c).$$

Exercise 4.4.5 *Define a_n for $n \geq 1$ by*

$$\sum_{n=1}^{\infty} \frac{a_n}{n^s} = \frac{1}{\zeta^2(s)}.$$

Prove that

$$\sum_{n \leq x} a_n = O\left(x \exp\left(-c(\log x)^{1/10}\right)\right)$$

for some positive constant c.

Exercise 4.4.6 *Prove that*

$$\sum_{n \leq x} \mu(n)d(n) = O\left(x \exp\left(-c(\log x)^{1/10}\right)\right)$$

for some constant $c > 0$.

Exercise 4.4.7 *If* $f(s) = \sum_{n=1}^{\infty} a_n/n^s$ *is a Dirichlet series converging absolutely for* $\sigma = \mathrm{Re}(s) = \sigma_a$, *show that*

$$\lim_{T \to \infty} \frac{1}{2T} \int_{-T}^{T} f(\sigma + it) m^{\sigma + it} dt = a_m.$$

Exercise 4.4.8 *Suppose*

$$f(s) := \sum_{n=1}^{\infty} a_n/n^s,$$

$$g(s) := \sum_{n=1}^{\infty} b_n/n^s,$$

and $f(s) = g(s)$ *in a half-plane of absolute convergence. Then prove that* $a_n = b_n$ *for all* n.

Exercise 4.4.9 *If*

$$f(s) = \sum_{n=1}^{\infty} a_n/n^s$$

converges absolutely for $\sigma = \mathrm{Re}(s) > \sigma_a$, *show that*

$$\lim_{T \to \infty} \frac{1}{2T} \int_{-T}^{T} |f(\sigma + it)|^2 dt = \sum_{n=1}^{\infty} \frac{|a_n|^2}{n^{2\sigma}}.$$

Exercise 4.4.10 *Let* $Q(x)$ *be the number of square-free numbers less than or equal to* x. *Show that*

$$Q(x) = \frac{x}{\zeta(2)} + O\left(x^{1/2} \exp\left(-c(\log x)^{1/10}\right)\right)$$

for some positive constant c.

Exercise 4.4.11 *Let* $\gamma(n) = \prod_{p|n} p$. *Show that*

$$\sum_{n \leq x} \frac{1}{n\gamma(n)} < \infty.$$

Exercise 4.4.12 *Show that*

$$\sum_{n\le x}\frac{n}{\phi(n)} \ll x.$$

Exercise 4.4.13 *Deduce by partial summation from the previous exercise that*

$$\sum_{n\le x}\frac{1}{\phi(n)} \ll \log x.$$

Exercise 4.4.14 *Prove that*

$$\sum_{n\le x}\frac{1}{\phi(n)} \sim c\log x$$

for some positive constant c.

Exercise 4.4.15 (Perron's formula) *Let* $f(s) = \sum_{n=1}^{\infty} a_n/n^s$ *be a Dirichlet series absolutely convergent for* $\mathrm{Re}(s) > 1$. *Show that for* x *not an integer and* $\sigma > 1$,

$$\sum_{n\le x} a_n$$

$$= \frac{1}{2\pi i}\int_{\sigma-iT}^{\sigma+iT} f(s)\frac{x^s}{s}ds + O\left(\sum_{n=1}^{\infty}\left(\frac{x}{n}\right)^{\sigma}|a_n|\min\left(1,\frac{1}{T|\log\frac{x}{n}|}\right)\right).$$

Exercise 4.4.16 *Suppose* $a_n = O(n^\epsilon)$ *for any* $\epsilon > 0$ *in the previous exercise. Show that for* x *not an integer,*

$$\sum_{n\le x} a_n = \frac{1}{2\pi i}\int_{\sigma-iT}^{\sigma+iT} f(s)\frac{x^s}{s}ds + O\left(\frac{x^{\sigma+\epsilon}}{T}\right).$$

Exercise 4.4.17 *Let* $f(s) = \sum_{n=1}^{\infty} a_n/n^s$, *with* $a_n = O(n^\epsilon)$. *Suppose that*

$$f(s) = \zeta(s)^k g(s),$$

where k *is a natural number and* $g(s)$ *is a Dirichlet series absolutely convergent in* $\mathrm{Re}(s) > 1-\delta$ *for some* $0 < \delta < 1$. *Show that*

$$\sum_{n\le x} a_n \sim g(1)x(\log x)^{k-1}/(k-1)!$$

as $x \to \infty$.

Exercise 4.4.18 *Let $\nu(n)$ denote the number of distinct prime factors of n. Show that*

$$\sum_{n \le x} 2^{\nu(n)} \sim \frac{x \log x}{\zeta(2)}$$

as $x \to \infty$.

5

Functional Equations

In this chapter we will derive the functional equations of $\zeta(s)$ and Dirichlet's $L(s,\chi)$. Our main tools will be the Poisson summation formula and the theory of Fourier transforms.

5.1 Poisson's Summation Formula

Let us recall Fejér's fundamental theorem concerning Fourier series. Let $f(x)$ be a function of a real variable that is bounded, measurable, and periodic with period 1. The Fourier coefficients of f are, by definition, given by

$$c_n = \int_0^1 f(x)e^{-2\pi inx}\,dx,$$

for each $n \in \mathbb{Z}$. The partial sums of the Fourier series of f are defined as

$$S_N(x) = \sum_{|n|\leq N} c_n e^{2\pi inx}.$$

Let $x_0 \in \mathbb{R}$ be such that the function $f(x)$ admits left and right limits:

$$f(x_0 \pm 0) = \lim_{h\to 0+} f(x_0 \pm h).$$

Then Fejér proved

$$\frac{f(x_0+0)+f(x_0-0)}{2} = \lim_{N\to\infty} \frac{S_0(x_0)+\cdots+S_N(x_0)}{N+1}.$$

If $f(x)$ is continuous at x_0, and the partial sums $S_N(x_0)$ converge, then

$$f(x_0) = c_0 + \sum_{n=1}^{\infty} \left(c_n e^{2\pi i n x_0} + c_{-n} e^{-2\pi i n x_0} \right).$$

When $f(x)$ is continuous and $\sum_{-\infty}^{\infty} |c_n| < \infty$, then the function is represented by the absolutely convergent Fourier series

$$f(x) = \sum_{-\infty}^{\infty} c_n e^{2\pi i n x}.$$

If $F(x)$ is continuous such that

$$\int_{-\infty}^{\infty} |F(x)| dx < \infty,$$

then we define its Fourier transform by

$$\hat{F}(u) = \int_{-\infty}^{\infty} F(x) e^{-2\pi i x u} dx.$$

It is also a continuous function of u. If

$$\int_{-\infty}^{\infty} |\hat{F}(u)| du < \infty,$$

then we have the Fourier inversion formula

$$F(x) = \int_{-\infty}^{\infty} \hat{F}(u) e^{2\pi i x u} du.$$

Thus, the Fourier transform of $\hat{F}(u)$ is $F(-x)$.

Exercise 5.1.1 *For* $\mathrm{Re}(c) > 0$, *let* $F(x) = e^{-c|x|}$. *Show that*

$$\hat{F}(u) = \frac{2c}{c^2 + 4\pi^2 u^2}.$$

Exercise 5.1.2 *For* $F(x) = e^{-\pi x^2}$, *show that* $\hat{F}(u) = e^{-\pi u^2}$.

Theorem 5.1.3 (Poisson summation formula) *Let $F \in L^1(\mathbb{R})$. Suppose that the series*

$$\sum_{n \in \mathbb{Z}} F(n+v)$$

converges absolutely and uniformly in v, and that

$$\sum_{m \in \mathbb{Z}} |\hat{F}(m)| < \infty.$$

Then

$$\sum_{n \in \mathbb{Z}} F(n+v) = \sum_{n \in \mathbb{Z}} \hat{F}(n) e^{2\pi i n v}.$$

Proof. The function

$$G(v) = \sum_{n \in \mathbb{Z}} F(n+v)$$

is a continuous function of v of period 1. The Fourier coefficients of G are given by

$$\begin{aligned} c_m &= \int_0^1 G(v) e^{-2\pi i m v} dv \\ &= \sum_{n \in \mathbb{Z}} \int_0^1 F(n+v) e^{-2\pi i m v} dv \\ &= \sum_{n \in \mathbb{Z}} \int_n^{n+1} F(x) e^{-2\pi i m x} dx \\ &= \int_{-\infty}^{\infty} F(x) e^{-2\pi i m x} dx = \hat{F}(m). \end{aligned}$$

Since $\sum_{m \in \mathbb{Z}} |\hat{F}(m)| < \infty$, we can represent G by its Fourier series

$$\sum_{n \in \mathbb{Z}} F(n+v) = \sum_{n \in \mathbb{Z}} \hat{F}(n) e^{2\pi i n v},$$

as desired. □

Corollary 5.1.4 *With F as above,*

$$\sum_{n \in \mathbb{Z}} F(n) = \sum_{n \in \mathbb{Z}} \hat{F}(n).$$

Proof. Set $v = 0$ in the theorem. □

Exercise 5.1.5 *With F as in Theorem 5.1.3, show that*

$$\sum_{n\in\mathbb{Z}} F\left(\frac{v+n}{t}\right) = \sum_{n\in\mathbb{Z}} |t|\hat{F}(nt)e^{2\pi intv}.$$

Exercise 5.1.6 *Show that*

$$\frac{e^c+1}{e^c-1} = \sum_{-\infty}^{\infty} \frac{2c}{c^2+4\pi^2n^2}.$$

Exercise 5.1.7 *Show that*

$$\sum_{n\in\mathbb{Z}} e^{-(n+\alpha)^2\pi/x} = x^{1/2}\sum_{n\in\mathbb{Z}} e^{-n^2\pi x+2\pi in\alpha}$$

for any $\alpha \in \mathbb{R}$, and $x > 0$.

Setting $\alpha = 0$ in the previous exercise gives the following theorem.

Theorem 5.1.8

$$\sum_{n\in\mathbb{Z}} e^{-n^2\pi/x} = x^{1/2}\sum_{n\in\mathbb{Z}} e^{-n^2\pi x}.$$

5.2 The Riemann Zeta Function

We will now derive the functional equation of the Riemann zeta function and its analytic continuation to the entire complex plane. To this end, we introduce the θ-function

$$\theta(z) = \sum_{n\in\mathbb{Z}} e^{\pi in^2 z}$$

for $z \in \mathbb{C}$, with $\mathrm{Im}(z) > 0$. If we put $z = iy$ and set $\omega(y) = \theta(iy)$, Theorem 5.1.8 gives us the functional equation:

$$\omega(1/x) = x^{1/2}\omega(x).$$

Riemann derives his functional equation from this fact. Recall that the Γ-function is given by the integral

$$\Gamma(s) = \int_0^\infty e^{-t}t^{s-1}dt,$$

valid for $\mathrm{Re}(s) > 0$.

Exercise 5.2.1 *Show that*

$$\Gamma(s+1) = s\Gamma(s)$$

for $\mathrm{Re}(s) > 0$ *and that this functional equation can be used to extend* $\Gamma(s)$ *as a meromorphic function for all* $s \in \mathbb{C}$ *with only simple poles at* $s = 0, -1, -2, \dots$.

Noting that

$$\Gamma\left(\frac{s}{2}\right) = \int_0^\infty e^{-t} t^{\frac{s}{2}-1} dt,$$

and putting $t = n^2\pi x$, we get

$$\pi^{-s/2}\Gamma\left(\frac{s}{2}\right)n^{-s} = \int_0^\infty x^{\frac{s}{2}-1} e^{-n^2\pi x} dx.$$

Hence, for $\sigma > 1$, we can sum both sides of the above equation over all positive integers n, to obtain

$$\pi^{-s/2}\Gamma\left(\frac{s}{2}\right)\zeta(s) = \int_0^\infty x^{\frac{s}{2}-1}\left(\sum_{n=1}^\infty e^{-n^2\pi x}\right)dx,$$

the inversion being justified by the absolute convergence of the right-hand side. Indeed, notice that

$$\sum_{n=1}^\infty e^{-n^2\pi x} \ll e^{-\pi x}.$$

Observing that

$$\sum_{n=1}^\infty e^{-n^2\pi x} = \frac{\omega(x)-1}{2},$$

we get

$$\pi^{-s/2}\Gamma\left(\frac{s}{2}\right)\zeta(s) = \int_0^\infty x^{\frac{s}{2}-1}\left(\frac{\omega(x)-1}{2}\right)dx.$$

Let us put

$$W(x) = \frac{\omega(x)-1}{2},$$

and write the right-hand side as

$$\int_0^\infty x^{\frac{s}{2}-1} W(x) dx.$$

We decompose this as

$$\int_1^\infty x^{\frac{s}{2}-1}W(x)dx + \int_0^1 x^{\frac{s}{2}-1}W(x)dx$$

and make the change of variables $x \mapsto 1/x$ in the second integral to get

$$\int_1^\infty x^{\frac{s}{2}}W(x)\frac{dx}{x} + \int_1^\infty W\Big(\frac{1}{x}\Big)x^{-\frac{s}{2}}\frac{dx}{x}.$$

Now,

$$W\Big(\frac{1}{x}\Big) = \frac{\omega(1/x)-1}{2} = \frac{x^{1/2}\omega(x)-1}{2} = -\frac{1}{2}+\frac{1}{2}x^{1/2}+x^{1/2}W(x)$$

by Theorem 5.1.8. Therefore,

$$\begin{aligned}\int_1^\infty W\Big(\frac{1}{x}\Big)x^{-\frac{s}{2}}\frac{dx}{x} &= \int_1^\infty \Big(-\frac{1}{2}+\frac{1}{2}x^{1/2}+x^{1/2}W(x)\Big)x^{-\frac{s}{2}}\frac{dx}{x} \\ &= -\frac{1}{s}+\frac{1}{s-1}+\int_1^\infty x^{\frac{1-s}{2}}W(x)\frac{dx}{x}.\end{aligned}$$

Putting this together proves that

$$\pi^{-s/2}\Gamma\Big(\frac{s}{2}\Big)\zeta(s) = \frac{1}{s(s-1)} + \int_1^\infty W(x)\left(x^{\frac{s}{2}}+x^{\frac{1-s}{2}}\right)\frac{dx}{x}$$

for $\mathrm{Re}(s) > 1$. However, the integral on the right-hand side converges absolutely for all $s \in \mathbb{C}$, since $W(x) = O(e^{-\pi x})$ as $x \to \infty$. This gives the analytic continuation and functional equation for $\zeta(s)$:

Theorem 5.2.2 *We have*

$$\pi^{-s/2}\Gamma\Big(\frac{s}{2}\Big)\zeta(s) = \frac{1}{s(s-1)} + \int_1^\infty W(x)\left(x^{\frac{s}{2}}+x^{\frac{1-s}{2}}\right)\frac{dx}{x}$$

for all $s \in \mathbb{C}$. Moreover, if we define

$$\xi(s) := s(s-1)\pi^{-s/2}\Gamma\Big(\frac{s}{2}\Big)\zeta(s),$$

then $\xi(s)$ is entire and $\xi(1-s) = \xi(s)$.

Exercise 5.2.3 *Show that $\zeta(s)$ has simple zeros at $s = -2n$, for n a positive integer.*

Exercise 5.2.4 *Prove that $\zeta(0) = -1/2$.*

Exercise 5.2.5 *Show that $\zeta(s) \neq 0$ for any real s satisfying $0 < s < 1$.*

5.3 Gauss Sums

For any character $\chi \pmod{q}$, the **Gauss sum** $\tau(\chi)$ is defined by

$$\tau(\chi) = \sum_{m=1}^{q} \chi(m) e\left(\frac{m}{q}\right),$$

where $e(t) = e^{2\pi i t}$. The Gauss sum plays a significant role in the functional equation of Dirichlet L-functions.

Before we proceed, we classify Dirichlet characters $\pmod{q}$ into two types: primitive and imprimitive. Let χ_0 denote the trivial character (mod q). If $d|q$ and ψ is a character (mod d), then $\chi_0\psi$ is a character (mod q). If d is a proper divisor of q characters (mod q) obtained in this way will be called imprimitive. Otherwise, we shall say the character is primitive.

Example 5.3.1 *If $(n, q) = 1$, then*

$$\chi(n)\tau(\overline{\chi}) = \sum_{m=1}^{q} \overline{\chi}(m) e\left(\frac{mn}{q}\right).$$

Solution. We have

$$\begin{aligned} \chi(n)\tau(\overline{\chi}) &= \sum_{m=1}^{q} \overline{\chi}(m)\chi(n) e\left(\frac{m}{q}\right) \\ &= \sum_{h=1}^{q} \overline{\chi}(h) e\left(\frac{nh}{q}\right) \end{aligned}$$

on putting $h \equiv mn^{-1} \pmod{q}$, which we can do, since $(n, q) = 1$. □

Exercise 5.3.2 *If χ is a primitive, nonprincipal character* (mod q), *show that*

$$\chi(n)\tau(\overline{\chi}) = \sum_{m=1}^{q} \overline{\chi}(m)e\Big(\frac{mn}{q}\Big)$$

if $(n, q) > 1$.

Theorem 5.3.3 *If χ is a primitive character* (mod q), *then* $|\tau(\chi)| = q^{1/2}$.

Proof. By Exercise 5.3.2,

$$\chi(n)\tau(\overline{\chi}) = \sum_{m=1}^{q} \overline{\chi}(m)e\Big(\frac{mn}{q}\Big).$$

Thus

$$|\chi(n)|^2|\tau(\chi)|^2 = \sum_{m_1=1}^{q}\sum_{m_2=1}^{q} \overline{\chi}(m_1)\chi(m_2)e\Big(\frac{n(m_1 - m_2)}{q}\Big).$$

Summing over n for $1 \le n \le q$ gives

$$\phi(q)|\tau(\chi)|^2 = q\phi(q),$$

so that $|\tau(\chi)|^2 = q$, as required. □

5.4 Dirichlet L-functions

The functional equation for a Dirichlet L-function $L(s, \chi)$ can be derived easily by means of the Poisson summation formula. The discussion splits according as χ is an **even** or **odd** character, that is, according as $\chi(-1) = 1$ or -1, respectively.

We discuss the even case first and relegate the odd case to the exercises. Thus, suppose $\chi(-1) = 1$. We have

$$\pi^{-s/2}q^{s/2}\Gamma\Big(\frac{s}{2}\Big)n^{-s} = \int_0^{\infty} e^{-n^2\pi x/q}x^{\frac{s}{2}}\frac{dx}{x}.$$

We multiply this equation by $\chi(n)$ and sum over n to get

$$\pi^{-s/2}q^{s/2}\Gamma\Big(\frac{s}{2}\Big)L(s, \chi) = \int_0^{\infty} x^{\frac{s}{2}}\Big(\sum_{n=1}^{\infty}\chi(n)e^{-n^2\pi x/q}\Big)\frac{dx}{x},$$

for $\text{Re}(s) > 1$. Since $\chi(-1) = 1$ and $\chi(0) = 0$, we rewrite this as

$$\frac{1}{2}\int_0^\infty x^{s/2}\theta(x,\chi)\frac{dx}{x},$$

where

$$\theta(x,\chi) = \sum_{n=-\infty}^{\infty} \chi(n)e^{-n^2\pi x/q}.$$

We can derive a functional equation for $\theta(x,\chi)$ by noting that upon multiplication of the Gauss sum $\tau(\overline{\chi})$, we get

$$\tau(\overline{\chi})\theta(x,\chi) = \sum_{m=1}^{q} \overline{\chi}(m) \sum_{n=-\infty}^{\infty} e^{-n^2\pi x/q+2\pi imn/q}.$$

By Exercise 5.1.7, the inner sum is equal to

$$(q/x)^{1/2} \sum_{n=-\infty}^{\infty} e^{-(n+m/q)^2\pi q/x},$$

so that

$$\begin{aligned}
\tau(\overline{\chi})\theta(x,\chi) &= (q/x)^{1/2} \sum_{m=1}^{q} \overline{\chi}(m) \sum_{n=-\infty}^{\infty} e^{-(nq+m)^2\pi/xq} \\
&= (q/x)^{1/2} \sum_{l=-\infty}^{\infty} \overline{\chi}(l)e^{-l^2\pi/xq} \\
&= (q/x)^{1/2}\theta(x^{-1},\overline{\chi}).
\end{aligned}$$

Thus, as before, we write the integral for $L(s,\chi)$ as

$$\frac{1}{2}\int_1^\infty x^{\frac{s}{2}}\theta(x,\chi)\frac{dx}{x} + \frac{1}{2}\int_1^\infty x^{-\frac{s}{2}}\theta(x^{-1},\chi)\frac{dx}{x}$$

$$= \frac{1}{2}\int_1^\infty x^{\frac{s}{2}}\theta(x,\chi)\frac{dx}{x} + \frac{q^{1/2}}{2\tau(\overline{\chi})}\int_1^\infty x^{\frac{1-s}{2}}\theta(x,\overline{\chi})\frac{dx}{x}.$$

The right-hand side is regular for all $s \in \mathbb{C}$, since $\theta(x,\chi) = O(e^{-\pi x})$. Also, if we replace s by $1-s$ and χ by $\overline{\chi}$, the expression becomes

$$\frac{q^{1/2}}{2\tau(\chi)}\int_1^\infty x^{\frac{s}{2}}\theta(x,\chi)\frac{dx}{x} + \frac{1}{2}\int_1^\infty x^{\frac{1-s}{2}}\theta(x,\overline{\chi})\frac{dx}{x},$$

which is the previous expression multiplied by $q^{1/2}/\tau(\chi)$, since $|\tau(\chi)|^2 = q$. This proves the following theorem.

Theorem 5.4.1 *Suppose* $\chi(-1) = 1$. *Set*

$$\xi(s, \chi) = \pi^{-s/2} q^{s/2} \Gamma\left(\frac{s}{2}\right) L(s, \chi).$$

Then $\xi(s, \chi)$ *in entire and*

$$\xi(s, \chi) = w_\chi \xi(1 - s, \overline{\chi}),$$

where $w_\chi = \tau(\chi)/\sqrt{q}$.

Exercise 5.4.2 *Suppose* $\chi(-1) = 1$. *Show that* $L(s, \chi)$ *has simple zeros at* $s = -2, -4, -6, \ldots$.

Below, we will derive the functional equation in the case $\chi(-1) = -1$. Note that the above argument fails because for now, $\theta(x, \chi)$ is identically zero.

Exercise 5.4.3 *Prove that*

$$\pi^{-(s+1)/2} q^{(s+1)/2} \Gamma\left(\frac{s+1}{2}\right) n^{-s} = \int_0^\infty n e^{-\pi n^2 x/q} x^{\frac{s+1}{2}} \frac{dx}{x}$$

and hence deduce that

$$\pi^{-\left(\frac{s+1}{2}\right)} q^{\left(\frac{s+1}{2}\right)} \Gamma\left(\frac{s+1}{2}\right) L(s, \chi) = \frac{1}{2} \int_0^\infty \theta_1(x, \chi) x^{\frac{s+1}{2}} \frac{dx}{x},$$

where

$$\theta_1(x, \chi) = \sum_{n=-\infty}^{\infty} n \chi(n) e^{-n^2 \pi x/q}.$$

Exercise 5.4.4 *Prove that*

$$\sum_{n=-\infty}^{\infty} n e^{-n^2 \pi x/q + 2\pi i m n/q} = i\,(q/x)^{3/2} \sum_{n=-\infty}^{\infty} \left(n + \frac{m}{q}\right) e^{-\pi (n + m/q)^2 q/x}.$$

Exercise 5.4.5 *Prove that for* $\chi(-1) = -1$, *if we set*

$$\xi(s, \chi) = \pi^{-s/2} q^{s/2} \Gamma\left(\frac{s+1}{2}\right) L(s, \chi),$$

then $\xi(s, \chi)$ *is entire and*

$$\xi(s, \chi) = w_\chi \xi(1 - s, \overline{\chi}),$$

where $w_\chi = \tau(\chi)/i q^{1/2}$.

5.5 Supplementary Problems

Exercise 5.5.1 *Let*

$$f(y) = \sum_{n=1}^{\infty} a_n e^{-2\pi n y}$$

converge for $y > 0$. *Suppose that for some* $w \in \mathbb{Z}$,

$$f(1/y) = (-1)^w y^r f(y),$$

and that $a_n = O(n^c)$ *for some constant* $c > 0$. *Let*

$$L_f(s) = \sum_{n=1}^{\infty} a_n n^{-s}.$$

Show that $(2\pi)^{-s}\Gamma(s)L_f(s)$ *extends to an entire function and satisfies the functional equation*

$$(2\pi)^{-s}\Gamma(s)L_f(s) = (-1)^w (2\pi)^{-(r-s)}\Gamma(r-s)L_f(r-s).$$

Exercise 5.5.2 *Let*

$$g(y) = \sum_{n=0}^{\infty} a_n e^{-2\pi n y}$$

converge for $y > 0$. *Suppose that for some* $w \in \mathbb{Z}$,

$$g(1/y) = (-1)^w y^r g(y)$$

and that $a_n = O(n^c)$ *for some constant* $c > 0$. *Let* $L_g(s) = \sum_{n=1}^{\infty} a_n n^{-s}$. *Show that* $(2\pi)^{-s}\Gamma(s)L_g(s)$ *extends to a meromorphic function with at most simple poles at* $s = 0$ *and* $s = r$ *and satisfies the functional equation*

$$(2\pi)^{-s}\Gamma(s)L_g(s) = (-1)^w (2\pi)^{r-s}\Gamma(r-s)L_g(r-s).$$

Exercise 5.5.3 *Let*

$$\Psi(x) = \begin{cases} x - [x] - \frac{1}{2} & \text{if } x \notin \mathbb{Z}, \\ 0 & \text{if } x \in \mathbb{Z}. \end{cases}$$

Show that

$$\left| \Psi(x) + \sum_{0<|m|\leq M} \frac{e(mx)}{2\pi i m} \right| \leq \frac{1}{2\pi M \|x\|},$$

where $e(t) = e^{2\pi i t}$ *and* $\|x\|$ *denotes the distance from* x *to the nearest integer.*

Exercise 5.5.4 *Let $f(x)$ be a differentiable function on $[0,1]$ satisfying $|f'(x)| \leq K$. Show that*

$$\left| \sum_{|m|\leq M} \int_0^1 f(x)e(mx)dx - \frac{f(0)+f(1)}{2} \right| \ll \frac{K \log M}{M}.$$

Deduce that

$$\sum_{-\infty}^{\infty} \int_0^1 f(x)e(mx)dx = \frac{f(0)+f(1)}{2}.$$

Exercise 5.5.5 *By using the previous exercise with $f(x) = x^2$, deduce that*

$$\sum_{m=1}^{\infty} \frac{1}{m^2} = \frac{\pi^2}{6}.$$

Exercise 5.5.6 (Pólya - Vinogradov inequality) *Let χ be a primitive character* $\bmod q$. *Show that for $q > 1$,*

$$\left| \sum_{n\leq x} \chi(n) \right| \ll q^{1/2} \log q.$$

Exercise 5.5.7 *Show that if χ is a primitive character* $(\bmod q)$, *then*

$$L(1,\chi) = \sum_{n\leq x} \frac{\chi(n)}{n} + O\left(\frac{q^{1/2}\log q}{x} \right)$$

for any $x \geq 1$ and $q > 1$.

Exercise 5.5.8 *Prove that*

$$\sum_{\chi\neq\chi_0} L(1,\chi) = \varphi(q) + O(q^{1/2}\log q),$$

where the summation is over all nontrivial characters $(\bmod q)$.

Exercise 5.5.9 *For any $s \in \mathbb{C}$ with* $\mathrm{Re}(s) > 0$, *show that for any $x \geq 1$,*

$$L(s,\chi) = \sum_{n\leq x} \frac{\chi(n)}{n^s} + O\left(\frac{|s|q^{1/2}\log q}{\sigma x^\sigma} \right),$$

where χ is a nontrivial character $\bmod q$ *and* $\sigma = \mathrm{Re}(s)$.

Exercise 5.5.10 *Prove that for any* $\sigma > 1/2$,

$$\sum_{\chi \neq \chi_0} L(\sigma, \chi) = \varphi(q) + O(q^{3/2-\sigma}),$$

where the sum is over all nontrivial characters $(\bmod q)$.

Exercise 5.5.11 *Let* $B_n(x)$ *denote the* n*th Bernoulli polynomial introduced in Chapter 2. For* $n \geq 2$*, show that*

$$\frac{B_n(x)}{n!} = \sum_{m \neq 0} \frac{e(mx)}{(2\pi i m)^n}.$$

Exercise 5.5.12 *Let* $f(x)$ *be differentiable on* $[A, B]$ *and satisfy for some constant* K, $|f'(x)| \leq K$ *for all* $x \in [A, B]$. *Show that*

$$\sum_{n=A}^{B}{}' f(n) = \sum_{m=-\infty}^{\infty} \int_A^B f(x) e(mx) dx,$$

where the dash on the summation means that the end-terms are replaced by $f(A)/2$ *and* $f(B)/2$. (Hint: Use Exercise 5.5.4.)

Exercise 5.5.13 *Apply the previous exercise to each of the functions* $f(x) = \cos(2\pi x^2/N)$ *and* $f(x) = \sin(2\pi x^2/N)$ *to deduce that*

$$S = \sum_{n=0}^{N-1} e\left(\frac{n^2}{N}\right) = \begin{cases} (1+i)N^{1/2} & \text{if } N \equiv 0 \pmod 4, \\ N^{1/2} & \text{if } N \equiv 1 \pmod 4, \\ 0 & \text{if } N \equiv 2 \pmod 4, \\ iN^{1/2} & \text{if } N \equiv 3 \pmod 4. \end{cases}$$

Exercise 5.5.14 *Let* χ *be a nontrivial quadratic character* $\bmod p$ *with* p *prime. Show that*

$$\tau(\chi) = \sum_{m=1}^{p-1} \chi(m) e\left(\frac{m}{p}\right) = \begin{cases} \sqrt{p} & \text{if } p \equiv 1 \pmod 4, \\ i\sqrt{p} & \text{if } p \equiv 3 \pmod 4. \end{cases}$$

From this result we can deduce the law of quadratic reciprocity as follows.

Let p and q be distinct odd primes. Let $\tau(\chi)$ be the Gauss sum for χ the quadratic character modulo p. Using the above formula and

$(a/q) \equiv a^{(q-1)/2} \pmod q$ we get

$$\begin{aligned}
\tau(\chi)^{q+1} &= (-1)^{(p-1)/2} p(\tau(\chi)^2)^{(q-1)/2} \\
&= (-1)^{(p-1)/2} p(-1)^{(p-1)(q-1)/4} p^{(q-1)/2} \\
&\equiv (-1)^{(p-1)/2+(p-1)(q-1)/4} p \left(\frac{p}{q}\right) \pmod q.
\end{aligned}$$

On the other hand, using the multinomial theorem, we obtain

$$\tau(\chi)^{q+1} = \tau(\chi)\tau(\chi)^q = \tau(\chi)\left(\sum_n \left(\frac{n}{p}\right) e(nq/p) + f(e(1/p))\right),$$

where $f(x)$ is a polynomial with integer coefficients divisible by q. Using Exercise 5.3.2 we get

$$\sum_n \left(\frac{n}{p}\right) e(nq/p) = \left(\frac{q}{p}\right) \tau(\chi).$$

So

$$\tau(\chi)^{q+1} = \left(\frac{q}{p}\right) (-1)^{(p-1)/2} p + h(e(1/p))$$

for another polynomial $h(x)$ with integer coefficients divisible by q. Collecting same powers of $e(1/p)$ and using the fact that $1, e(1/p), e(2/p), \ldots, e((p-2)/p)$ are linearly independent, since $1 + x + x^2 + \cdots + x^{p-1}$ is irreducible (see, for example, [EM, p. 37 and p. 183]) we get

$$\left(\frac{q}{p}\right) p(-1)^{(p-1)/2} \equiv (-1)^{(p-1)/2+(p-1)(q-1)/4} p \left(\frac{p}{q}\right) \pmod q,$$

from which it follows easily that

$$\left(\frac{q}{p}\right) = \left(\frac{p}{q}\right) (-1)^{(p-1)(q-1)/4}.$$

Exercise 5.5.15 *Let $\phi(s) = (2\pi)^{-s}\Gamma(s)\zeta(s)\zeta(s+1)$. Show that $\phi(-s) = \phi(s)$.*

Exercise 5.5.16 *Show that $\phi(s)$ in Exercise 5.5.15 has a double pole at $s = 0$ and simple poles at $s = \pm 1$. Show further that $\text{Res}_{s=1}\phi(s) = \pi/12$ and $\text{Res}_{s=-1}\phi(s) = -\pi/12$.*

Exercise 5.5.17 *Show that if* $\sigma(n) = \sum_{d|n} d$, *then*

$$\sum_{n=1}^{\infty} \frac{\sigma(n)}{n^{s+1}} = \zeta(s)\zeta(s+1),$$

and that

$$\sum_{n=1}^{\infty} \frac{\sigma(n)}{n} e^{-nx} = \frac{1}{2\pi i} \int_{2-i\infty}^{2+i\infty} x^{-s}\Gamma(s)\zeta(s)\zeta(s+1)ds.$$

Exercise 5.5.18 *Show that*

$$\sum_{n=1}^{\infty} \frac{\sigma(n)}{n} e^{-2\pi nx} = \frac{\pi}{12x} - \frac{\pi x}{12} + \frac{1}{2}\log x + \sum_{n=1}^{\infty} \frac{\sigma(n)}{n} e^{-2\pi n/x}.$$

Exercise 5.5.19 *For* a *and* b *coprime integers and* $b > 0$, *define*

$$C\left(\frac{a}{b}\right) = \sum_{j=0}^{b-1} e^{2\pi i j^2 a/b}.$$

Let q *be prime and* $(p, q) = 1$. *Show that*

$$\lim_{t \to 0} \sqrt{t}\theta\left(t + \frac{2pi}{q}\right) = \frac{1}{q} C\left(-\frac{p}{q}\right).$$

Exercise 5.5.20 *Let* $r = p/q$. *Show that*

$$\lim_{t \to 0} \sqrt{\frac{t}{t+2ir}}\,\theta\left(\frac{1}{t+2ir}\right) = \frac{(1-i)}{4\sqrt{pq}} C\left(\frac{q}{4p}\right),$$

with notation as in the previous exercise.

Exercise 5.5.21 *Deduce from the previous exercise the law of quadratic reciprocity*

$$\left(\frac{p}{q}\right)\left(\frac{q}{p}\right) = (-1)^{\frac{p-1}{2}\cdot\frac{q-1}{2}}$$

for odd primes p *and* q, *and where* $\left(\frac{a}{b}\right)$ *denotes the Legendre symbol.*

Exercise 5.5.22 *Suppose that* $f(s)$ *is an entire function satisfying the functional equation*

$$A^s\Gamma(s)f(s) = A^{1-s}\Gamma(1-s)f(1-s).$$

Show that if $f(1/2) \neq 0$, *then*

$$f'\left(\frac{1}{2}\right) = -f(1/2)\left(\log A + \frac{\Gamma'(1/2)}{\Gamma(1/2)}\right).$$

6

Hadamard Products

An entire function $f(z)$ is said to be of **finite order** if for some $\alpha \geq 0$, we have

$$f(z) = O\left(e^{|z|^\alpha}\right)$$

as $|z| \to \infty$. If $\alpha = 0$, then $f(z)$ is constant by Liouville's theorem. The infimum of the numbers α such that the above estimate holds is called the **order** of $f(z)$.

In the 1890s, Hadamard developed the theory of entire functions of finite order. He showed that, very much like polynomials, they can be factored into an infinite product over the zeros of $f(z)$.

In this chapter we will derive this factorization theorem of Hadamard for entire functions of order 1 and then apply it to derive a wider zero free region for $\zeta(s)$.

6.1 Jensen's Theorem

Let $f(z)$ be an entire function of finite order β. Jensen's theorem relates β to the distribution of the zeros of $f(z)$.

Example 6.1.1 *Show that an entire function $f(z)$ of finite order β without any zeros must be of the form $f(z) = e^{g(z)}$, where $g(z)$ is a polynomial and $\beta = \deg g$.*

Solution.

Let $h(z) = \log f(z) - \log f(0)$. Then $h(z)$ is entire, since $f(z)$ has no zeros. Also, for any $\epsilon > 0$,

$$\operatorname{Re} h(z) = \log |f(z)| \ll R^{\beta+\epsilon}.$$

Writing

$$h(z) = \sum_{n=0}^{\infty} (a_n + ib_n) z^n$$

with $a_n, b_n \in \mathbb{R}$, we see that for $z = Re^{i\theta}$,

$$\operatorname{Re}(h(z)) = \sum_{n=0}^{\infty} a_n R^n \cos n\theta - \sum_{n=0}^{\infty} b_n R^n \sin n\theta.$$

By Fourier analysis, we get

$$|a_n| R^n \ll \int_0^{2\pi} \left| \operatorname{Re}\left(h\left(Re^{i\theta}\right)\right) \right| d\theta.$$

Since $h(0) = 0$, we have $a_0 = 0$, and therefore

$$\int_0^{2\pi} \operatorname{Re}\left(h\left(Re^{i\theta}\right)\right) d\theta = 0.$$

Observe that for $x \in \mathbb{R}$, we have

$$|x| + x = \begin{cases} 2x & \text{if} \quad x \geq 0, \\ 0 & \text{if} \quad x < 0. \end{cases}$$

Hence

$$\begin{aligned} |a_n| R^n &\ll \int_0^{2\pi} \left\{ |\operatorname{Re}(h(Re^{i\theta}))| + \operatorname{Re} h(Re^{i\theta}) \right\} d\theta \\ &\ll R^{\beta+\epsilon}. \end{aligned}$$

Letting $R \to \infty$ yields $a_n = 0$ if $n > \beta$. □

Notice that in this example the same result holds if the estimate

$$|f(z)| \ll e^{R_i^{\beta+\epsilon}}$$

holds for $|z| = R_i$ and R_i is a sequence tending to infinity.

Theorem 6.1.2 (Jensen's theorem) *Let $f(z)$ be an entire function of order β such that $f(0) \neq 0$. If $z_1, z_2, \ldots, z_n$ are the zeros of $f(z)$ in $|z| < R$, counted with multiplicity, then*

$$\frac{1}{2\pi}\int_0^{2\pi} \log|f(Re^{i\theta})|d\theta = \log|f(0)| + \log\left(\frac{R^n}{|z_1|\cdots|z_n|}\right).$$

Proof. We may assume, without loss of generality, that $f(0) = 1$. Also, it is clear that if the theorem is true for functions g and h, that it is also true for the product gh. Thus, it suffices to prove it for functions with either no zero or one zero in $|z| < R$. Indeed, if f has no zeros in $|z| < R$, the right-hand side is zero. The left-hand side is

$$\frac{1}{2\pi i}\int_{|z|=R} (\log f(z))\frac{dz}{z},$$

which by Cauchy's theorem is zero. Taking real parts gives the desired result.

If f has one zero $z = z_1$ in $|z| < R$, we consider the contour $|z| = R$ taken in the counterclockwise direction and cut it from z_1 to the boundary. We deform the contour so that we go around z_1 in a clockwise direction along a circle of radius ϵ (say). Then, by Cauchy's theorem with $g(z) = \log f(z)$,

$$0 = \frac{1}{2\pi i}\int_C g(z)\frac{dz}{z}$$

where C is the contour given above.

Since the argument changes by $-2\pi i$ when $g(z)$ goes around the zero $z = z_1$, we see that as $\epsilon \to 0$, we deduce

$$\frac{1}{2\pi}\int_0^{2\pi} \log|f(Re^{i\theta})|d\theta = \log\frac{R}{|z_1|},$$

as desired. This completes the proof. □

An alternative proof of Jensen's theorem can be given that avoids the use of cutting the plane. One considers

$$f(z) = \frac{R(z - z_1)}{R^2 - \overline{z_1}z}.$$

Then $f(z)$ is regular for $|z| \leq R$. Moreover, $|f(z)| = 1$ on $|z| = R$, and $|f(0)| = |z_1|/R$, as a simple calculation shows. Jensen's theorem is easily verified for this choice of f. But any holomorphic function on $|z| \leq R$ can be written as a function with no zeros in $|z| \leq R$ and a product of functions of the form

$$\frac{R(z - z_i)}{R^2 - \overline{z_i} z}.$$

Now Jensen's theorem easily follows.

Corollary 6.1.3 *Let f be as in Theorem 6.1.2. Then*

$$\log\left(\frac{R^n}{|z_1| \cdots |z_n|}\right) \leq \max_{|z|=R} \log|f(z)| - \log|f(0)|.$$

Proof. This is clear from Jensen's theorem. □

Now define $n_f(r) := n(r)$ to be the number of zeros of f in $|z| \leq r$.

Exercise 6.1.4 *Show that*

$$\int_0^R \frac{n(r)dr}{r} \leq \max_{|z|=R} \log|f(z)| - \log|f(0)|,$$

with f as in Jensen's theorem.

Exercise 6.1.5 *If $f(z)$ is of order β, show that $n_f(r) = O(r^{\beta+\epsilon})$, for any $\epsilon > 0$.*

Exercise 6.1.6 *Let $f(z)$ be an entire function of order β. Show that*

$$\sum_{n=1}^{\infty} |z_n|^{-\beta-\epsilon}$$

converges for any $\epsilon > 0$ (Here, we have indexed the zeros z_i so that $|z_1| \leq |z_2| \leq \cdots$).

6.2 Entire Functions of Order 1

We will now derive a factorization theorem for entire functions of order 1. A similar result holds for entire functions of higher order, and we relegate their study to the supplementary problems.

Theorem 6.2.1 *Let $f(z)$ be an entire function of order 1 with zeros $z_1, z_2, \ldots$ arranged so that $|z_1| \leq |z_2| \leq \cdots$ and repeated with appropriate multiplicity. Then f can be written as*

$$f(z) = e^{A+Bz} \prod_{n=1}^{\infty} \left(1 - \frac{z}{z_n}\right) e^{z/z_n},$$

where A and B are constants.

Proof. The product

$$P(z) = \prod_{n=1}^{\infty} \left(1 - \frac{z}{z_n}\right) e^{z/z_n}$$

converges absolutely for all z, since

$$(1 - z)e^{z} = 1 - z^2 + \cdots$$

and by Exercise 6.1.6. Thus, $P(z)$ represents an entire function. If we write

$$f(z) = P(z)F(z),$$

then $F(z)$ is an entire function without zeros. If F were of finite order, we could conclude by Example 6.1.1 that $F(z) = e^{g(z)}$, where $g(z)$ is a polynomial.

By the remark after Example 6.1.1, it suffices to show that

$$|F(z)| \ll e^{R_i^{1+\epsilon}}$$

to deduce that $F(z) = e^{g(z)}$ where $g(z)$ is of the form $A + Bz$ for certain constants A and B.

To this end, we will choose R_i satisfying

$$\Big| R_i - |z_n| \Big| > |z_n|^{-2}$$

for all n. This can be done, since the total measure of the intervals $(|z_n| - |z_n|^{-2}, |z_n| + |z_n|^{-2})$ is bounded by

$$2 \sum_{n=1}^{\infty} |z_n|^{-2} < \infty,$$

since $f(z)$ has order 1.

We write
$$P(z) = P_1(z)P_2(z)P_3(z),$$
where in P_1, $|z_n| < \frac{1}{2}R_i$, in P_2, $\frac{1}{2}R_i \le |z_n| \le 2R_i$, and in P_3, $|z_n| > 2R_i$. For the factors of P_1 we have for $|z| = R_i$,

$$\left|\left(1 - \frac{z}{z_n}\right) e^{z/z_n}\right| \ge \left(\left|\frac{z}{z_n}\right| - 1\right) e^{-|z|/|z_n|} > e^{-R_i/|z_n|}.$$

Since
$$\sum_{|z_n|<\frac{1}{2}R} |z_n|^{-1} \le \left(\frac{1}{2}R\right)^{\epsilon} \sum_{n=1}^{\infty} |z_n|^{-1-\epsilon},$$

we get
$$|P_1(z)| > \exp(-R_i^{1+\epsilon}).$$

For $P_2(z)$,

$$\left|\left(1 - \frac{z}{z_n}\right) e^{z/z_n}\right| \ge e^{-2}|z - z_n|/2R_i \gg R_i^{-3}$$

by the way we have chosen R_i.

Since $n(R_i) = O(R_i^{1+\epsilon})$, we get

$$|P_2(z)| \gg (R_i^{-3})^{R_i^{1+\epsilon}} \ge \exp(-c_1 R_i^{1+2\epsilon}).$$

Finally, for $P_3(z)$, we have $|z/z_n| < 1/2$ so that

$$\left|\left(1 - \frac{z}{z_n}\right) e^{z/z_n}\right| \ge e^{-c_2 R_i^2/|z_n|^2}$$

and
$$\sum_{|z_n|>2R} |z_n|^{-2} < (2R)^{-1+\epsilon} \sum_{n=1}^{\infty} |z_n|^{-1-\epsilon}.$$

Thus, on $|z| = R_i$ we have

$$|P(z)| > \exp(-R^{1+3\epsilon}),$$

so that
$$|F(z)| < \exp(R^{1+4\epsilon}).$$

Hence, $F(z) = e^{g(z)}$, where $g(z)$ is a polynomial of degree at most 1. □

6.3 The Gamma Function

We will prove that $1/\Gamma(z)$ is an entire function of order 1 and derive its Hadamard factorization.

Exercise 6.3.1 *Show that*

$$\int_0^\infty \frac{v^{x-1}dv}{1+v} = \frac{\pi}{\sin \pi x}$$

for $0 < x < 1$.

Exercise 6.3.2 *Show that*

$$\Gamma(x)\Gamma(y) = 2\Gamma(x+y)\int_0^{\pi/2} (\cos\theta)^{2x-1}(\sin\theta)^{2y-1}d\theta$$

for $x, y > 0$.

Exercise 6.3.3 *Show that*

$$\Gamma(x)\Gamma(y) = \Gamma(x+y)\int_0^1 \lambda^{x-1}(1-\lambda)^{y-1}d\lambda.$$

(The integral is denoted $B(x,y)$ and called the beta function.)

Exercise 6.3.4 *Prove that*

$$\Gamma(x)\Gamma(1-x) = \frac{\pi}{\sin \pi x}$$

for $0 < x < 1$.

Exercise 6.3.5 *Prove that*

$$\Gamma\left(\frac{1}{2}\right) = \sqrt{\pi}.$$

Exercise 6.3.6 (Legendre's duplication formula) *Show that*

$$\Gamma(2x)\Gamma\left(\frac{1}{2}\right) = 2^{2x-1}\Gamma(x)\Gamma\left(x+\frac{1}{2}\right)$$

for $x > 0$.

Exercise 6.3.7 *Let c be a positive constant. Show that as $x \to \infty$,*

$$\Gamma(x+c) \sim x^c \Gamma(x).$$

Exercise 6.3.8 (Stirling's formula) *Show that*

$$\Gamma(x) \sim e^{-x} x^{x-1/2} \sqrt{2\pi}$$

as $x \to \infty$.

Exercise 6.3.9 *Show that $1/\Gamma(z)$ is an entire function with simple zeros at $z = 0, -1, -2, \dots$.*

Exercise 6.3.10 *Show that for some constant K,*

$$\frac{\Gamma'(z)}{\Gamma(z)} = \int_0^1 \left\{1 - (1-t)^{z-1}\right\} \frac{dt}{t} - K.$$

Exercise 6.3.11 *Show that for z not a negative integer,*

$$\frac{\Gamma'(z)}{\Gamma(z)} = \sum_{n=0}^{\infty} \left(\frac{1}{n+1} - \frac{1}{n+z} \right) - K$$

for some constant K.

Exercise 6.3.12 *Derive the Hadamard factorization of $1/\Gamma(z)$:*

$$\frac{1}{\Gamma(z)} = e^{\gamma z} z \prod_{n=1}^{\infty} \left(1 + \frac{z}{n}\right) e^{-z/n},$$

where γ denotes Euler's constant.

Exercise 6.3.13 *Show that*

$$\log \Gamma(z) = \left(z - \frac{1}{2}\right) \log z - z + \frac{1}{2} \log 2\pi + \int_0^{\infty} \frac{[u] - u + \frac{1}{2}}{u+z} du.$$

Exercise 6.3.14 *For any $\delta > 0$, show that*

$$\log \Gamma(z) = \left(z - \frac{1}{2}\right) \log z - z + \frac{1}{2} \log 2\pi + O\left(\frac{1}{|z|}\right)$$

uniformly for $-\pi + \delta \le \arg z \le \pi - \delta$.

Exercise 6.3.15 *If σ is fixed, and $|t| \to \infty$, show that*

$$|\Gamma(\sigma + it)| \sim e^{-\frac{1}{2}\pi|t|}|t|^{\sigma-\frac{1}{2}}\sqrt{2\pi}.$$

Exercise 6.3.16 *Show that $1/\Gamma(z)$ is of order 1.*

Exercise 6.3.17 *Show that*

$$\frac{\Gamma'(z)}{\Gamma(z)} = \log z + O\Big(\frac{1}{|z|}\Big)$$

for $|z| \to \infty$ in the sector $-\pi + \delta < \arg z < \pi - \delta$ for any fixed $\delta > 0$.

6.4 Infinite Products for $\xi(s)$ and $\xi(s,\chi)$

In this section we will establish that $\xi(s)$ and $\xi(s,\chi)$ are entire functions of order 1. Then we will derive their Hadamard factorizations.

Recall that

$$\xi(s) = \frac{1}{2}s(s-1)\pi^{-s/2}\Gamma\Big(\frac{s}{2}\Big)\zeta(s)$$

and that when χ is a primitive character (mod q),

$$\xi(s,\chi) = (q/\pi)^{\frac{s+a}{2}}\Gamma\Big(\frac{s+a}{2}\Big)L(s,\chi),$$

where $a = 0$ or 1 according as $\chi(-1) = 1$ or -1.

Exercise 6.4.1 *Show that for some constant c,*

$$|\xi(s)| < \exp(c|s|\log|s|)$$

as $|s| \to \infty$. Conclude that $\xi(s)$ has order 1.

Exercise 6.4.2 *Prove that $\zeta(s)$ has infinitely many zeros in $0 \le \mathrm{Re}(s) \le 1$.*

Exercise 6.4.3 *Show that*

$$\xi(s) = e^{A+Bs}\prod_{\rho}\Big(1 - \frac{s}{\rho}\Big)e^{s/\rho},$$

where the product is over the nontrivial zeroes of $\zeta(s)$ in the region $0 \le \mathrm{Re}(s) \le 1$ and $A = -\log 2$, $B = -\gamma/2 - 1 + \frac{1}{2}\log 4\pi$, where γ is Euler's constant.

Exercise 6.4.4 *Let χ be a primitive character* (mod q). *Show that $\xi(s,\chi)$ is an entire function of order* 1.

Exercise 6.4.5 *Show that $L(s,\chi)$ has infinitely many zeros in $0 \le \mathrm{Re}(s) \le 1$ and that*

$$\xi(s,\chi) = e^{A+Bs}\prod_{\rho}\left(1-\frac{s}{\rho}\right)e^{s/\rho},$$

where the product is over the nontrivial zeros of $L(s,\chi)$.

Exercise 6.4.6 *For A and B occurring in the previous exercise, show that*

$$e^{A} = \xi(0,\chi)$$

and that

$$\mathrm{Re}(B) = -\sum_{\rho}\mathrm{Re}\left(\frac{1}{\rho}\right),$$

where the sum is over the nontrivial zeros ρ of $L(s,\chi)$.

6.5 Zero-Free Regions for $\zeta(s)$ and $L(s,\chi)$

In Exercise 3.2.5 we proved the nonvanishing of $\zeta(s)$ for $\mathrm{Re}(s) = 1$. A similar deduction was made for $L(s,\chi)$ in Exercise 3.2.12. Using the Hadamard factorization for $\xi(s)$ and $\xi(s,\chi)$, we will derive a wider zero-free region.

The starting point is

$$-\mathrm{Re}\left(\frac{\zeta'(s)}{\zeta(s)}\right) = \sum_{n=1}^{\infty}\frac{\Lambda(n)\cos(t\log n)}{n^{\sigma}},$$

where, following custom, we write $s = \sigma + it$.

Exercise 6.5.1 *Show that*

$$-3\frac{\zeta'(\sigma)}{\zeta(\sigma)} - 4\,\mathrm{Re}\left(\frac{\zeta'(\sigma+it)}{\zeta(\sigma+it)}\right) - \mathrm{Re}\left(\frac{\zeta'(\sigma+2it)}{\zeta(\sigma+2it)}\right) \ge 0$$

for $t \in \mathbb{R}$ and $\sigma > 1$.

Exercise 6.5.2 *For $1 < \sigma < 2$, show that*

$$-\frac{\zeta'(\sigma)}{\zeta(\sigma)} < \frac{1}{\sigma - 1} + A$$

for some constant A.

Exercise 6.5.3 *Prove that*

$$-\operatorname{Re}\left(\frac{\zeta'(s)}{\zeta(s)}\right) < A \log |t| - \sum_{\rho} \operatorname{Re}\left(\frac{1}{s-\rho} + \frac{1}{\rho}\right)$$

for $1 \leq \sigma \leq 2$ and $|t| \geq 2$.

Exercise 6.5.4 *Show that*

$$\operatorname{Re}\left(\frac{1}{s-\rho} + \frac{1}{\rho}\right) \geq 0.$$

Deduce that

$$-\operatorname{Re}\left(\frac{\zeta'(s)}{\zeta(s)}\right) < A \log |t|$$

for $1 \leq \sigma \leq 2$, $|t| \geq 2$.

Exercise 6.5.5 *Let $\rho = \beta + i\gamma$ be any nontrivial zero of $\zeta(s)$. Show that for $|t| \geq 2$,*

$$-\operatorname{Re}\left(\frac{\zeta'(\sigma + it)}{\zeta(\sigma + it)}\right) < A \log |t| - \frac{1}{\sigma - \beta}.$$

Theorem 6.5.6 *There exists a constant $c > 0$ such that $\zeta(s)$ has no zero in the region*

$$\sigma \geq 1 - \frac{c}{\log |t|}, \qquad |t| \geq 2.$$

Proof. By Exercise 6.5.5,

$$-\operatorname{Re}\left(\frac{\zeta'(\sigma + it)}{\zeta(\sigma + it)}\right) < A_1 \log |t| - \frac{1}{\sigma - \beta}.$$

We also know, by Exercises 6.5.2 and 6.5.4, that

$$-\frac{\zeta'(\sigma)}{\zeta(\sigma)} < \frac{1}{\sigma - 1} + A_2$$

and

$$-\operatorname{Re}\left(\frac{\zeta'(\sigma + 2it)}{\zeta(\sigma + 2it)}\right) < A_3 \log |t|.$$

Inserting these inequalities into

$$-3\frac{\zeta'(\sigma)}{\zeta(\sigma)} - 4\,\mathrm{Re}\left(\frac{\zeta'(\sigma+it)}{\zeta(\sigma+it)}\right) - \mathrm{Re}\left(\frac{\zeta'(\sigma+2it)}{\zeta(\sigma+2it)}\right) \geq 0$$

(Exercise 6.5.1), we obtain

$$\frac{4}{\sigma-\beta} < \frac{3}{\sigma-1} + A\log|t|$$

for some constant A. Taking $\sigma = 1 + \delta/\log|t|$ gives

$$\beta < 1 + \frac{\sigma}{\log|t|} - \frac{4\sigma}{(3+A\delta)\log|t|},$$

so that if δ is sufficiently small, we get

$$\beta < 1 - \frac{c}{\log|t|}$$

for some suitable positive constant c. □

Corollary 6.5.7 *There exists a constant $c > 0$ such that $\zeta(s)$ has no zero in the region*

$$\sigma \geq 1 - \frac{c}{\log(|t|+2)}.$$

Proof. The region $\sigma \geq 1$, $|t| \leq 2$ contains no zeros of $\zeta(s)$. Thus, there must be a constant $c_1 > 0$ such that $\zeta(s)$ has no zeros in $\sigma \geq 1 - c_1$ and $|t| \leq 2$. Combining such a region with the zero-free region provided by the theorem gives the result. □

Exercise 6.5.8 *Show that*

$$-\mathrm{Re}\left(\frac{\zeta'(s)}{\zeta(s)}\right) < \mathrm{Re}\left(\frac{1}{s-1}\right) + c_1\log(|t|+2)$$

for some constant $c_1 > 0$ and $\sigma > 1$.

In the following exercises we will derive analogous results for the Dirichlet L-functions $L(s,\chi)$.

Exercise 6.5.9 *Suppose that χ is a primitive character* $\pmod q$ *satisfying $\chi^2 \neq \chi_0$. Show that there is a constant $c > 0$ such that $L(s,\chi)$ has no zero in the region*

$$\sigma > 1 - \frac{c}{\log(q|t|+2)}.$$

Exercise 6.5.10 *Show that the previous result remains valid when χ is a nonreal imprimitive character.*

We now proceed to extend the previous results for $\chi^2 = \chi_0$. Let us first observe that

$$\left|\frac{L'(s,\chi_0)}{L(s,\chi_0)} - \frac{\zeta'(s)}{\zeta(s)}\right| \le \log q$$

for $\sigma > 1$. By Exercise 6.5.8,

$$-\operatorname{Re}\left(\frac{\zeta'(s)}{\zeta(s)}\right) < \operatorname{Re}\left(\frac{1}{s-1}\right) + c_1 \log(|t|+2).$$

Hence, if $\chi^2 = \chi_0$,

$$-\operatorname{Re}\left(\frac{L'(\sigma+2it,\chi^2)}{L(\sigma+2it,\chi^2)}\right) < \operatorname{Re}\left(\frac{1}{\sigma-1+2it}\right) + c_2 \log(q(|t|+2)).$$

When we insert this estimate into our previous calculations, we obtain

$$\frac{4}{\sigma-\beta} < \frac{3}{\sigma-1} + \operatorname{Re}\left(\frac{1}{\sigma-1+2i\gamma}\right) + c_3 \log(q(|\gamma|+2)).$$

Let us write $\mathcal{L}$ for $\log(q(|t|+2))$. Taking $\sigma = 1+\delta/\mathcal{L}$ and assuming $\gamma > \delta/\mathcal{L}$ gives

$$\frac{4}{\sigma-\beta} < \frac{3\mathcal{L}}{\delta} + \frac{\mathcal{L}}{5\delta} + c_3\mathcal{L},$$

so that

$$\beta < 1 - \frac{4-c_3\delta}{16+5c_3\delta}\frac{\delta}{\mathcal{L}}.$$

Hence if δ is sufficiently small in relation to c_3, we get the following:

Theorem 6.5.11 *There exists an absolute constant $c > 0$ such that if $0 < \delta < c$ and χ is a real, nonprincipal character* $(\bmod\, q)$, *$L(s,\chi)$ has no zeros in the region*

$$\sigma > 1 - \frac{\delta}{4\mathcal{L}}$$

and

$$|t| > \frac{\delta}{\log q},$$

where $\mathcal{L} = \log(q(|t|+2))$.

The case where $|t| < \delta/\log q$ still needs to be considered. We will show that for suitable δ (independent of q) there is at most one zero in the region and this zero is simple and real. Such a zero, if it exists, is called a **Siegel zero** in the literature.

Theorem 6.5.12 *There exists a positive absolute constant c such that if $0 < \delta < c$, then $L(s,\chi)$ has no zeros in the region*

$$\delta > 1 - \frac{c}{\log q(|t|+2)}$$

except possibly if χ is real and nonprincipal, in which case there is at most one simple, real zero in the region.

Proof. We need only consider the case where χ is real and nonprincipal and $|\gamma| < \delta/\log q$. First suppose there are two complex zeros in the region. We have

$$-\frac{L'(\sigma,\chi)}{L(\sigma,\chi)} < c_1 \log q - \sum_\rho \frac{1}{\sigma-\rho},$$

the sum over the zeros being real, since they occur in complex conjugate pairs. If $\beta \pm i\gamma$ are zeros of $L(s,\chi)$, with $\gamma \neq 0$, then

$$-\frac{L'(\sigma,\chi)}{L(\sigma,\chi)} < c_1 \log q - \frac{2(\sigma-\beta)}{(\sigma-\beta)^2+\gamma^2}.$$

Also,

$$-\frac{L'(\sigma,\chi)}{L(\sigma,\chi)} = \sum_{n=1}^{\infty} \frac{\chi(n)\Lambda(n)}{n^\sigma} \geq -\sum_{n=1}^{\infty} \frac{\Lambda(n)}{n^\sigma} = \frac{\zeta'(\sigma)}{\zeta(\sigma)} > -\frac{1}{\sigma-1} - c_0$$

for some constant c_0. Thus

$$-\frac{1}{\sigma-1} < c_2 \log q - \frac{2(\sigma-\beta)}{(\sigma-\beta)^2+\gamma^2},$$

and taking $\sigma = 1 + 2\delta/\log q$ gives

$$-\frac{1}{\sigma-1} < c_2 \log q - \frac{8}{5(\sigma-\beta)}$$

because

$$|\gamma| < \frac{\delta}{\log q} = \frac{1}{2}(\sigma-1) < \frac{1}{2}(\sigma-\beta).$$

Therefore,

$$\beta < 1 - \frac{\delta}{\log q}$$

for a sufficiently small δ. The argument for two real zeros or a double real zero is the same. This completes the proof. □

6.6 Supplementary Problems

Exercise 6.6.1 *Prove that* $\Gamma(s)$ *has poles only at* $s = 0, -1, -2, \ldots,$ *and that these are simple, with*

$$\mathrm{Res}_{s=-k}\Gamma(s) = (-1)^k/k!.$$

Exercise 6.6.2 *Show that*

$$e^{-1/x} = \frac{1}{2\pi i}\int_{(\sigma)} x^s\Gamma(s)ds,$$

for any $\sigma > 1$ *and* $x \geq 1$.

Exercise 6.6.3 *Let* $f(s) = \sum_{n=1}^{\infty} a_n/n^s$ *be an absolutely convergent Dirichlet series in the half-plane* $\mathrm{Re}(s) > 1$. *Show that*

$$\sum_{n=1}^{\infty} a_n e^{-n/x} = \frac{1}{2\pi i}\int_{(\sigma)} f(s)x^s\Gamma(s)ds$$

for any $\sigma > 1$.

Exercise 6.6.4 *Prove that*

$$\sin z = z\prod_{n=1}^{\infty}\left(1 - \frac{z^2}{n^2\pi^2}\right).$$

Exercise 6.6.5 *Using the previous exercise, deduce that*

$$\sum_{n=1}^{\infty}\frac{1}{n^2} = \frac{\pi^2}{6}.$$

7
Explicit Formulas

In this chapter our goal is to derive the explicit formula

$$\psi(x) = x - \sum_{\rho} \frac{x^{\rho}}{\rho} - \frac{\zeta'(0)}{\zeta(0)} - \frac{1}{2}\log\left(1 - x^{-2}\right),$$

where the sum is over the nontrivial zeros ρ of $\zeta(s)$. The method will then be used to derive the result

$$\psi(x) = x + O\left(x^{1/2}\log^{2} x\right)$$

assuming the Riemann hypothesis. A similar result can be obtained for primes in arithmetic progressions.

7.1 Counting Zeros

If $f(z)$ is analytic in C and non-vanishing on C, then the integral

$$\frac{1}{2\pi i}\int_{C} \frac{f'}{f}(z)dz$$

is equal to the number of zeros of f inside C, counted with multiplicity. This is easily seen by Cauchy's theorem.

Since

$$\frac{d}{dz}\log f(z) = \frac{f'(z)}{f(z)},$$

we have

$$\int_C \frac{f'}{f}(z)dz = \Delta_C \log f(z),$$

where Δ_C denotes the variation of $\log f(z)$ around the contour C. Also,

$$\log f(z) = \log|f(z)| + i\arg f(z),$$

and $\log|f|$ is single-valued. Thus, the formula can be rewritten as

$$\frac{1}{2\pi}\Delta_C \arg f(z).$$

Exercise 7.1.1 *Let L be the two line segments formed by the line joining 2 to $2 + iT$ and then $\frac{1}{2} + iT$. Show that*

$$\Delta_L \arg(s-1) = \frac{\pi}{2} + O\left(\frac{1}{T}\right).$$

Exercise 7.1.2 *With L as in the previous exercise, show that*

$$\Delta_L \arg \pi^{-s/2} = -\frac{T}{2}\log\pi.$$

Exercise 7.1.3 *With L as in the previous exercise, show that*

$$\Delta_L \arg\Gamma\left(\frac{s}{2}+1\right) = \frac{T}{2}\log\frac{T}{2} - \frac{T}{2} + \frac{3}{8}\pi + O\left(\frac{1}{T}\right).$$

Exercise 7.1.4 *Show that*

$$\sum_\rho \frac{1}{1+(T-\gamma)^2} = O(\log T),$$

where the sum is over the nontrivial zeros $\rho = \beta + i\gamma$ of $\zeta(s)$.

Exercise 7.1.5 *Let $N(T)$ be the number of zeros of $\zeta(s)$ with $0 < \mathrm{Im}(s) \le T$. Show that*

$$N(T+1) - N(T) = O(\log T).$$

Exercise 7.1.6 *Let $s = \sigma + it$ with t unequal to an ordinate of a zero. Show that for large $|t|$ and $-1 \le \sigma \le 2$,*

$$\frac{\zeta'(s)}{\zeta(s)} = \sum_{\rho}{}' \frac{1}{s-\rho} + O(\log|t|),$$

where the dash on the summation indicates that it is limited to those ρ for which $|t-\gamma| < 1$.

Theorem 7.1.7 *Let $N(T)$ be the number of zeros of $\zeta(s)$ in the rectangle $0 < \sigma < 1, 0 < t < T$. Then*

$$N(T) = \frac{T}{2\pi}\log\frac{T}{2\pi} - \frac{T}{2\pi} + \frac{7}{8} + S(T) + O\left(\frac{1}{T}\right),$$

where

$$\pi S(T) = \Delta_L \arg\zeta(s)$$

and L denotes the path of line segments joining 2 to $2+iT$ and then to $\frac{1}{2}+iT$. We also have

$$S(T) = O(\log T).$$

Proof. Let R be the rectangle with vertices 2, $2+iT$, $-1+iT$, and -1, traversed in the counterclockwise direction. Then

$$2\pi N(T) = \Delta_R \arg\xi(s).$$

There is no change in the argument as s goes from -1 to 2. Also, the change when s moves from $\frac{1}{2}+iT$ to $-1+iT$ and then to -1 is equal to the change as s moves from 2 to $2+iT$ and then to $\frac{1}{2}+iT$, since

$$\xi(\sigma+it) = \xi(1-\sigma-it) = \overline{\xi(1-\sigma+it)}.$$

Hence $\pi N(T) = \Delta_L \arg\xi(s)$, where L denotes the path of line segments joining 2 to $2+iT$ and then to $\frac{1}{2}+iT$. By Exercises 7.1.1 and 7.1.3, we deduce

$$N(T) = \frac{T}{2\pi}\log\frac{T}{2\pi} - \frac{T}{2\pi} + \frac{7}{8} + S(T) + O\left(\frac{1}{T}\right),$$

where

$$\pi S(T) = \Delta_L \arg\zeta(s).$$

Now, the variation of $\zeta(s)$ along $\sigma = 2$ is bounded, since $\log\zeta(s)$ is bounded there. Thus,

$$\pi S(T) = O(1) - \int_{\frac{1}{2}+iT}^{2+iT} \text{Im}\left(\frac{\zeta'(s)}{\zeta(s)}\right) ds.$$

We now apply Exercise 7.1.6, which says that

$$\frac{\zeta'(s)}{\zeta(s)} = \sum_{\rho}{}' \frac{1}{s-\rho} + O(\log t),$$

where the dash on the summation means $|\,\text{Im}(s-\rho)| < 1$. Observing that

$$\int_{\frac{1}{2}+iT}^{2+iT} \text{Im}\left(\frac{1}{s-\rho}\right) ds = \Delta \arg(s-\rho)$$

is at most π and noting that the number of terms in the sum above is $O(\log|t|)$ by Exercise 7.1.5 gives us

$$S(T) = O(\log T).$$

This completes the proof. □

7.2 Explicit Formula for $\psi(x)$

Our main tool in deriving the explicit formula for $\psi(x)$ will be Theorem 4.1.4. Recall that this theorem says that

$$I(x, R) = \frac{1}{2\pi i} \int_{c-iR}^{c+iR} \frac{x^s}{s} ds$$

satisfies

$$|I(x, R) - \delta(x)| < \begin{cases} x^c \min(1, R^{-1}|\log x|^{-1}) & \text{if} \quad x \neq 1, \\ \dfrac{c}{R} & \text{if} \quad x = 1, \end{cases}$$

where

$$\delta(x) = \begin{cases} 0 & \text{if} \quad 0 < x < 1, \\ 1/2 & \text{if} \quad x = 1, \\ 1 & \text{if} \quad x > 1. \end{cases}$$

Exercise 7.2.1 *Show that if x is not a prime power and $x > 1$, then*

$$\psi(x) = \frac{1}{2\pi i}\int_{c-iR}^{c+iR} -\frac{\zeta'(s)}{\zeta(s)}\frac{x^s}{s}ds$$

$$+O\left(\sum_{n=1}^{\infty}\Lambda(n)\left(\frac{x}{n}\right)^c \min\left(1, R^{-1}\left|\log\frac{x}{n}\right|^{-1}\right)\right).$$

Exercise 7.2.2 *Prove that if x is not an integer, then*

$$\sum_{\frac{1}{2}x<n<2x}\left|\log\frac{x}{n}\right|^{-1} = O\left(\frac{x}{\|x\|}\log x\right),$$

where $\|x\|$ denotes the distance of x to the nearest integer.

Exercise 7.2.3 *By choosing $c = 1 + \frac{1}{\log x}$ in the penultimate exercise, deduce that*

$$\psi(x) = \frac{1}{2\pi i}\int_{c-iR}^{c+iR} -\frac{\zeta'(s)}{\zeta(s)}\frac{x^s}{s}ds + O\left(\frac{x\log^2 x}{R}\right)$$

if $x - \frac{1}{2}$ is a positive integer.

Exercise 7.2.4 *Let C be the rectangle with vertices $c-iR$, $c+iR$, $-U+iR$, $-U-iR$, where $c = 1+1/\log x$, and U is an odd positive integer. Show that*

$$\frac{1}{2\pi i}\int_C -\frac{\zeta'(s)}{\zeta(s)}\frac{x^s}{s}ds = x - \sum_{|\gamma|<R}\frac{x^\rho}{\rho} - \frac{\zeta'(0)}{\zeta(0)} + \sum_{0<2m<U}\frac{x^{-2m}}{2m},$$

where we are writing the nontrivial zeros of $\zeta(s)$ as $\rho = \beta + i\gamma$. (R is chosen so that it is not the ordinate of any zero of $\zeta(s)$.)

Exercise 7.2.5 *Recall that the number of zeros $\rho = \beta + i\gamma$ satisfying $|\gamma - R| < 1$ is $O(\log R)$. Show that we can ensure $|\gamma - R| \gg (\log R)^{-1}$ by varying R by a bounded amount.*

Exercise 7.2.6 *Let U be a positive odd number. Prove that*

$$|\zeta'(s)/\zeta(s)| \ll (\log 2|s|)$$

for $-U \leq \sigma \leq -1$, provided that we exclude circles of a fixed positive radius around the trivial zeros $s = -2, -4, \ldots$ of $\zeta(s)$.

Exercise 7.2.7 *In Exercise 7.2.4, letting $U \to \infty$ along the odd numbers and $R \to \infty$ appropriately (that is, as in Exercise 7.2.5) prove that*

$$\psi(x) = x - \sum_{\rho} \frac{x^\rho}{\rho} - \frac{\zeta'(0)}{\zeta(0)} + \frac{1}{2}\log(1 - x^{-2}),$$

whenever x is half more than an integer.

We use these ideas to prove the following result:

Theorem 7.2.8 *For some constant $c_1 > 0$,*

$$\psi(x) = x + O\left(x \exp\left(-c_1\sqrt{\log x}\right)\right)$$

Proof. By the solution to Exercise 7.2.7, we know that

$$\psi(x) = x - \sum_{|\rho|<R} \frac{x^\rho}{\rho} - \frac{\zeta'(0)}{\zeta(0)} + \frac{1}{2}\log\left(1 - x^{-2}\right) + O\left(\frac{x\log^2 x}{R} + \frac{x\log^2 R}{R\log x}\right).$$

By Theorem 6.5.6, we have $\operatorname{Re}(\rho) = \beta < 1 - \frac{c}{\log R}$, so that the sum over the zeros is

$$x\exp\left(-\frac{c\log x}{\log R}\right)\sum_{|\gamma|<R}\frac{1}{|\rho|}.$$

By partial summation and Theorem 7.1.7 we have

$$\sum_{|\gamma|<R}\frac{1}{|\rho|} \ll \int_1^R \frac{\log t}{t}dt \ll \log^2 R.$$

The optimal choice for R satisfies

$$\log R = c_2(\log x)^{1/2}$$

for some appropriate constant c_2. It is now easily verified that this gives the desired result. □

Exercise 7.2.9 *Assuming the Riemann hypothesis, show that*

$$\psi(x) = x + O\left(x^{1/2}\log^2 x\right)$$

as $x \to \infty$.

Exercise 7.2.10 *Show that if*

$$\psi(x) = x + O\left(x^{1/2}\log^2 x\right)$$

then $\zeta(s)$ has no zeros for $\operatorname{Re}(s) > 1/2$.

7.3 Weil's Explicit Formula

The general philosophy of explicit formulas is to relate the sum of a suitable function over prime powers to the sum of the Fourier transform of that function over the zeros of the zeta function. The same philosophy applies to any function of the Selberg class (see Chapter 8). Here, we develop it only for the zeta function. In many applications, such formulas are useful in establishing subtle information on the distribution of prime numbers by exploiting information about the zeros of $\zeta(s)$ or vice versa.

Lemma 7.3.1 *Let $\epsilon > 0$ and let $h(s)$ be holomorphic in the strip $-\frac{1}{2} - \epsilon \leq \mathrm{Re}(s) \leq \frac{1}{2} + \epsilon$ and satisfy $h(s) = h(-s)$, $h(s) = O(|s|^{-1-\epsilon})$ as $|s| \to \infty$. Then*

$$\frac{1}{2\pi i} \int_{(\frac{1}{2}+\epsilon)} \frac{\xi'(\frac{1}{2}+s)}{\xi(\frac{1}{2}+s)} h(s) ds = \sum_{\gamma} h(i\gamma),$$

where $\xi(s) = s(s-1)\pi^{-s/2}\Gamma(s/2)\zeta(s)$, and the summation is over all γ such that $\frac{1}{2} + i\gamma$ is a zero of $\zeta(s)$ with $\mathrm{Im}(i\gamma) > 0$.

Proof. Recall that $\xi(s)$ is an entire function of order 1 and has the factorization

$$\frac{1}{2} e^{Bs} \prod_{\rho} \left(1 - \frac{s}{\rho}\right) e^{s/\rho},$$

where the product is over the nontrivial zeros $\rho = \frac{1}{2} + i\gamma$ of $\zeta(s)$ in $0 \leq \mathrm{Re}(s) \leq 1$ (Exercise 6.4.3).

Thus

$$\begin{aligned} \frac{\xi'(s)}{\xi(s)} &= B + \sum_{\rho} \left(\frac{1}{s-\rho} + \frac{1}{\rho} \right) \\ &= B + \sum_{\rho} \frac{s}{\rho(s-\rho)}. \end{aligned}$$

By the argument in Exercise 7.1.6, we see that

$$\frac{\xi'(s)}{\xi(s)} = \sum_{\rho}{}' \frac{s}{\rho(s-\rho)} + O(\log(|t|+1)),$$

where the dash on the summation means $|\mathrm{Im}(s-\rho)| < 1$ and $t = \mathrm{Im}(s)$. For any given T we can vary T by a bounded amount to ensure that $|\gamma - t| \geq (\log T)^{-1}$ by the argument in Exercise 7.2.5. Thus the summation is $O(|s| \log |s|)$ for $\mathrm{Im}(s) = T$.

Thus, by the hypothesis on $h(s)$, we can always find arbitrarily large $T > 0$ such that

$$\frac{\xi'(\frac{1}{2}+s)}{\xi(\frac{1}{2}+s)} h(s) = O(|s|^{-\epsilon})$$

for $s = \sigma + iT$ and $-1-\epsilon \leq \sigma \leq 1+\epsilon$. Now let R_T be the closed rectangular contour described by traversing the vertices $\frac{1}{2}+\epsilon - iT$, $\frac{1}{2}+\epsilon+iT$, $-\frac{1}{2}-\epsilon+iT$, and $-\frac{1}{2}-\epsilon-iT$. Since the zeros of $\zeta(s)$ occur in pairs $1/2 \pm i\gamma$, it follows by Cauchy's theorem that

$$\frac{1}{2\pi i}\int_{R_T} \frac{\xi'(\frac{1}{2}+s)}{\xi(\frac{1}{2}+s)} h(s)ds = 2 \sum_{0<\mathrm{Im}(i\gamma)<T} h(i\gamma).$$

Since

$$\lim_{T\to\infty} \frac{\xi'(\sigma+iT)}{\xi(\sigma+iT)} h(\sigma+iT) = \lim_{T\to\infty} O(T^{-\epsilon}) = 0$$

for $-\frac{1}{2}-\epsilon \leq \sigma \leq \frac{1}{2}+\epsilon$, it follows that the horizontal integrals tend to 0 as $T \to \infty$. By the functional equation $\xi(s) = \xi(1-s)$, we have

$$\frac{\xi'(\frac{1}{2}+s)}{\xi(\frac{1}{2}+s)} = -\frac{\xi'(\frac{1}{2}-s)}{\xi(\frac{1}{2}-s)},$$

so that the vertical line integrals are equal to

$$\frac{1}{2\pi i}\int_{\frac{1}{2}+\epsilon-iT}^{\frac{1}{2}+\epsilon+iT} \frac{\xi'(\frac{1}{2}+s)}{\xi(\frac{1}{2}+s)} h(s)ds.$$

Now

$$\frac{\xi'(\frac{1}{2}+s)}{\xi(\frac{1}{2}+s)} = \frac{1}{s-1/2} - \frac{1}{2}\log\pi + \frac{\Gamma'(s/2+3/4)}{2\Gamma(s/2+3/4)} + \frac{\zeta'(s+1/2)}{\zeta(s+1/2)}.$$

On the vertical line $\mathrm{Re}(s) = \frac{1}{2}+\epsilon$, the quantity

$$\frac{1}{s-1/2} + \frac{1}{2}\log\pi + \frac{\zeta'(s+1/2)}{\zeta(s+1/2)}$$

is bounded by

$$1/\epsilon + \frac{1}{2}\log\pi + \frac{\zeta'(1+\epsilon)}{\zeta(1+\epsilon)}.$$

Also,

$$\frac{\Gamma'(s/2+3/4)}{\Gamma(s/2+3/4)}$$

is bounded according to Exercise 6.3.17 by $O(\log(|s|+1))$. Since $h(s) = O(|s|^{-1-\epsilon})$, the above integral converges absolutely. Letting $T \to \infty$ establishes the lemma. □

Theorem 7.3.2 (Weil's explicit formula) *Assume that $h(s)$ satisfies the conditions of Lemma 7.3.1. In addition, assume that $h(it) = h_0(t/2\pi)$ is a real-valued function for $t \in \mathbb{R}$ whose Fourier transform*

$$\hat{h}_0(y) = \int_{-\infty}^{\infty} h_0(t) e^{-2\pi i t y}\, dt$$

satisfies the bound

$$\hat{h}_0(y) = O(e^{-(\frac{1}{2}+\epsilon)y})$$

for fixed $\epsilon > 0$ as $y \to \infty$. Then we have

$$\sum_{\gamma} h(i\gamma) + \sum_{n=1}^{\infty} \frac{\Lambda(n)}{\sqrt{n}} \hat{h}_0(\log n)$$

$$= h\left(\frac{1}{2}\right) - \frac{1}{2}(\log \pi)\hat{h}_0(0) + \int_{-\infty}^{\infty} \frac{\Gamma'(1/4+i\pi t)}{\Gamma(1/4+i\pi t)} h_0(t)\,dt,$$

where the first sum is over all zeros $1/2 + i\gamma$ satisfying $\mathrm{Im}(i\gamma) > 0$, and $\Lambda(n)$ is the von Mangoldt function, so that the second sum is over all prime powers.

Remark. The growth conditions on h and $\hat{h}_0$ ensure that the integrals and sums in the formula converge absolutely.

Proof. Recall that

$$\frac{\xi'(\frac{1}{2}+s)}{\xi(\frac{1}{2}+s)}$$

$$= \frac{1}{s+1/2} + \frac{1}{s-1/2} - \frac{1}{2}\log \pi + \frac{\Gamma'(1/4+s/2)}{\Gamma(1/4+s/2)} - \sum_{n=1}^{\infty} \frac{\Lambda(n)}{n^{s+1/2}},$$

so that inserting this into Lemma 7.3.1 we see that

$$\frac{1}{2\pi i}\int_{(\frac{1}{2}+\epsilon)}\left\{\frac{1}{s+1/2}+\frac{1}{s-1/2}-\frac{1}{2}\log\pi\right.$$

$$\left.+\frac{\Gamma'(1/4+s/2)}{\Gamma(1/4+s/2)}-\sum_{n=1}^{\infty}\frac{\Lambda(n)}{n^{s+1/2}}\right\}h(s)ds=\sum_{\gamma}h(i\gamma).$$

Observe that by the growth condition on h,

$$\frac{1}{2\pi i}\int_{(\frac{1}{2}+\epsilon)}h(s)n^{-s}ds=\frac{1}{2\pi i}\int_{(0)}h(s)n^{-s}ds$$

by moving the line of integration to the purely imaginary axis. Thus

$$\begin{aligned}\frac{1}{2\pi i}\int_{(\frac{1}{2}+\epsilon)}h(s)n^{-s}ds &= \frac{1}{2\pi}\int_{-\infty}^{\infty}h(it)e^{-it\log n}dt\\ &= \frac{1}{2\pi}\int_{-\infty}^{\infty}h_0(t/2\pi)e^{-it\log n}dt\\ &= \frac{1}{2\pi}\int_{-\infty}^{\infty}h_0(t)e^{-2\pi it\log n}dt\\ &= \hat{h}_0(\log n).\end{aligned}$$

Similarly, we can also move the other integrals to $\mathrm{Re}(s)=0$, which gives rise to the other terms of the formula. This completes the proof. □

7.4 Supplementary Problems

Exercise 7.4.1 *Using the method of Exercise 6.5.3, prove that for* $1\leq\sigma\leq 2$, $|t|\geq 2$,

$$-\mathrm{Re}\left(\frac{L'(s,\chi)}{L(s,\chi)}\right)<A_1\log q(|t|+2)-\sum_{\rho}\mathrm{Re}\left(\frac{1}{s-\rho}\right),$$

where A_1 *is an absolute constant, and the summation is over all zeros* ρ *of* $L(s,\chi)$, *and* χ *is a primitive Dirichlet character* $(\mathrm{mod}\, q)$. (Of course, $s=\sigma+it$, as usual.)

Exercise 7.4.2 *Let χ be a primitive Dirichlet character $(\bmod\, q)$. If $\rho = \beta + i\gamma$ runs through the nontrivial zeros of $L(s,\chi)$, then show that for any real t,*

$$\sum_\rho \frac{1}{1+(t-\gamma)^2} = O(\log q(|t|+2)).$$

Exercise 7.4.3 *With χ a primitive character $(\bmod\, q)$ and t not coinciding with the ordinate of a zero, show that for $-3/2 \le \sigma \le 5/2$, $|t| \ge 2$,*

$$\frac{L'}{L}(s,\chi) = \sum_\rho{}' \frac{1}{s-\rho} + O(\log q(|t|+2)),$$

where the dash on the sum means we sum over $\rho = \beta + i\gamma$ for which $|t-\gamma| < 1$.

Exercise 7.4.4 *Let χ be a primitive Dirichlet character $(\bmod\, q)$. Let $N(T,\chi)$ be the number of zeros of $L(s,\chi)$ in the rectangle $0 < \sigma < 1$, $|t| < T$. Show that*

$$N(T,\chi) = \frac{T}{\pi}\log\frac{qT}{2\pi} - \frac{T}{2\pi} + O(\log qT)$$

for $T \ge 2$.

Exercise 7.4.5 *Let χ be a primitive Dirichlet character $(\bmod\, q)$. If x is not a prime power and $\chi(-1) = -1$, derive the explicit formula*

$$\begin{aligned} \psi(x,\chi) &:= \sum_{n\le x}\chi(n)\Lambda(n) \\ &= -\sum_\rho \frac{x^\rho}{\rho} - \frac{L'(0,\chi)}{L(0,\chi)} + \sum_{m=1}^{\infty}\frac{x^{1-2m}}{2m-1}, \end{aligned}$$

where the first sum on the right-hand side is over the nontrivial zeros of $L(s,\chi)$.

Exercise 7.4.6 *Let χ be a primitive Dirichlet character $(\bmod\, q)$. If x is not a prime power and $\chi(-1) = 1$, derive the explicit formula*

$$\psi(x,\chi) = -\sum_\rho \frac{x^\rho}{\rho} - \log x - b(\chi) - \frac{1}{2}\log(1-x^{-2}),$$

where $b(\chi) = \lim_{s\to 0}\left(\frac{L'(s,\chi)}{L(s,\chi)} - \frac{1}{s}\right)$, and the sum on the right-hand side is over the nontrivial zeros of $L(s,\chi)$.

Exercise 7.4.7 *Let χ be a primitive Dirichlet character* $(\bmod\, q)$ *and set $a = 0$ or 1 according as $\chi(-1) = 1$ or -1. If $x - 1/2$ is a positive integer, show that*

$$\begin{aligned}\psi(x,\chi) &= -\sum_{|\gamma|<R} \frac{x^\rho}{\rho} - (1-a)(\log x + b(\chi)) \\ &\quad + \sum_{m=1}^{\infty} \frac{x^{a-2m}}{2m-a} + O\left(\frac{x \log^2 qxR}{R}\right),\end{aligned}$$

where the first summation is over zeros $\rho = \beta + i\gamma$ and R is chosen greater than or equal to 2 so as not to coincide with the ordinate of any zero of $L(s,\chi)$.

Exercise 7.4.8 *If we assume that all the nontrivial zeros of $L(s,\chi)$ lie on* $\mathrm{Re}(s) = 1/2$ *(the generalized Riemann hypothesis), prove that*

$$\psi(x,\chi) = O(x^{1/2} \log^2 qx).$$

Exercise 7.4.9 *Let*

$$\psi(x,q,a) = \sum_{\substack{n \le x \\ n \equiv a (\bmod q)}} \Lambda(n).$$

Show that the generalized Riemann hypothesis implies

$$\psi(x,q,a) = \frac{x}{\phi(q)} + O\left(x^{1/2} \log^2 qx\right)$$

when $(a,q) = 1$.

Exercise 7.4.10 *Assuming the generalized Riemann hypothesis, show that there is always a prime $p \ll q^2 \log^4 q$ satisfying $p \equiv a \pmod q$ whenever $(a,q) = 1$.*

Exercise 7.4.11 *Show that if q is prime, then*

$$\frac{\varphi(q-1)}{q-1} \sum_{d \mid q-1} \frac{\mu(d)}{\varphi(d)} \sum_{o(\chi)=d} \chi(a) = \begin{cases} 1 & \text{if } a \text{ has order } q-1 \\ 0 & \text{otherwise,} \end{cases}$$

where the inner sum is over characters $\chi \bmod q$ whose order is d.

Exercise 7.4.12 *Let q be prime and assume the generalized Riemann hypothesis. For q sufficiently large, show that there is always a prime $p < q$ such that p is a primitive root* $\pmod q$*,*

Exercise 7.4.13 *Let q be a prime. Show that the smallest primitive root* $\bmod q$ *is* $O\left(2^{\nu(q-1)} q^{1/2} \log q\right)$, *where $\nu(q-1)$ is the number of distinct prime factors of $q-1$.*

Exercise 7.4.14 *Let q be a prime and assume the generalized Riemann hypothesis. Show that there is always a prime-power primitive root satisfying the bound* $O\left(4^{\nu(q-1)} \log^4 q\right)$.

Exercise 7.4.15 *Let q be prime and assume the generalized Riemann hypothesis. Show that the least quadratic nonresidue* $\pmod q$ *is* $O(\log^4 q)$.

Exercise 7.4.16 *Let q be prime and assume the generalized Riemann hypothesis. Show that the least prime quadratic residue* $\pmod q$ *is* $O(\log^4 q)$.

Exercise 7.4.17 *Prove that for $n > 1$,*

$$\lim_{T\to\infty} \frac{1}{T} \sum_{|\gamma| \leq T} n^{\rho} = -\frac{\Lambda(n)}{\pi},$$

where the summation is over zeros $\rho = \beta + i\gamma$, $\beta \in \mathbb{R}$, of the Riemann zeta function.

8

The Selberg Class

The Selberg class S consists of functions $F(s)$ of a complex variable s satisfying the following properties:

1. (Dirichlet series): For $\mathrm{Re}(s) > 1$,

$$F(s) = \sum_{n=1}^{\infty} \frac{a_n}{n^s}$$

   where $a_1 = 1$. (We will write $a_n(F) = a_n$ for the coefficients of the Dirichlet series of F.)

2. (Analytic continuation): For some integer $m \geq 0$, $(s-1)^m F(s)$ extends to an entire function of finite order.

3. (Functional equation): There are numbers $Q > 0$, $\alpha_i > 0$, $r_i \in \mathbb{C}$ with $\mathrm{Re}(r_i) \geq 0$ such that

$$\Phi(s) = Q^s \prod_{i=1}^{d} \Gamma(\alpha_i s + r_i) F(s)$$

   satisfies the functional equation

$$\Phi(s) = w\overline{\Phi}(1-s),$$

   where w is a complex number with $|w| = 1$ and $\overline{\Phi}(s) = \overline{\Phi(\overline{s})}$.

4. (Euler product): For $\mathrm{Re}(s) > 1$,

$$F(s) = \prod_p F_p(s),$$

where

$$F_p(s) = \exp\Big(\sum_{k=1}^{\infty} \frac{b_{p^k}}{p^{ks}}\Big)$$

and $b_{p^k} = O(p^{k\theta})$ for some $\theta < 1/2$, and p denotes a prime number here. We shall write $b_p(F) = b_p$.

5. (Ramanujan hypothesis): For any fixed $\epsilon > 0$,

$$a_n = O(n^{\epsilon}),$$

where the implied constant may depend upon ϵ.

A prototypical example of an element of $\mathcal{S}$ is, of course, the Riemann zeta function. But more exemplary is the Ramanujan zeta function

$$L_{\Delta}(s) = \sum_{n=1}^{\infty} \frac{\tau_n}{n^s},$$

where $\tau_n = \tau(n)/n^{11/2}$ and τ is defined by the infinite product

$$\sum_{n=1}^{\infty} \tau(n) q^n = q \prod_{n=1}^{\infty} (1 - q^n)^{24}.$$

Ramanujan established properties (1), (2), and (3) and conjectured (4) and (5). Property (4) was proved by Mordell and (5) by Deligne.

8.1 The Phragmén - Lindelöf Theorem

We discuss an important theorem that allows us to estimate the growth of a function in the region $a \le \mathrm{Re}(s) \le b$ from its behaviour on $\mathrm{Re}(s) = a$ and $\mathrm{Re}(s) = b$. We first recall the maximum modulus principle.

Exercise 8.1.1 *Let $f(z)$ be an analytic function, regular in a region R and on the boundary ∂R, which we assume to be a simple closed contour. If $|f(z)| \le M$ on ∂R, show that $|f(z)| \le M$ for all $z \in R$.*

Exercise 8.1.2 (The maximum modulus principle) *If f is as in the previous exercise, show that $|f(z)| < M$ for all interior points $z \in R$, unless f is constant.*

Theorem 8.1.3 (Phragmén - Lindelöf) *Suppose that $f(s)$ is entire in the region*

$$S(a,b) = \{s \in \mathbb{C} : a \leq \mathrm{Re}(s) \leq b\}$$

and that as $|t| \to \infty$,

$$|f(s)| = O\left(e^{|t|^\alpha}\right)$$

for some $\alpha \geq 1$. If $f(s)$ is bounded on the two vertical lines $\mathrm{Re}(s) = a$ and $\mathrm{Re}(s) = b$, then $f(s)$ is bounded in $S(a,b)$.

Proof. We first select an integer $m > \alpha$, $m \equiv 2 \pmod 4$. Since $\arg s \to \pi/2$ as $t \to \infty$, we can choose T_1 sufficiently large so that

$$|\arg s - \pi/2| < \pi/4m.$$

Then for $|\mathrm{Im}(s)| \geq T_1$, we find that $\arg s = \pi/2 - \delta = \theta$ (say) satisfies

$$\cos m\theta = -\cos m\delta < -1/\sqrt{2}.$$

Therefore, if we consider

$$g_\epsilon(s) = e^{\epsilon s^m} f(s),$$

then

$$|g_\epsilon(s)| \leq K e^{|t|^\alpha} e^{-\epsilon |s|^m/\sqrt{2}}.$$

Thus, $|g_\epsilon(s)| \to 0$ as $|t| \to \infty$. Let B be the maximum of $f(s)$ in the region

$$a \leq \mathrm{Re}(s) \leq b, \quad 0 \leq |\mathrm{Im}(s)| \leq T_1.$$

Let T_2 be chosen such that

$$|g_\epsilon(s)| \leq B$$

for $|\mathrm{Im}(s)| \geq T_2$. Thus,

$$|f(s)| \leq B e^{-\epsilon |s|^m \cos(m \arg s)} \leq B e^{\epsilon |s|^m}$$

for $|\mathrm{Im}(s)| \geq T_2$. Applying the maximum modulus principle to the region

$$a \leq \mathrm{Re}(s) \leq b, \quad 0 \leq |\mathrm{Im}(s)| \leq T_2,$$

we find that $|f(s)| \leq B e^{\epsilon |s|^m}$. This estimate holds for all s in $S(a,b)$. Letting $\epsilon \to 0$ yields the result. □

Corollary 8.1.4 *Suppose that $f(s)$ is entire in $S(a,b)$ and that $|f(s)| = O(e^{|t|^\alpha})$ for some $\alpha \geq 1$ as $|t| \to \infty$. If $f(s)$ is $O(|t|^A)$ on the two vertical lines $\mathrm{Re}(s) = a$ and $\mathrm{Re}(s) = b$, then $f(s) = O(|t|^A)$ in $S(a,b)$.*

Proof. We apply the theorem to the function $g(s) = f(s)/(s-u)^A$, where $u > b$. Then g is bounded on the two vertical strips, and the result follows. □

Exercise 8.1.5 *Show that for any entire function $F \in \mathcal{S}$, we have*

$$F(s) = O\left(|t|^A\right),$$

for some $A > 0$, in the region $0 \leq \mathrm{Re}(s) \leq 1$.

8.2 Basic Properties

We begin by stating the following theorem of Selberg:

Theorem 8.2.1 (Selberg) *For any $F \in \mathcal{S}$, let $N_F(T)$ be the number of zeros ρ of $F(s)$ satisfying $0 \leq \mathrm{Im}(\rho) \leq T$, counted with multiplicity. Then*

$$N_F(T) \sim \left(2\sum_{i=1}^{d} \alpha_i\right) \frac{T \log T}{2\pi}$$

as $T \to \infty$.

Proof. This is easily derived by the method used to count zeros of $\zeta(s)$ and $L(s,\chi)$ as in Theorem 7.1.7 and Exercise 7.4.4. □

Clearly, the functional equation for $F \in \mathcal{S}$ is not unique, by virtue of Legendre's duplication formula. However, the above theorem shows that the sum of the α_i's is well-defined. Accordingly, we define the degree of F by

$$\deg F := 2\sum_{i=1}^{d} \alpha_i.$$

Lemma 8.2.2 (Conrey and Ghosh) *If $F \in \mathcal{S}$ and $\deg F = 0$, then $F = 1$.*

Proof. We follow [CG]. A Dirichlet series can be viewed as a power series in the infinitely many variables p^{-s} as we range over primes p. Thus, if $\deg F = 0$, we can write our functional equation as

$$\sum_{n=1}^{\infty} a_n \left(\frac{Q^2}{n}\right)^s = wQ \sum_{n=1}^{\infty} \frac{\overline{a_n}}{n} n^s,$$

where $|w| = 1$.

Thus, if $a_n \neq 0$ for some n, then Q^2/n is an integer. Hence Q^2 is an integer. Moreover, $a_n \neq 0$ implies $n|Q^2$, so that our Dirichlet series is really a Dirichlet polynomial. Therefore, if $Q^2 = 1$, then $F = 1$, and we are done. So, let us suppose $q := Q^2 > 1$. Since $a_1 = 1$, comparing the Q^{2s} term in the functional equation above gives $|a_q| = Q$. Since a_n is multiplicative, we must have for some prime power $p^r || q$ that $|a_{p^r}| \geq p^{r/2}$. Now consider the p-Euler factor

$$F_p(s) = \sum_{j=0}^{r} \frac{a_{p^j}}{p^{js}}$$

with logarithm

$$\log F_p(s) = \sum_{j=0}^{\infty} \frac{b_{p^j}}{p^{js}}.$$

Viewing these as power series in $x = p^{-s}$, we write

$$P(x) = \sum_{j=0}^{r} A_j x^j,$$

$$\log P(x) = \sum_{j=0}^{\infty} B_j x^j,$$

where $A_j = a_{p^j}$, $B_j = b_{p^j}$. Since $a_1 = 1$, we can factor

$$P(x) = \prod_{j=1}^{r} (1 - R_i x),$$

so that

$$B_j = -\sum_{i=1}^{r} \frac{R_i^j}{j}.$$

We also know that

$$\prod_{i=1}^{r} |R_i| \geq p^{r/2},$$

so that

$$\max_{1\le i\le r} |R_i| \ge p^{1/2}.$$

But

$$|b_{p^j}|^{1/j} = |B_j|^{1/j} = \left|\sum_{i=1}^{r} \frac{R_i^j}{j}\right|^{1/j}$$

tends to $\max_{1\le i\le r} |R_i|$ as $j \to \infty$, which is greater than or equal to $p^{1/2}$. This contradicts the condition that $b_n = O(n^\theta)$ with $\theta < 1/2$. Therefore, $Q = 1$ and hence $F = 1$. □

We can now prove the following basic result:

Theorem 8.2.3 (Selberg) *If $F \in \mathcal{S}$ and F is of positive degree, then* $\deg F \ge 1$.

Proof. We follow [CG]. Consider the identity

$$\sum_{n=1}^{\infty} a_n e^{-nx} = \frac{1}{2\pi i}\int_{(2)} F(s)x^{-s}\Gamma(s)ds.$$

Because of the Phragmen - Lindelöf principle and the functional equation, we find that $F(s)$ has polynomial growth in $|\operatorname{Im}(s)|$ in any vertical strip. Thus, moving the line of integration to the left, and taking into account the possible pole at $s = 1$ of $F(s)$ as well as the poles of $\Gamma(s)$ at $s = 0, -1, -2, \ldots$, we obtain

$$\sum_{n=1}^{\infty} a_n e^{-nx} = \frac{P(\log x)}{x} + \sum_{n=0}^{\infty} \frac{F(-n)(-1)^n x^n}{n!},$$

where P is a polynomial. The functional equation relates $F(-n)$ to $F(n+1)$ with a product of gamma functions. If $0 < \deg F < 1$, we find by Stirling's formula that the sum on the right-hand side converges for all x. Moreover, $P(\log x)$ is analytic in $\mathbb{C}\backslash\{x \le 0 : x \in \mathbb{R}\}$. Hence the left-hand side is analytic in $\mathbb{C}\backslash\{x \le 0 : x \in \mathbb{R}\}$. But since the left-hand side is periodic with period $2\pi i$, we find that

$$f(z) = \sum_{n=1}^{\infty} a_n e^{-nz}$$

is entire. Thus, for any x,

$$a_n e^{-nx} = \int_0^{2\pi} f(x+iy)e^{iny}dy \ll n^{-2}$$

by integrating by parts. Choosing $x = 1/n$ gives $a_n = O(1/n^2)$. Hence the Dirichlet series

$$F(s) = \sum_{n=1}^{\infty} \frac{a_n}{n^s}$$

converges absolutely for $\operatorname{Re} s > -1$. However, relating $F(-1/2+it)$ to $F(3/2-it)$ by the functional equation and using Stirling's formula, we find that $F(-1/2+it)$ is not bounded. This contradiction forces $\deg F \geq 1$. □

An element $F \in \mathcal{S}$ will be called **primitive** if $F \neq 1$ and $F = F_1F_2$ with $F_1, F_2 \in \mathcal{S}$ implies $F_1 = 1$ or $F_2 = 1$. Thus, a primitive function cannot be factored nontrivially in $\mathcal{S}$.

Exercise 8.2.4 *Show that*

$$\deg F_1F_2 = \deg F_1 + \deg F_2.$$

Exercise 8.2.5 *If $F \in \mathcal{S}$ has degree 1, show that it is primitive.*

Exercise 8.2.6 *Show that any $F \in \mathcal{S}$, $F \neq 1$, can be written as a product of primitive functions.*

Exercise 8.2.7 *Show that the Riemann zeta function is a primitive function.*

Exercise 8.2.8 *If χ is a primitive character $\pmod q$, show that $L(s,\chi)$ is a primitive function of $\mathcal{S}$.*

Exercise 8.2.9 *If $F \in \mathcal{S}$ and $\epsilon > 0$ is fixed, show that $|a_n| \leq c(\epsilon)n^{\epsilon}$ implies that*

$$|b_{p^k}| \leq c(\epsilon)(2^k - 1)p^{k\epsilon}/k.$$

Exercise 8.2.10 *Prove the asymmetric form of the functional equation for $\zeta(s)$:*

$$\zeta(1-s) = 2^{1-s}\pi^{-s}\Big(\cos\frac{s\pi}{2}\Big)\Gamma(s)\zeta(s).$$

Exercise 8.2.11 *Show that for $k \in \mathbb{N}$,*

$$|\zeta(-k)| \leq Ck!/(2\pi)^k$$

for some absolute constant C.

Exercise 8.2.12 *Show that*

$$\sum_{n=1}^{\infty} e^{-nx} = x^{-1} + \sum_{k=0}^{\infty} \frac{\zeta(-k)(-x)^k}{k!}.$$

Deduce that for $k = 2, 3, \ldots$

$$\zeta(1-k) = -B_k/k$$

and $\zeta(0) = -1/2$, *where* B_k *denotes the kth Bernoulli number.*

Exercise 8.2.13 *Let* χ *be a primitive Dirichlet character* $(\bmod\, q)$ *satisfying* $\chi(-1) = 1$. *Prove that*

$$L(1-s, \overline{\chi}) = \sqrt{\frac{2}{\pi}} \frac{q^{1/2}}{\tau(\chi)} \left(\frac{2\pi}{q}\right)^{1/2-s} \left(\cos \frac{\pi s}{2}\right) \Gamma(s) L(s, \chi),$$

where $\tau(\chi)$ *denotes the Gauss sum.*

Exercise 8.2.14 *Let* χ *be a primitive character* $(\bmod\, q)$, *satisfying* $\chi(-1) = 1$. *Show that for* $k \in \mathbb{N}$,

$$|L(-k, \chi)| \leq C k! (q/2\pi)^k$$

for some constant $C = O(\sqrt{q})$.

Exercise 8.2.15 *Let* χ *be a primitive character* $(\bmod\, q)$, *satisfying* $\chi(-1) = 1$. *Show that*

$$L(1-s, \overline{\chi}) = -(2\pi)^{-1/2} \frac{i q^{1/2}}{\tau(\chi)} \left(\frac{2\pi}{q}\right)^{1/2-s} \left(\sin \frac{\pi s}{2}\right) \Gamma(s+1) L(s, \chi).$$

Exercise 8.2.16 *Let* χ *be a primitive Dirichlet character* $(\bmod\, q)$ *satisfying* $\chi(-1) = -1$. *Show that for* $k \in \mathbb{N}$,

$$|L(-k, \chi)| \leq C (k+1)! (q/2\pi)^k$$

for some constant $C = O(\sqrt{q})$.

Exercise 8.2.17 *Prove that*

$$\sum_{n=1}^{\infty} \chi(n) e^{-nx} = \sum_{k=0}^{\infty} \frac{L(-k, \chi)(-x)^k}{k!}.$$

Deduce that for $n \geq 1$,

$$L(1-n, \chi) = -B_{n,\chi}/n,$$

where

$$B_{n,\chi} = q^{n-1} \sum_{a=1}^{q} \chi(a) b_n\left(\frac{a}{q}\right),$$

with $b_n(x)$ *denoting the nth Bernoulli polynomial.*

8.3 Selberg's Conjectures

We have seen in the previous section that $\zeta(s)$ and Dirichlet's L-functions $L(s, \chi)$ are primitive since they are of degree 1. Selberg [S] conjectures that as $x \to \infty$:
(a) for any primitive function F,

$$\sum_{p \leq x} \frac{|a_p(F)|^2}{p} = \log \log x + O(1);$$

(b) for two distinct primitive functions F and G,

$$\sum_{p \leq x} \frac{a_p(F)\overline{a_p(G)}}{p} = O(1).$$

We have also seen that any function of $\mathcal{S}$ can be factored into primitive functions. Two of the important consequences of conjectures (a) and (b) are contained in the following exercises.

Exercise 8.3.1 *Assuming (a) and (b), prove that any function* $F \in \mathcal{S}$ *can be factored uniquely as a product of primitive functions.*

Exercise 8.3.2 *Suppose* $F, G \in \mathcal{S}$ *and* $a_p(F) = a_p(G)$ *for all but finitely many primes* p. *Assuming (a) and (b), prove that* $F = G$.

This exercise shows that a form of "strong multiplicity one" holds for the Selberg class. It is possible to prove a slightly stronger version of this fact without assuming (a) and (b). This is the goal of the exercises below.

Exercise 8.3.3 *If* $F(s) = \sum_{n=1}^{\infty} a_n n^{-s}$ *and* $\sigma = \text{Re}(s) > \sigma_a(F)$, *the abscissa of absolute convergence of* F, *then prove that*

$$\lim_{T\to\infty} \frac{1}{2T} \int_{-T}^{T} F(\sigma+it) y^{\sigma+it}\, dt = \begin{cases} a_n(F) & \text{if} \quad n = y, \\ 0 & \text{otherwise,} \end{cases}$$

for any real y.

Exercise 8.3.4 *Prove that*

$$\frac{1}{2\pi i} \int_{(c)} \frac{y^s ds}{(\alpha s + \beta)^2} = \begin{cases} \alpha^{-2} y^{-\beta/\alpha} \log y & \text{if} \quad y > 1, \\ 0 & \text{if} \quad 0 \leq y \leq 1, \end{cases}$$

for $c > 0$ *and* $\alpha, \beta > 0$.

Exercise 8.3.5 *Let* $f(s)$ *be a meromorphic function on* $\mathbb{C}$, *analytic for* $\text{Re}(s) \geq \frac{1}{2}$, *and nonvanishing there. Suppose that* $\log f(s)$ *is a Dirichlet series and that* $f(s)$ *satisfies the functional equation*

$$H(s) = w\overline{H}(1-s),$$

where w *is a complex number of absolute value* 1, *and*

$$H(s) = A^s \frac{\prod_{i=1}^{d_1} \Gamma(\alpha_i s + \beta_i)}{\prod_{i=1}^{d_2} \Gamma(\gamma_i s + \delta_i)} f(s)$$

with certain A, α_i, $\gamma_i > 0$ *and* $\text{Re}(\beta_i)$, $\text{Re}(\delta_i) \geq 0$. *Show that* $f(s)$ *is constant.*

Exercise 8.3.6 *Let* $F, G \in \mathcal{S}$. *Suppose* $a_p(F) = a_p(G)$, $a_{p^2}(F) = a_{p^2}(G)$ *for all but finitely many primes* p. *Show that* $F = G$.

Exercise 8.3.7 *Assume Selberg's conjectures (a) and (b). If* $F \in \mathcal{S}$ *has a pole of order* m *at* $s = 1$, *show that* $F(s)/\zeta(s)^m$ *is entire.*

Exercise 8.3.8 *Assume Selberg's conjectures (a) and (b). Show that for any* $F \in \mathcal{S}$, *there are no zeros on* $\text{Re}(s) = 1$.

8.4 Supplementary Problems

Exercise 8.4.1 *Verify that the primitive functions $\zeta(s)$ and $L(s,\chi)$, where χ is a primitive character $(\bmod q)$, satisfy Selberg's conjectures (a) and (b).*

Exercise 8.4.2 *For each F, G in S, define*

$$(F \otimes G)(s) = \prod_p H_p(s),$$

where

$$H_p(s) = \exp\left(\sum_{k=1}^{\infty} k b_{p^k}(F) b_{p^k}(G) p^{-ks}\right).$$

If $F_p(s) = \det(1 - A_p p^{-s})^{-1}$ and $G_p(s) = \det(1 - B_p p^{-s})^{-1}$ for certain nonsingular matrices A_p and B_p, show that

$$H_p(s) = \det(1 - (A_p \otimes B_p)p^{-s})^{-1}.$$

Exercise 8.4.3 *With notation as in the previous exercise, show that if $F, G \in \mathcal{S}$, then $F \otimes G$ converges absolutely for $\mathrm{Re}(s) > 1$.*

Exercise 8.4.4 *If $F \in \mathcal{S}$ and $F \otimes F$ extends to an analytic function for $\mathrm{Re}(s) \geq 1/2$, except for a simple pole at $s = 1$, we will say that F is $\otimes$-simple. Prove that a $\otimes$-simple function has at most a simple pole at $s = 1$.*

Exercise 8.4.5 *If $F \in \mathcal{S}$ and*

$$F = F_1^{e_1} F_2^{e_2} \cdots F_k^{e_k}$$

is a factorization of F into distinct primitive functions, show that

$$\sum_{p \leq x} \frac{|a_p(F)|^2}{p} = (e_1^2 + e_2^2 + \cdots + e_k^2) \log\log x + O(1),$$

assuming Selberg's conjectures (a) and (b).

Exercise 8.4.6 *If $F \in \mathcal{S}$ and $F \otimes \bar{F} \in \mathcal{S}$ show that F is $\otimes$-simple if and only if F is primitive, assuming Selberg's conjectures (a) and (b).*

Exercise 8.4.7 *If $F \in \mathcal{S}$ is $\otimes$-simple and entire, prove that $F(1 + it) \neq 0$ for all $t \in \mathbb{R}$.*

Exercise 8.4.8 *Let $F \in \mathcal{S}$ and write*

$$-\frac{F'}{F}(s) = \sum_{n=1}^{\infty} \Lambda_F(n) n^{-s}.$$

For $T > 1$ and $n \in \mathbb{N}$, $n > 1$, show that

$$\sum_{|\gamma| \leq T} n^{\rho} = -\frac{T}{\pi}\Lambda_F(n) + O\left(n^{3/2} \log^2 T\right),$$

where $\rho = \beta + i\gamma$, $\beta > 0$, runs over the nontrivial zeros of $F(s)$.

Exercise 8.4.9 *Suppose $F, G \in \mathcal{S}$. Let*

$$Z_F(T) = \{\rho = \beta + i\gamma, \beta > 0, F(\rho) = 0 \, and \, |\gamma| \leq T\}.$$

Suppose that as $T \to \infty$,

$$|Z_F(T) \Delta Z_G(T)| = o(T),$$

where Δ denotes the symmetric difference $A \Delta B = (A \setminus B) \cup (B \setminus A)$. Show that $F = G$.

9
Sieve Methods

9.1 The Sieve of Eratosthenes

The basic principle of a sieve method is the following: Given a finite set of natural numbers, estimate its size (from above and below) given information about the image of the set $\bmod p$ for a given set of primes p. For example, let S be the set of primes in the interval $[\sqrt{x}, x]$. We know that for each prime $p \leq \sqrt{x}$, the image of $S \bmod p$ fails to contain the zero residue class. Given this information, the estimation of S from above and below gives us estimates for $\pi(x) - \pi(\sqrt{x})$.

The oldest method to attack this question is the sieve of Eratosthenes (300 B.C.). It was formally written in the following form by Legendre in the eighteenth century.

Example 9.1.1 (Eratosthenes-Legendre) *Let P_z be the product of the primes $p \leq z$, and $\pi(x, z)$ the number of $n \leq x$ that are not divisible by any prime $p \leq z$. Then*

$$\pi(x, z) = \sum_{d|P_z} \mu(d) \left[\frac{x}{d}\right].$$

Solution. Clearly,

$$\begin{aligned}\pi(x,z) &= \sum_{n\le x}\sum_{d|(n,P_z)} \mu(d) \\ &= \sum_{d|P_z}\mu(d)\sum_{\substack{n\le x\\ d|n}} 1 = \sum_{d|P_z}\mu(d)\left[\frac{x}{d}\right]\end{aligned}$$

as required. □

We saw in Exercise 1.5.10 that

$$\pi(x,z) = x\prod_{p\le z}\left(1-\frac{1}{p}\right) + O(2^z),$$

and in Exercise 1.5.11 that

$$\prod_{p\le z}\left(1-\frac{1}{p}\right)^{-1} \gg \log z.$$

This gives the estimate (Exercise 1.5.12)

$$\pi(x,z) \ll \frac{x}{\log z} + O(2^z).$$

Choosing $z = \log x$, we obtain

$$\pi(x) = O\left(\frac{x}{\log\log x}\right).$$

Exercise 9.1.2 *Prove that there is a constant c such that*

$$\prod_{p\le z}\left(1-\frac{1}{p}\right) = \frac{e^{-c}}{\log z}\left(1+O\left(\frac{1}{\log z}\right)\right).$$

There is a famous theorem of Mertens that shows that the constant c in the previous exercise is Euler's constant γ, given by

$$\gamma = \lim_{t\to\infty}\left(\sum_{n\le t}\frac{1}{n} - \log t\right).$$

This is proved in the following way. For $\sigma > 0$, we have

$$\zeta(1+\sigma) = \sum_{n=1}^{\infty}\frac{1}{n^{1+\sigma}}.$$

Now consider

$$\begin{aligned} f(\sigma) &= \log \zeta(1+\sigma) - \sum_p \frac{1}{p^{1+\sigma}} \\ &= -\sum_p \left\{ \log\left(1 - \frac{1}{p^{1+\sigma}}\right) + \frac{1}{p^{1+\sigma}} \right\}. \end{aligned}$$

In Exercise 9.1.2 it was proved that $c = c_0 + c_1$, where

$$c_0 = -\sum_p \left\{ \log\left(1 - \frac{1}{p}\right) + \frac{1}{p} \right\}$$

and

$$c_1 = \lim_{z\to\infty} \left(\sum_{p\le z} \frac{1}{p} - \log\log z \right).$$

Hence $c_0 = \lim_{\sigma\to 0} f(\sigma)$. It is clear that as $\sigma \to 0^+$, $\log \zeta(1+\sigma) = \log \frac{1}{\sigma} + O(\sigma)$. Now, as $\sigma \to 0^+$, $\log(1 - e^{-\sigma}) = \log \sigma + O(\sigma)$, so that as $\sigma \to 0^+$,

$$\begin{aligned} \log \zeta(1+\sigma) &= -\log(1 - e^{-\sigma}) + O(\sigma) \\ &= \sum_{n=1}^{\infty} \frac{e^{-\sigma n}}{n} + O(\sigma). \end{aligned}$$

Put $H(t) = \sum_{n\le t} \frac{1}{n}$ and $P(t) = \sum_{p\le t} \frac{1}{p}$. By partial summation,

$$\begin{aligned} \sum_p \frac{1}{p^{1+\sigma}} &= \sigma \int_1^{\infty} \frac{P(u)}{u^{1+\sigma}} du \\ &= \sigma \int_0^{\infty} P(e^t) e^{-\sigma t} dt. \end{aligned}$$

Similarly,

$$\log \zeta(1+\sigma) = \sigma \int_0^{\infty} e^{-\sigma t} H(t) dt + O(\sigma)$$

as $\sigma \to 0^+$. Hence,

$$f(\sigma) = \sigma \int_0^{\infty} e^{-\sigma t} \Big(H(t) - P(e^t) \Big) dt + O(\sigma).$$

Since $H(t) = \log t + \gamma + O(1/t)$ by Example 2.1.10, and $P(e^t) = \log t + c_1 + O(1/t)$, we deduce

$$\begin{aligned} f(\sigma) &= \sigma \int_0^\infty e^{-\sigma t}\Big(\gamma - c_1 + O\Big(\frac{1}{t+1}\Big)\Big)dt + O(\sigma) \\ &= (\gamma - c_1) + \sigma \int_0^\infty e^{-\sigma t} O\Big(\frac{1}{t+1}\Big)dt + O(\sigma). \end{aligned}$$

An easy integration by parts shows that the integrand is $O(\sigma)$, so that $f(0) = c_0 = \gamma - c_1$. This proves the following theorem:

Theorem 9.1.3 (Mertens)

$$V(z) := \prod_{p \le z}\Big(1 - \frac{1}{p}\Big) = \frac{e^{-\gamma}}{\log z}\Big(1 + O\Big(\frac{1}{\log z}\Big)\Big).$$

Exercise 9.1.4 *For $z \le \log x$, prove that*

$$\pi(x, z) = (1 + o(1))\frac{xe^{-\gamma}}{\log z}$$

whenever $z = z(x) \to \infty$ as $x \to \infty$.

We now define $\Phi(x, z)$ to be the number of $n \le x$ all of whose prime factors are less than or equal to z. This function, along with $\pi(x, z)$, plays an important role in sieve problems.

Exercise 9.1.5 (Rankin's trick) *Prove that*

$$\Phi(x, z) \le x^\delta \prod_{p \le z}\Big(1 - \frac{1}{p^\delta}\Big)^{-1}$$

for any $\delta > 0$.

Exercise 9.1.6 *Choose $\delta = 1 - \frac{1}{\log z}$ in the previous exercise to deduce that*

$$\Phi(x, z) \ll x(\log z)\exp\Big(-\frac{\log x}{\log z}\Big).$$

Exercise 9.1.7 *Prove that*

$$\pi(x, z) = x\sum_{\substack{d | P_z \\ d \le x}} \frac{\mu(d)}{d} + O\Big(x(\log z)\exp\Big(-\frac{\log x}{\log z}\Big)\Big)$$

for $z = z(x) \to \infty$ as $x \to \infty$.

Exercise 9.1.8 *Prove that*

$$\sum_{\substack{d|P_z \\ d\leq x}} \frac{\mu(d)}{d} = \prod_{p\leq z}\left(1-\frac{1}{p}\right) + O\left((\log z)^2 \exp\left(-\frac{\log x}{\log z}\right)\right),$$

with $z = z(x) \to \infty$ *as* $x \to \infty$.

Exercise 9.1.9 *Prove that*

$$\pi(x, z) = xV(z) + O\left(x(\log z)^2 \exp\left(-\frac{\log x}{\log z}\right)\right),$$

where

$$V(z) = \prod_{p\leq z}\left(1-\frac{1}{p}\right)$$

and $z = z(x) \to \infty$ *as* $x \to \infty$.

Exercise 9.1.10 *Prove that*

$$\pi(x) \ll \frac{x}{\log x} \log\log x$$

by setting $\log z = \epsilon \log x/\log\log x$, *for some sufficiently small* ϵ, *in the previous exercise.*

Exercise 9.1.11 *For any* $A > 0$, *show that*

$$\pi(x, (\log x)^A) \sim \frac{xe^{-\gamma}}{A\log\log x}$$

as $x \to \infty$.

The estimate of Exercise 9.1.9 for $\pi(x)$ will be seen to be as good as the one obtained by the elementary Brun sieve of the next section. Let A be any set of natural numbers and let $\mathcal{P}$ be a set of primes. To each prime $p \in \mathcal{P}$, let there be $\omega(p)$ distinguished residue classes $\bmod\, p$. Let A_p denote the set of elements of A belonging to at least one of these distinguished classes $\bmod\, p$. For any square-free number d composed of primes $p \in \mathcal{P}$, let

$$A_d = \cap_{p|d} A_p.$$

We denote by $S(A, \mathcal{P}, z)$ the number of elements of

$$A \setminus \cup_{p\in\mathcal{P}, p\leq z} A_p.$$

Let $\omega(d) = \prod_{p|d} \omega(p)$, and $P(z) = \prod_{p\leq z, p\in\mathcal{P}} p$.

Exercise 9.1.12 *Suppose that*

$$\sum_{\substack{p \leq z \\ p \in \mathcal{P}}} \frac{\omega(p) \log p}{p} \leq \kappa \log z + O(1).$$

Show that

$$F_\omega(t, z) := \sum_{\substack{d \leq t \\ d | P(z)}} \omega(d)$$

is bounded by $O\Big(t(\log z)^\kappa \exp\Big(-\frac{\log t}{\log z}\Big)\Big)$.

Exercise 9.1.13 *Let C be a constant. With the same hypothesis as in the previous exercise, show that*

$$\sum_{\substack{d | P(z) \\ d > Cx}} \frac{\omega(d)}{d} = O\Big((\log z)^{\kappa+1} \exp\Big(-\frac{\log x}{\log z}\Big)\Big).$$

We are now ready to prove our version of the sieve of Eratosthenes. We follow [MS]. We suppose there is an X such that

$$|A_d| = \frac{X\omega(d)}{d} + R_d$$

with $R_d = O(\omega(d))$. We also assume

$$\sum_{\substack{p \leq z \\ p \in \mathcal{P}}} \frac{\omega(p) \log p}{p} \leq \kappa \log z + O(1)$$

and set

$$W(z) = \prod_{\substack{p \leq z \\ p \in \mathcal{P}}} \left(1 - \frac{\omega(p)}{p}\right).$$

Exercise 9.1.14 (Sieve of Eratosthenes) *Suppose there is a constant $C > 0$ such that $|A_d| = 0$ for $d > Cx$. Then*

$$S(A, \mathcal{P}, z) = XW(z) + O\Big(x(\log z)^{\kappa+1} \exp\Big(-\frac{\log x}{\log z}\Big)\Big).$$

9.2 Brun's Elementary Sieve

By comparing coeffients of x^r on both sides of the identity

$$(1-x)^{-1}(1-x)^{\nu} = (1-x)^{\nu-1}$$

we deduce

$$\sum_{k\le r}(-1)^k\binom{\nu}{k} = (-1)^r\binom{\nu-1}{r}.$$

This implies that

$$\sum_{\substack{d|n \\ \nu(d)\le r}} \mu(d) = (-1)^r\binom{\nu(n)-1}{r},$$

where $\nu(n)$ is the number of prime factors of n. This observation is the basis of Brun's elementary sieve. Namely, let

$$\mu_r(d) = \begin{cases} \mu(d) & \text{if} \quad \nu(d)\le r \\ 0 & \text{if} \quad \nu(d) > r. \end{cases}$$

Then setting

$$\psi_r(n) = \sum_{d|n}\mu_r(d),$$

we find that if r is even, $\sum_{d|n}\mu(d) \le \psi_r(n)$ and if r is odd, $\sum_{d|n}\mu(d) \ge \psi_r(n)$. Thus

$$\sum_{d|n}\mu(d) = \psi_r(n) + O\Big(\sum_{\substack{d|n \\ \nu(d)=r+1}} |\mu(d)|\Big).$$

Exercise 9.2.1 *Show that for r even,*

$$\pi(x,z) \le x\sum_{d|P_z}\frac{\mu_r(d)}{d} + O(z^r).$$

We now turn our attention to

$$\sum_{d|P_z}\frac{\mu_r(d)}{d}.$$

By Möbius inversion,

$$\mu_r(d) = \sum_{\delta | d} \mu(d/\delta)\psi_r(\delta),$$

so that

$$\begin{aligned}
\sum_{d|P_z} \frac{\mu_r(d)}{d} &= \sum_{d|P_z} \frac{1}{d} \sum_{\delta|d} \mu(d/\delta)\psi_r(\delta) \\
&= \sum_{\delta|P_z} \frac{\psi_r(\delta)}{\delta} \sum_{d|P_z/\delta} \frac{\mu(d)}{d} \\
&= V(z) \sum_{\delta|P_z} \frac{\psi_r(\delta)}{\phi(\delta)},
\end{aligned}$$

where $V(z)$ is as in the previous section and ϕ denotes Euler's function. Let us note that

$$\sum_{d|P_z} \frac{\mu_r(d)}{d} = V(z) + V(z) \sum_{\substack{\delta|P_z \\ \delta>1}} \frac{\psi_r(\delta)}{\phi(\delta)}.$$

We now want to estimate the last sum. Observe that

$$\psi_r(\delta) \leq \binom{\nu(\delta)-1}{r},$$

so that the sum under consideration is bounded by

$$\begin{aligned}
\sum_{\substack{\delta|P_z \\ \delta>1}} \binom{\nu(\delta)-1}{r} \frac{1}{\phi(\delta)} &\leq \sum_{r \leq m \leq \pi(z)} \binom{m}{r} \frac{1}{m!} \Big(\sum_{p \leq z} \frac{1}{p-1} \Big)^m \\
&\leq \frac{1}{r!} (\log\log z + c_1)^r \exp(\log\log z + c_1),
\end{aligned}$$

where we have utilized the elementary estimate

$$\sum_{p \leq z} \frac{1}{p} < \log\log z + c_1$$

for some constant c_1. Since $e^r \geq \frac{r^r}{r!}$, we can write $1/r! \leq (e/r)^r$, and thus

$$V(z) \sum_{\substack{\delta|P_z \\ \delta>1}} \frac{\psi_r(\delta)}{\phi(\delta)} \leq c_2 \exp(r - r\log r + r\log\Lambda),$$

where $\Lambda = \log\log z + c_1$, and we have used the estimate

$$V(z) \ll \frac{1}{\log z}.$$

The idea is to choose r so that the $r \log r$ term dominates. This will minimize the error term. Indeed, choosing r to be the nearest even integer to $\alpha \log x / \log z$, with $\alpha < 1$, gives an error term of

$$O\Big(x \exp\Big(-c_3 \frac{\log x}{\log z}\Big)\Big)$$

for some constant c_3, and we impose

$$\frac{\alpha \log x}{\log z} > 2(\log\log z + c_1)$$

to ensure that the error term is sufficiently small. This proves the following theorem:

Theorem 9.2.2 *There is a constant $c_4 > 0$ such that for*

$$\log z < c_4 \log x / \log\log x,$$

we have

$$\pi(x, z) \leq xV(z) + O\Big(x \exp\Big(-c_3 \frac{\log x}{\log z}\Big)\Big).$$

Remark. Observe that this is comparable to the estimate obtained earlier by using the sieve of Eratosthenes combined with the careful application of Rankin's trick (Exercises 9.1.8 and 9.1.9).

Also note that Theorem 9.2.2 gives us the upper bound

$$\pi(x) \ll \frac{x}{\log x}(\log\log x),$$

which is comparable to the estimate we obtained in Exercise 9.1.10.

Brun used his method described above to deduce that the number of primes $p \leq x$ such that $p + 2$ is also prime is bounded by

$$\ll \frac{x}{(\log x)^2}(\log\log x)^2.$$

From this, it is easy to deduce by partial summation that

$$\sum{}' \frac{1}{p} < \infty,$$

where p is such that $p+2$ is prime, a result that created a sensation at the time it was proved by Brun.

Let A be a finite set of natural numbers, $\mathcal{P}$ a set of primes. For square-free d composed of primes from $\mathcal{P}$, let A_d be the set of elements of A divisible by d. For some $\omega(d)$ multiplicative, suppose

$$|A_d| = \frac{\omega(d)}{d}|A| + R_d.$$

Let $S(A, \mathcal{P}, z)$ denote the number of elements of A coprime to

$$P(z) = \prod_{\substack{p \le z \\ p \in \mathcal{P}}} p.$$

As above

$$\begin{aligned} S(A, \mathcal{P}, z) &= \sum_{n \in A} \sum_{d|(n, P(z))} \mu(d) \\ &= \sum_{n \in A} \Bigl(\psi_r(n, P(z)) + O\Bigl(\sum_{\substack{d|(n,P(z)) \\ \nu(d)=r+1}} 1\Bigr)\Bigr) \\ &= \sum_{d|P(z)} \mu_r(d) \left(\frac{\omega(d)}{d}|A| + R_d\right) + O\left(\frac{|A|z^r}{r!}\right). \end{aligned}$$

We make the hypothesis $|R_d| \le \omega(d)$. Then

$$\begin{aligned} S(A, \mathcal{P}, z) &= |A| \sum_{d|P(z)} \frac{\mu_r(d)\omega(d)}{d} + O\left(\frac{|A|z^r}{r!}\right) \\ &+ O\Bigl(1 + \sum_{p<z} \omega(p)\Bigr)^r. \end{aligned}$$

Exercise 9.2.3 *Show that*

$$\sum_{d|P(z)} \frac{\mu_r(d)\omega(d)}{d} = \prod_{\substack{p \le z \\ p \in \mathcal{P}}} \left(1 - \frac{\omega(p)}{p}\right) \sum_{\delta|P(z)} \frac{\psi_r(\delta)\omega(\delta)}{\Omega(\delta)},$$

where $\Omega(\delta) = \prod_{p|\delta}(p - \omega(p))$.

Exercise 9.2.4 *Suppose that $\omega(p) \leq c$, and that*

$$\sum_{\substack{p \leq z \\ p \in \mathcal{P}}} \frac{\omega(p)}{p} \leq c_1 \log\log z + c_2$$

for some constants c, c_1, and c_2. Show that there are constants c_3, c_4, and c_5 such that

$$\sum_{\substack{\delta | P(z) \\ \delta > 1}} \frac{\psi_r(\delta)\omega(\delta)}{\Omega(\delta)} \ll \frac{1}{r!}(c_3 \log\log z + c_4)^r (\log z)^{c_5}.$$

We can put these inequalities together in the following form:

Theorem 9.2.5 (Brun's elementary sieve) *Suppose that $\omega(p) \leq c$ and that*

$$\sum_{\substack{p \leq z \\ p \in \mathcal{P}}} \frac{\omega(p)}{p} \leq c_1 \log\log z + c_2$$

for some constants c, c_1, and c_2. Suppose further that $R_d = O(\omega(d))$. Then there are constants c_3 and c_4 such that

$$\begin{aligned} S(A, \mathcal{P}, z) &= |A| \prod_{\substack{p \leq z \\ p \in \mathcal{P}}} \left(1 - \frac{\omega(p)}{p}\right) + O\left(\frac{|A| z^r}{r!}\right) + O(z^r) \\ &+ O\left(\frac{|A| \cdot (c_1 \log\log z + c_4)^r}{r!} (\log z)^{c_3}\right) \end{aligned}$$

for any even number r.

To make this amenable for applications, we use the inequality

$$\frac{1}{r!} \leq \left(\frac{e}{r}\right)^r$$

to obtain

$$S(A, \mathcal{P}, z) = |A|(W(z) + O(\exp(-r \log r + r \log z + r))) + O(z^r).$$

Our intention now is to make the $r \log r$ term dominate so that we can get a small error term in the above result. Suppose that $|A| \ll x$. We choose r to be the nearest even integer to

$$\eta \log\log z$$

for some $\eta = \eta(x, z)$ soon to be specified. With this choice of r the error term becomes

$$\ll x \exp(-\eta(\log \eta) \log\log z + \eta(\log z) \log\log z + \eta \log\log z) + z^{\eta \log\log z}.$$

If we choose

$$\eta = \frac{\alpha \log x}{(\log z)(\log\log z)},$$

for some $\alpha < 1$, the error term is

$$\ll x \exp\left(-c_3 \frac{\log x}{\log z}\right),$$

for some positive constant c_3. In particular, there is a constant $c > 0$ such that for

$$\log z < \frac{c \log x}{\log\log x},$$

we have

$$S(A, \mathcal{P}, z) = |A| W(z) + O\left(x \exp\left(-c_3 \frac{\log x}{\log\log x}\right)\right). \qquad (9.1)$$

Exercise 9.2.6 *Show that the number of primes $p \leq x$ such that $p + 2$ is also prime is $\ll x(\log\log x)^2/(\log x)^2$.*

Exercise 9.2.7 (Brun, 1915) *Show that*

$$\sum{}' \frac{1}{p} < \infty,$$

where the dash on the sum means we sum over primes p such that $p + 2$ is also prime.

9.3 Selberg's Sieve

The key idea of Selberg is to replace the use of the Möbius function that appears in Brun's sieve as well as the sieve of Eratosthenes by another sequence optimally chosen so as to minimize the resulting estimates. The method is best illustrated by the example below.

Let $\lambda_1 = 1$, and let us set $\lambda_d = 0$ for $d > z$. Let us now consider the problem of estimating $\pi(x, z)$.

Exercise 9.3.1 *Let $P_z = \prod_{p \leq z} p$ be the product of the primes $p \leq z$. Show that*

$$\pi(x, z) \leq \sum_{n \leq x} \Big(\sum_{d|(n, P_z)} \lambda_d \Big)^2,$$

for any sequence λ_d of real numbers satisfying $\lambda_1 = 1$.

Exercise 9.3.2 *Show that if $|\lambda_d| \leq 1$, then*

$$\pi(x, z) \leq \sum_{d_1, d_2 \leq z} \frac{\lambda_{d_1} \lambda_{d_2}}{[d_1, d_2]} x + O(z^2),$$

where $[d_1, d_2]$ is the least common multiple of d_1 and d_2.

The main idea is to notice that we have a quadratic form on the right-hand side, given by

$$\sum_{d_1, d_2 \leq z} \frac{\lambda_{d_1} \lambda_{d_2}}{[d_1, d_2]},$$

and we seek to minimize it. We will show that there is a choice of λ_d's such that $|\lambda_d| \leq 1$, as required in Exercise 9.3.2. It should also be noted that the error term here is $O(z^2)$ in contrast to $O(2^z)$, which we obtained in the simplest form of the sieve of Eratosthenes.

Exercise 9.3.3 *Prove that*

$$[d_1, d_2](d_1, d_2) = d_1 d_2,$$

where (d_1, d_2) is the greatest common divisor of d_1 and d_2.

Exercise 9.3.4 *Show that*

$$\sum_{d_1, d_2 \leq z} \frac{\lambda_{d_1} \lambda_{d_2}}{[d_1, d_2]} = \sum_{\delta \leq z} \phi(\delta) \Big(\sum_{\substack{\delta | d \\ d \leq z}} \frac{\lambda_d}{d} \Big)^2.$$

We now use the method of Lagrange multipliers to minimize the quadratic form of the previous exercise.

Exercise 9.3.5 *If*

$$u_\delta = \sum_{\substack{\delta | d \\ d \leq z}} \frac{\lambda_d}{d}$$

show that

$$\frac{\lambda_\delta}{\delta} = \sum_{\delta | d} \mu(d/\delta) u_d.$$

(Note that $u_\delta = 0$ for $\delta > z$, since $\lambda_d = 0$ for $d > z$.)

Exercise 9.3.6 *Show that if $\lambda_1 = 1$, then*

$$\sum_{d_1, d_2 \leq z} \frac{\lambda_{d_1} \lambda_{d_2}}{[d_1, d_2]}$$

attains the minimum value $1/V(z)$, where

$$V(z) = \sum_{d \leq z} \frac{\mu^2(d)}{\phi(d)}.$$

By Exercise 9.3.4, we must minimize

$$\sum_{\delta \leq z} \phi(\delta) \Bigl(\sum_{\substack{\delta | d \\ d \leq z}} \frac{\lambda_d}{d} \Bigr)^2$$

subject to the constraint $\lambda_1 = 1$.

Exercise 9.3.7 *Show that for the choice of*

$$u_\delta = \mu(\delta)/(\phi(\delta) V(z)),$$

we have $|\lambda_d| \leq 1$.

This leads to the following problem:

Exercise 9.3.8 *Show that*

$$\pi(x, z) \leq \frac{x}{V(z)} + O(z^2).$$

Deduce that $\pi(x) = O\left(x/\log x\right)$ by setting $z = x^{1/2-\epsilon}$.

Exercise 9.3.9 *Let f be a multiplicative function. Show that*

$$f([d_1, d_2]) f((d_1, d_2)) = f(d_1) f(d_2).$$

Let $\mathcal{P}$ be a set of primes. Suppose that we are given a sequence of integers $A = \{a_n\}_{n=1}^{\infty}$ and we would like to count the number $N(x,z)$ of $n \leq x$ such that $(a_n, P(z)) = 1$ where $P(z)$ is the product of the primes $p \leq z, p \in \mathcal{P}$. We now derive a more formal version of Selberg's sieve. For convenience, we write $N(d)$ for the number of $n \leq x$ such that $d|a_n$, and assume

$$N(d) = \frac{X}{f(d)} + R_d$$

for some multiplicative function f and some X.

Theorem 9.3.10 (Selberg's sieve, 1947)

$$N(x,z) \leq \frac{X}{U(z)} + O\Big(\sum_{d_1,d_2 \leq z} |R_{[d_1,d_2]}| \Big),$$

where

$$U(z) = \sum_{d \leq z} \frac{\mu^2(d)}{f_1(d)}$$

and

$$f(n) = \sum_{d|n} f_1(d).$$

Proof. We have

$$N(x,z) \leq \sum_{\substack{n \leq x \\ a_n \in A}} \Big(\sum_{d|(a_n,P(z))} \lambda_d \Big)^2,$$

where $\lambda_1 = 1$ and λ_d are real numbers to be chosen. We will set $\lambda_d = 0$ for $d > z$. Expanding the right-hand side of the above inequality, we get

$$\begin{aligned} N(x,z) &\leq \sum_{d_1,d_2 \leq z} \lambda_{d_1}\lambda_{d_2} \Big(\sum_{\substack{d_1,d_2|a_n \\ n \leq x}} 1 \Big) \\ &\leq X \sum_{d_1,d_2 \leq z} \frac{\lambda_{d_1}\lambda_{d_2}}{f([d_1,d_2])} + O\Big(\sum_{d_1,d_2 \leq z} |\lambda_{d_1}||\lambda_{d_2}||R_{[d_1,d_2]}| \Big). \end{aligned}$$

By Exercise 9.3.8, we have

$$f([d_1,d_2]) = f(d_1)f(d_2)/f((d_1,d_2)).$$

Hence, the first sum can be rewritten as

$$\sum_{d_1,d_2\le z}\frac{\lambda_{d_1}\lambda_{d_2}}{f(d_1)f(d_2)}f((d_1,d_2)) = \sum_{d_1,d_2\le z}\frac{\lambda_{d_1}\lambda_{d_2}}{f(d_1)f(d_2)}\sum_{\delta|d_1,d_2} f_1(\delta).$$

Rearranging, we get

$$\sum_{\delta\le z} f_1(\delta)\Big(\sum_{\substack{\delta|d\\ d\le z}}\frac{\lambda_d}{f(d)}\Big)^2,$$

which we seek to minimize subject to the condition $\lambda_1 = 1$. As before, we set

$$u_\delta = \sum_{\substack{\delta|d\\ d\le z}}\frac{\lambda_d}{f(d)}.$$

By Möbius inversion (Exercise 1.5.16),

$$\frac{\lambda_\delta}{f(\delta)} = \sum_{\delta|d}\mu(d/\delta)u_d.$$

Thus, we must minimize

$$\sum_{\delta\le z} f_1(\delta)u_\delta^2$$

subject to the condition

$$1 = \sum_d \mu(d)u_d.$$

By the Lagrange multiplier method,

$$2f_1(\delta)u_\delta = \lambda\mu(\delta)$$

for some scalar λ. Thus,

$$u_\delta = \frac{\lambda\mu(\delta)}{2f_1(\delta)},$$

so that

$$\frac{\lambda}{2}\sum_{\delta\le z}\frac{\mu^2(\delta)}{f_1(\delta)} = 1.$$

Therefore, the minimum is

$$\sum_{\delta \leq z} f_1(\delta) \frac{\lambda^2 \mu^2(\delta)}{4 f_1^2(\delta)} = \frac{1}{U(z)}.$$

In addition, we have

$$\begin{aligned} \frac{\lambda_d}{f(d)} &= \sum_t \mu(t) u_{dt} \\ &= \frac{\mu(d)}{f_1(d)} \sum_{\substack{(d,t)=1 \\ t \leq z/d}} \frac{\mu^2(t)}{f_1(t) U(z)}. \end{aligned}$$

Hence,

$$U(z)\lambda_d = \mu(d) \frac{f(d)}{f_1(d)} \sum_{\substack{(d,t)=1 \\ t \leq z/d}} \frac{\mu^2(t)}{f_1(t)}.$$

Now, for d square-free,

$$\begin{aligned} \frac{f(d)}{f_1(d)} &= \prod_{p|d} \frac{f(p)}{f_1(p)} = \prod_{p|d} \frac{f(p)}{f(p)-1} \\ &= \prod_{p|d} \left(1 + \frac{1}{f(p)-1}\right). \end{aligned}$$

Therefore,

$$U(z)\lambda_d = \mu(d) \left(\sum_{\delta | d} \frac{\mu^2(\delta)}{f_1(\delta)} \right) \left(\sum_{\substack{(d,t)=1 \\ t \leq z/d}} \frac{\mu^2(t)}{f_1(t)} \right),$$

from which we see that $|\lambda_d| \leq 1$. Hence, the error term is

$$O\left(\sum_{[d_1,d_2] \leq z} \left| R_{[d_1,d_2]} \right| \right).$$

We have therefore proved

$$N(x,z) \leq \frac{X}{U(z)} + O\Big(\sum_{[d_1,d_2] \leq z} |R_{[d_1,d_2]}| \Big),$$

as desired. □

Exercise 9.3.11 *Show that*

$$U(z) \geq \sum_{\delta \leq z} \frac{1}{\tilde{f}(\delta)},$$

where $\tilde{f}(n)$ is the completely multiplicative function defined by $\tilde{f}(p) = f(p)$.

Exercise 9.3.12 *Let $\pi_2(x)$ denote the number of twin primes $p \leq x$. Using Selberg's sieve, show that*

$$\pi_2(x) = O\Big(\frac{x}{\log^2 x}\Big).$$

Exercise 9.3.13 (The Brun - Titchmarsh theorem) *For $(a, k) = 1$, and $k \leq x$, show that*

$$\pi(x, k, a) \leq \frac{(2+\epsilon)x}{\varphi(k)\log(2x/k)}$$

for $x > x_0(\epsilon)$, where $\pi(x, k, a)$ denotes the number of primes less than x which are congruent to $a \pmod k$.

Exercise 9.3.14 (Titchmarsh divisor problem) *Show that*

$$\sum_{p \leq x} d(p-1) = O(x),$$

where the sum is over primes and $d(n)$ denotes the divisor function.

9.4 Supplementary Problems

Exercise 9.4.1 *Show that*

$$\sum_{\substack{p \leq x \\ p \equiv 1 \pmod k}} \frac{1}{p} \ll \frac{\log\log x + \log k}{\varphi(k)},$$

where the implied constant is absolute.

Exercise 9.4.2 *Suppose that P is a set of primes such that*

$$\sum_{p \in P} \frac{1}{p} = +\infty.$$

Show that the number of $n \leq x$ not divisible by any prime $p \in P$ is $o(x)$ as $x \to \infty$.

Exercise 9.4.3 *Show that the number of solutions of* $[d_1, d_2] \leq z$ *is* $O(z(\log z)^3)$.

Exercise 9.4.4 *Prove that*

$$\sum_{p \leq x/2} \frac{1}{p \log(x/p)} = O\left(\frac{\log \log x}{\log x}\right),$$

where the summation is over prime numbers.

Exercise 9.4.5 *Let* $\pi_k(x)$ *denote the number of* $n \leq x$ *with* k *prime factors (not necessarily distinct). Using the sieve of Eratosthenes, show that*

$$\pi_k(x) \leq \frac{x(A \log \log x + B)^k}{k! \log x}$$

for some constants A *and* B.

Exercise 9.4.6 *Let* a *be an even integer. Show that the number of primes* $p \leq x$ *such that* $p + a$ *is also prime is*

$$\ll \frac{x}{(\log x)^2} \prod_{p|a} \left(1 + \frac{1}{p}\right),$$

where the implied constant is absolute.

Exercise 9.4.7 *Let* k *be a positive even integer greater than* 1. *Show that the number of primes* $p \leq x$ *such that* $kp + 1$ *is also prime is*

$$\ll \frac{x}{(\log x)^2} \prod_{p|k} \left(1 + \frac{1}{p}\right).$$

Exercise 9.4.8 *Let* k *be even and satisfy* $2 \leq k < x$. *The number of primes* $p \leq x$ *such that* $p - 1 = kq$ *with* q *prime is*

$$\ll \frac{x}{\varphi(k) \log^2(x/k)}.$$

Exercise 9.4.9 *Let* n *be a natural number. Show that the number of solutions of the equation* $[a, b] = n$ *is* $d(n^2)$, *where* $d(n)$ *is the number of divisors of* n.

Exercise 9.4.10 *Show that the error term in Theorem 9.3.10 can be replaced by*

$$O\Big(\sum_{a < z^2} d(a^2)|R_a|\Big).$$

Exercise 9.4.11 *Show that*

$$\sum_{p \leq x} \frac{p-1}{\varphi(p-1)} = O\left(\frac{x}{\log x}\right),$$

where the summation is over prime numbers.

Exercise 9.4.12 *Prove that*

$$\prod_{r<p\leq x} \left(1 - \frac{r}{p}\right) \ll \frac{1}{(\log x)^r}.$$

Exercise 9.4.13 *Prove that for some constant $c > 0$, we have*

$$\sum_{n \leq x} \frac{d(n^2)}{\varphi(n)} = c(\log x)^3 + O(\log^2 x).$$

Exercise 9.4.14 *Let $d(n)$ denote the number of divisors of n. Show that*

$$\sum_{p \leq x} d^2(p-1) = O(x \log^2 x \log\log x),$$

where the summation is over prime numbers.

Exercise 9.4.15 *Show that the result in the previous exercise can be improved to $O(x \log^2 x)$ by noting that $d^2(n) \leq d_4(n)$, where $d_4(n)$ is the number of ways of writing n as a product of four natural numbers.*

10
p-adic Methods

10.1 Ostrowski's Theorem

Recall that a metric on a set X is a map $d : X \times X \to \mathbb{R}_+$ such that

1. $d(x, y) = 0 \Leftrightarrow x = y$;
2. $d(x, y) = d(y, x)$;
3. $d(x, y) \leq d(x, z) + d(z, y) \ \forall z \in X$.

Property (3) is called the triangle inequality. The pair (X, d) is then called a metric space, with metric d.

A **norm** on a field F is a map $|| \cdot || : F \to \mathbb{R}_+$ such that
(1) $||x|| = 0 \Leftrightarrow x = 0$;
(2) $||xy|| = ||x|| ||y||$;
(3) $||x + y|| \leq ||x|| + ||y||$ (triangle inequality).

Exercise 10.1.1 *If F is a field with norm $||\cdot||$, show that $d(x, y) = ||x-y||$ defines a metric on F.*

The well-known norm on the field of rational numbers is, of course, the usual absolute value $| \cdot |$. The induced metric $|x - y|$ is the usual distance function on the real line. But there are other norms that we can define on $\mathbb{Q}$ that give rise to other metrics and "new" notions of distance. For each prime p and any rational number $x \neq 0$, we can write $x = p^{\nu_p(x)} x_1$ where x_1 is a rational number coprime to p (that is, when x_1 is written in lowest terms, neither the numerator nor the denominator is divisible by p). Define a norm $|\cdot|_p$ by

$$|x|_p = p^{-\nu_p(x)}$$

for $x \neq 0$ and for $x = 0$, $|0|_p = 0$.

Exercise 10.1.2 *Show that $|\cdot|_p$ is a norm on $\mathbb{Q}$.*

A norm satisfying

$$||x+y|| \leq \max(||x||, ||y||)$$

is called a nonarchimedean norm (or a finite valuation). The solution of Exercise 10.1.2 shows that the p-adic metric $|\cdot|_p$ is nonarchimedean. A metric that is not nonarchimedean is called Archimedean (or an infinite valuation).

Exercise 10.1.3 *Show that the usual absolute value on $\mathbb{Q}$ is archimedean.*

The celebrated theorem of Ostrowski states that essentially the only norms we can define on $\mathbb{Q}$ are the p-adic norms and the usual absolute value. To make this precise, we need the notion of equivalence of two norms.

Given a metric space X, we can discuss the notion of a Cauchy sequence. This is any sequence $\{a_n\}_{n=1}^{\infty}$ of elements of X such that given any $\epsilon > 0$, there exists an N (depending only on ϵ) such that $d(a_m, a_n) < \epsilon$ for $m, n > N$.

Two metrics d_1, d_2 on X are said to be **equivalent** if every sequence that is Cauchy with respect to d_1 is also Cauchy with respect to d_2. Two norms on a field are said to be **equivalent** if they induce equivalent metrics.

Exercise 10.1.4 *If $0 < c < 1$ and p is prime, define*

$$||x|| = \begin{cases} c^{\nu_p(x)} & \text{if } x \neq 0, \\ 0 & \text{if } x = 0 \end{cases}$$

for all rational numbers x. Show that $||\cdot||$ is equivalent to $|\cdot|_p$ on $\mathbb{Q}$.

The usual absolute value on $\mathbb{Q}$ we will denote by $|\cdot|_\infty$ to distinguish it from the p-adic metrics. Note that we can always define a "trivial" norm by setting $||0|| = 0$, and $||x|| = 1$ for $x \neq 0$. We also note that $||-x|| = ||x||$ follows from the axioms.

Theorem 10.1.5 (Ostrowski) *Every nontrivial norm $||\cdot||$ on $\mathbb{Q}$ is equivalent to $|\cdot|_p$ for some prime p or $|\cdot|_\infty$.*

Proof. Case (i): Suppose there is a natural number n such that $||n|| > 1$. Let n_0 be the least such n. We know that $n_0 > 1$, so we can write $||n_0|| = n_0^\alpha$ for some positive α. Write any natural number n in base n_0:

$$n = a_0 + a_1 n_0 + \cdots + a_s n_0^s, \quad 0 \leq a_i < n_0,$$

and $a_s \neq 0$. Then, by the triangle inequality,

$$\begin{aligned} ||n|| &\leq ||a_0|| + ||a_1 n_0|| + \cdots + ||a_s n_0^s|| \\ &\leq ||a_0|| + ||a_1|| n_0^\alpha + \cdots + ||a_s|| n_0^{\alpha s}. \end{aligned}$$

Since all the a_i are less than n_0, we have $||a_i|| \leq 1$. Hence,

$$\begin{aligned} ||n|| &\leq 1 + n_0^\alpha + \cdots + n_0^{\alpha s} \\ &\leq n_0^{\alpha s}\left(1 + \frac{1}{n_0^\alpha} + \cdots\right). \end{aligned}$$

Since $n > n_0^s$, we deduce $||n|| \leq C n^\alpha$ for some constant C and for all natural numbers n. Thus, $||n^N|| \leq C n^{N\alpha}$, so that $||n|| \leq C^{1/N} n^\alpha$. Letting $N \to \infty$ gives $||n|| \leq n^\alpha$ for all natural numbers n. We can also get the reverse inequality as follows: since $n_0^{s+1} > n \geq n_0^s$, we have

$$\begin{aligned} ||n_0^{s+1}|| &= ||n + n_0^{s+1} - n|| \\ &\leq ||n|| + ||n_0^{s+1} - n||, \end{aligned}$$

so that

$$\begin{aligned} ||n|| &\geq ||n_0^{s+1}|| - ||n_0^{s+1} - n|| \\ &\geq n_0^{(s+1)\alpha} - (n_0^{s+1} - n)^\alpha. \end{aligned}$$

Thus,

$$||n|| \geq n_0^{(s+1)\alpha} - (n_0^{s+1} - n_0^s)^\alpha,$$

since $n \geq n_0^s$, so that

$$\begin{aligned} ||n|| &\geq n_0^{(s+1)\alpha}\left(1-\left(1-\frac{1}{n_0}\right)^{\alpha}\right) \\ &\geq C_1 n^{\alpha} \end{aligned}$$

for some constant C_1. Repeating the previous argument gives $||n|| \geq n^{\alpha}$ and therefore $||n|| = n^{\alpha}$ for all natural numbers n. Thus, $||\cdot||$ is equivalent to $|\cdot|_{\infty}$.

Case (ii): Suppose that $||n|| \leq 1$ for all natural numbers. Since the norm is nontrivial, there is an n such that $||n|| < 1$. Let n_0 be the least such n. Then n_0 must be prime, for if $n_0 = ab$, then $||n_0|| = ||a||||b|| < 1$ implies $||a|| < 1$ and $||b|| < 1$, contrary to the choice of n_0. Say $n_0 = p$. If q is a prime not equal to p, then we claim $||q|| = 1$. Indeed, if not, then $||q|| < 1$, and for sufficiently large N, $||q^N|| < 1/2$. Similarly, for sufficiently large M, $||p^M|| < 1/2$. Since p^M, q^N are coprime, we can find integers a and b such that $ap^M + bq^N = 1$. Hence

$$\begin{aligned} 1 &= ||ap^M + bq^N|| \leq ||a||||p^M|| + ||b||||q^N|| \\ &< 1/2 + 1/2 = 1, \end{aligned}$$

a contradiction. Therefore, $||q|| = 1$. Now write $C = ||p||$. Since any natural number can be written uniquely as a product of prime powers, we get

$$||n|| = C^{\nu_p(n)}.$$

By Exercise 10.1.4, this metric is equivalent to $|\cdot|_p$, which completes the proof. □

Exercise 10.1.6 *Let F be a field with norm $||\cdot||$ satisfying*

$$||x+y|| \leq \max(||x||, ||y||).$$

If $a \in F$, and $r > 0$, let $B(a,r)$ be the open disk, $\{x \in F : ||x-a|| < r\}$. Show that $B(a,r) = B(b,r)$ for any $b \in B(a,r)$. (This result says that every point of the disk is a "center" of the disc.)

Exercise 10.1.7 *Let F be a field with $||\cdot||$. Let R be the set of all Cauchy sequences $\{a_n\}_{n=1}^{\infty}$. Define addition and multiplication of sequences point-*

wise: that is,

$$\{a_n\}_{n=1}^{\infty} + \{b_n\}_{n=1}^{\infty} = \{a_n + b_n\}_{n=1}^{\infty},$$

$$\{a_n\}_{n=1}^{\infty} \times \{b_n\}_{n=1}^{\infty} = \{a_n b_n\}_{n=1}^{\infty}.$$

Show that $(R, +, \times)$ is a commutative ring. Show further that the subset R consisting of null Cauchy sequences (namely those satisfying $||a_n|| \to 0$ as $n \to \infty$) forms a maximal ideal m.

We can embed our field F in R by the map $a \mapsto (a, a, \ldots)$, which is clearly a Cauchy sequence. Since m is a maximal ideal, R/m is a field. R/m is called the completion of F with respect to $||\cdot||$. In the case of $F = \mathbb{Q}$ with norm $|\cdot|_p$, the completion is called the field of p-adic numbers, and denoted by $\mathbb{Q}_p$.

We can extend the concept of norm to $\mathbb{Q}_p$ by setting

$$|a|_p = \lim_{n\to\infty} |a_n|_p$$

for any Cauchy sequence $a = \{a_n\}_{n=1}^{\infty}$. It is easily seen that this is well-defined.

Theorem 10.1.8 *$\mathbb{Q}_p$ is complete with respect to $|\cdot|_p$.*

Proof. Let $\{a^{(j)}\}_{j=1}^{\infty}$ be a Cauchy sequence of equivalence classes in $\mathbb{Q}_p$. We must show that there is a Cauchy sequence to which it converges. We write $a^{(j)} = \{a_n^{(j)}\}_{n=1}^{\infty}$ and set $s = \{a_j^{(j)}\}_{j=1}^{\infty}$, the "diagonal" sequence. First, observe that s is a Cauchy sequence, since $\{a^{(j)}\}_{j=1}^{\infty}$ is Cauchy, so that given $\epsilon > 0$, there is an $N(\epsilon)$ such that for $j, k \geq N(\epsilon)$, we have $|a^{(j)} - a^{(k)}|_p < \epsilon$. This means that for $j, k, n \geq N_1(\epsilon)$ for some $N_1(\epsilon)$, we have

$$\left|a_n^{(j)} - a_n^{(k)}\right|_p < \epsilon.$$

In particular,

$$\left|a_j^{(j)} - a_k^{(k)}\right|_p \leq \max\left(\left|a_j^{(j)} - a_k^{(j)}\right|_p, \left|a_k^{(j)} - a_k^{(k)}\right|_p\right)$$

for $j, k \geq N_1(\epsilon)$. Therefore, s is a Cauchy sequence. We now show that $\lim_{j\to\infty} a^{(j)} = s$. That is, given $\epsilon > 0$, we must show that there is an $N_2(\epsilon)$ such that for $j \geq N_2(\epsilon)$, we have

$$|a^{(j)} - s|_p < \epsilon.$$

This means that we must show that for some $N_3(\epsilon)$ and $j, n \geq N_3(\epsilon)$, we have $|a_n^{(j)} - a_n^{(n)}|_p < \epsilon$. But this is clear from the above for $N_3(\epsilon) = N_1(\epsilon)$. □

When we complete $\mathbb{Q}$ with respect to the usual absolute value $|\cdot|_\infty$, we get the real number field $\mathbb{R}$, which is complete. When we complete $\mathbb{Q}$ with respect to $|\cdot|_p$ we get $\mathbb{Q}_p$, which we just proved to be complete. It is this point of view that motivates p-adic analysis. Real analysis is seen to be the special case of only one completion of $\mathbb{Q}$. As we shall see, it is fruitful to develop p-adic analysis on an equal footing. When applied to the context of number theory, we get an important theme of p-adic analytic number theory, which is playing a central role in the modern perspective.

Exercise 10.1.9 *Show that*

$$\mathbb{Z}_p = \{x \in \mathbb{Q}_p : |x|_p \leq 1\}$$

is a ring. (This ring is called the ring of p-adic integers.)

Exercise 10.1.10 *Given $x \in \mathbb{Q}$ satisfying $|x|_p \leq 1$, and any natural number i, show that $|x - a_i|_p \leq p^{-i}$. Moreover, we can choose a_i satisfying $0 \leq a_i < p^i$.*

Just as it is impractical to think of real numbers as Cauchy sequences, it is impractical to think of elements of $\mathbb{Q}_p$ as Cauchy sequences. It is better to think of them as formal series

$$\sum_{n=-N}^{\infty} b_n p^n, \quad 0 \leq b_n \leq p-1,$$

as the following theorem shows.

Theorem 10.1.11 *Every equivalence class s in $\mathbb{Q}_p$ for which $|s|_p \leq 1$ has exactly one representative Cauchy sequence $\{a_i\}_{i=1}^\infty$ satisfying $0 \leq a_i < p^i$ and $a_i \equiv a_{i+1} (\operatorname{mod} p^i)$ for $i = 1, 2, 3, \ldots$.*

Proof. The uniqueness is clear, for if $\{a_i'\}_{i=1}^\infty$ is another such sequence, we have $a_i \equiv a_i' (\operatorname{mod} p^i)$, which forces $a_i = a_i'$. Now let $\{c_i\}_{i=1}^\infty$ be a Cauchy sequence of $\mathbb{Q}_p$ in s. Then for each j, there is an $N(j)$ such that

$$|c_i - c_k|_p \leq p^{-j}$$

for $i, k \geq N(j)$. Without loss of generality, we may take $N(j) \geq j$. Since $|s|_p \leq 1$, we have $|c_i|_p \leq 1$ for $i \geq N(1)$ because

$$\begin{aligned} |c_i|_p &\leq \max(|c_k|_p, |c_i - c_k|_p) \\ &\leq \max(|c_k|_p, 1/p), \end{aligned}$$

so that by choosing a sufficiently large k we are ensured that $|c_k|_p \leq 1$, since $|s|_p = \lim_{k\to\infty} |c_k|_p \leq 1$. By Exercise 10.1.10, we can find a sequence of integers a_j such that

$$|a_j - c_{N(j)}|_p \leq p^{-j},$$

with $0 \leq a_j < p^j$. The claim is that $\{a_j\}_{j=1}^{\infty}$ is the required sequence. First observe that by the triangle inequality,

$$\begin{aligned} |a_{j+1} - a_j|_p &\leq \max(\left|a_{j+1} - c_{N(j+1)}\right|_p, \left|c_{N(j+1)} - c_{N(j)}\right|_p, \\ &\qquad \left|c_{N(j)} - a_j\right|_p) \\ &\leq \max\left(p^{-j-1}, p^{-j}, p^{-j}\right) = p^{-j}, \end{aligned}$$

so that

$$a_j \equiv a_{j+1} (\operatorname{mod} p^j),$$

for $i = 1, 2, \ldots$. Second, for any j, and $i \geq N(j)$, we have

$$\begin{aligned} |a_i - c_i|_p &\leq \max(|a_i - a_j|_p, |a_j - c_{N(j)}|_p, |c_{N(j)} - c_j|_p) \\ &\leq \max(p^{-j}, p^{-j}, p^{-j}) = p^{-j} \end{aligned}$$

so that $\lim_{i\to\infty} |a_i - c_i|_p = 0$. □

The above theorem says that $\mathbb{Z}$ is dense in $\mathbb{Z}_p$, the ring of p-adic integers. Now writing each a_i of Theorem 10.1.11 in base p, we see that

$$a_i = b_0 + b_1 p + \cdots + b_{i-1} p^{i-1},$$

where $0 \leq b_i < p$. The condition $a_i \equiv a_{i+1} \pmod{p^i}$ means that

$$a_{i+1} = b_0 + b_1 p + \cdots + b_{i-1} p^{i-1} + b_i p^i$$

in base p. Therefore, every element of $\mathbb{Z}_p$ can be written as $\sum_{n=0}^{\infty} b_n p^n$, $0 \leq b_n < p$. If $x \in \mathbb{Q}_p$, we can always multiply x by an appropriate power of p (say p^N) so that $|p^N x|_p \leq 1$. Then, we can expand $p^N x$ as above to deduce that every p-adic number has a unique expansion as $\sum_{n=-N}^{\infty} b_n p^n, 0 \leq b_n \leq p-1$.

It is useful to observe the analogy with Laurent series and the field of meromorphic functions of a complex variable. At each point $z \in \mathbb{C}$, the meromorphic function has a Laurent expansion, which is unique. Thus, if a rational number has denominator divisible by p, we can think of it as having a "pole" at p. This analogy has been a guiding force for much of the development in p-adic theory.

Exercise 10.1.12 *Show that the p-adic series*

$$\sum_{n=1}^{\infty} c_n, \quad c_n \in \mathbb{Q}_p,$$

converges if and only if $|c_n|_p \to 0$.

Thus convergence of infinite series is easily verified. Note, however, that the analogue of Exercise 10.1.12 is **not** true for the real numbers, as the example of the harmonic series shows.

Exercise 10.1.13 *Show that*

$$\sum_{n=1}^{\infty} n!$$

converges in $\mathbb{Q}_p$.

Exercise 10.1.14 *Show that*

$$\sum_{n=1}^{\infty} n \cdot n! = -1$$

in $\mathbb{Q}_p$.

Exercise 10.1.15 *Show that the power series*

$$\sum_{n=0}^{\infty} \frac{x^n}{n!}$$

converges in the disk $|x|_p < p^{-\frac{1}{p-1}}$.

Exercise 10.1.16 (Product formula) *Prove that for $x \in \mathbb{Q}$, $x \neq 0$,*

$$\prod_p |x|_p = 1$$

where the product is taken over all primes p including ∞.

Exercise 10.1.17 *Prove that for any natural number n and a finite prime p,*

$$|n|_p \geq \frac{1}{|n|_\infty}.$$

10.2 Hensel's Lemma

In many ways $\mathbb{Q}_p$ is analogous to $\mathbb{R}$. For example, $\mathbb{R}$ is not algebraically closed. The exercises below show that $\mathbb{Q}_p$ is not algebraically closed. However, by adjoining $i = \sqrt{-1}$ to $\mathbb{R}$, we get the field of complex numbers, which is algebraically closed. In contrast, the algebraic closure $\overline{\mathbb{Q}}_p$ of $\mathbb{Q}_p$ is not of finite degree over $\mathbb{Q}$. Moreover, $\mathbb{C}$ is complete with respect to the extension of the usual norm of $\mathbb{R}$. Unfortunately, $\overline{\mathbb{Q}}_p$ is not complete with respect to the extension of the p-adic norm. So after completing it (via the usual method of Cauchy sequences) we get a still larger field, usually denoted by $\mathbb{C}_p$, and it turns out to be both algebraically closed and complete. It is this field $\mathbb{C}_p$ that is the p-adic analogue of the field of complex numbers. Very little is known about it. The topic of rigid analytic spaces in the literature refers to its study, which we will **not** cover in this chapter. We confine much of our study to $\mathbb{Q}_p$.

Exercise 10.2.1 *Show that $x^2 = 7$ has no solution in $\mathbb{Q}_5$.*

Example 10.2.2 *Show that $x^2 = 6$ has a solution in $\mathbb{Q}_5$.*

Solution. The equation $x^2 \equiv 6 \pmod 5$ has a solution (namely $x \equiv 1 \pmod 5$). We will show inductively that $x^2 \equiv 6 \pmod{5^n}$ has a solution for every $n \geq 1$. Suppose

$$x_n^2 \equiv 6 \pmod{5^n}.$$

We want to find $x_{n+1}^2 \equiv 6 \pmod{5^{n+1}}$. Write $x_{n+1} = 5^n t + x_n$. So we must have

$$(5^n t + x_n)^2 \equiv 6 \pmod{5^{n+1}},$$

which means that $2 \cdot 5^n t x_n + x_n^2 \equiv 6 (\bmod 5^{n+1})$. This reduces to

$$2tx_n + \frac{x_n^2 - 6}{5^n} \equiv 0 \pmod 5,$$

so that we can clearly solve for t. The method produces a sequence of integers $\{x_n\}_{n=1}^{\infty}$ such that $x_n^2 \equiv 6 \pmod{5^n}$ and $x_{n+1} \equiv x_n (\bmod 5^n)$. The sequence is therefore Cauchy and its limit x (which exists in $\mathbb{Q}_p$ by completeness) satisfies $x^2 = 6$. □

The method suggested by the previous example is quite general. It is the main idea behind Hensel's lemma which is the following theorem.

Theorem 10.2.3 *Let $f(x) \in \mathbb{Z}_p[x]$ be a polynomial with coefficients in $\mathbb{Z}_p$. Write $f'(x)$ for its formal derivative. If $f(x) \equiv 0 \pmod p$ has a solution a_0 satisfying $f'(a_0) \not\equiv 0 \pmod p$, then there is a unique p-adic integer a such that $f(a) = 0$ and $a \equiv a_0 \pmod p$.*

Proof. We imitate the construction suggested by the example. Suppose

$$f(x) \equiv 0 \pmod{p^n}$$

has a solution a_n. We claim that there is a unique solution

$$a_{n+1} \left(\bmod p^{n+1}\right)$$

such that

$$f(a_{n+1}) \equiv 0 \left(\bmod p^{n+1}\right)$$

and $a_{n+1} \equiv a_n (\bmod p^n)$. Indeed, writing $a_{n+1} = p^n t + a_n$, we require $f(p^n t + a_n) \equiv 0 \pmod{p^{n+1}}$. We write $f(x) = \sum_i c_i x^i$, so that

$$\begin{aligned} f(p^n t + a_n) &= \sum_i c_i (a_n + p^n t)^i \\ &\equiv \sum_i c_i (a_n^i + i a_n^{i-1} p^n t) \pmod{p^{n+1}} \\ &\equiv f(a_n) + p^n t f'(a_n) \pmod{p^{n+1}}. \end{aligned}$$

We need to solve for t in the congruence

$$p^n t f'(a_n) + f(a_n) \equiv 0 \pmod{p^{n+1}}.$$

Since $f(a_n) \equiv 0 \pmod{p^n}$, this reduces to

$$tf'(a_n) \equiv -(f(a_n)/p^n) \pmod{p},$$

which has a unique solution $\pmod{p}$, since $f'(a_n) \not\equiv 0 \pmod{p}$, because $a_n \equiv a_0 (\bmod\, p)$. This proves the claim. As before, $\{a_n\}_{n=1}^{\infty}$ is a Cauchy sequence, whose limit is the required solution. Since a_{n+1} is a unique lifting $\pmod{p^{n+1}}$ of $a_n \pmod{p^n}$, the uniqueness of the solution is now clear. □

Exercise 10.2.4 *Let $f(x) \in \mathbb{Z}_p[x]$. Suppose for some N and $a_0 \in \mathbb{Z}_p$ we have $f(a_0) \equiv 0 \pmod{p^{2N+1}}$, $f'(a_0) \equiv 0 \pmod{p^N}$ but $f'(a_0) \not\equiv 0 \pmod{p^{N+1}}$. Show that there is a unique $a \in \mathbb{Z}_p$ such that $f(a) = 0$ and $a \equiv a_0 \pmod{p^{N+1}}$.*

Exercise 10.2.5 *For any prime p, and any positive integer m coprime to p, show that there exists a primitive mth root of unity in $\mathbb{Q}_p$ if and only if $m|(p-1)$.*

Exercise 10.2.6 *Show that the set of $(p-1)$st roots of unity in $\mathbb{Q}_p$ is a cyclic group of order $(p-1)$.*

Remark. The previous exercise shows the existence of p-adic numbers $\omega_0, \omega_1, \ldots, \omega_{p-1}$ that are roots of the polynomial $x^p - x = 0$ such that $\omega_i \equiv i \pmod{p}$. These roots are called the "Teichmüller representatives."

Exercise 10.2.7 (Polynomial form of Hensel's Lemma) *Suppose $f(x) \in \mathbb{Z}_p[x]$ and that there exist $g_1, h_1 \in (\mathbb{Z}/p\mathbb{Z})[x]$ such that*

$$f(x) \equiv g_1(x)h_1(x) \pmod{p},$$

with $(g_1, h_1) = 1$, $g_1(x)$ monic. Then there exist polynomials $g(x), h(x) \in \mathbb{Z}_p[x]$ such that $g(x)$ is monic, $f(x) = g(x)h(x)$, and $g(x) \equiv g_1(x) \pmod{p}$, $h(x) \equiv h_1(x) \pmod{p}$.

We now consider $\overline{\mathbb{Q}_p}$, the algebraic closure of $\mathbb{Q}_p$. The p-adic norm extends uniquely to $\overline{\mathbb{Q}_p}$ in the obvious way, which we will also denote by $|\cdot|_p$. Indeed, if $K/\mathbb{Q}_p$ is a finite extension of degree n, we have for $x \in K$,

$$|x|_p = (|N_{K/\mathbb{Q}_p}(x)|_p)^{1/n}.$$

Theorem 10.2.8 *$|\cdot|_p$ is a nonarchimedean norm on K.*

Proof. It is clear that $|x|_p = 0$ if and only if $x = 0$. It is also clear that $|xy|_p = |x|_p|y|_p$, since the norm is multiplicative. To prove that

$$|x + y|_p \leq \max(|x|_p, |y|_p)$$

we see (upon dividing by y) that it suffices to prove for $\alpha \in K$,

$$|\alpha + 1|_p \leq \max(|\alpha|_p, 1).$$

It is easily seen that this follows if we can show

$$|\alpha|_p \leq 1 \Rightarrow |\alpha - 1|_p \leq 1.$$

That is, we must show

$$|N_{K/\mathbb{Q}_p}(\alpha)|_p \leq 1 \Rightarrow |N_{K/\mathbb{Q}_p}(\alpha - 1)|_p \leq 1.$$

This reduces to showing

$$N_{K/\mathbb{Q}_p}(\alpha) \in \mathbb{Z}_p \Rightarrow N_{K/\mathbb{Q}_p}(\alpha - 1) \in \mathbb{Z}_p.$$

It is now necessary to use a little bit of commutative algebra. Clearly, $\mathbb{Q}_p(\alpha) = \mathbb{Q}_p(\alpha - 1)$. Now let

$$f(x) = x^n + a_{n-1}x^{n-1} + \cdots + a_1 x + a_0$$

be the minimal polynomial for α. The minimal polynomial for $\alpha - 1$ is clearly

$$f(x+1) = x^n + (a_{n-1} + n)x^{n-1} + \cdots + (1 + a_{n-1} + \cdots + a_1 + a_0).$$

Now $N_{K/\mathbb{Q}_p}(\alpha) = (-1)^n a_0$ and

$$N_{K/\mathbb{Q}_p}(\alpha - 1) = (-1)^n(1 + a_{n-1} + \cdots + a_1 + a_0).$$

We now use the polynomial form of Hensel's lemma. If all the coefficients of $f(x)$ are in $\mathbb{Z}_p$, we are done. So, assume that

$$f(x) = x^n + a_{n-1}x^{n-1} + \cdots + a_1 x + a_0$$

is such that $a_0 \in \mathbb{Z}_p$ but some $a_i \notin \mathbb{Z}_p$. Choose m to be the smallest exponent such that $p^m a_i \in \mathbb{Z}_p$ for all i and now "clear denominators":

$$g(x) = p^m f(x) = b_n x^n + b_{n-1}x^{n-1} + \cdots + b_1 x + b_0$$

with $b_i = p^m a_i$. Since $f(x)$ is monic, $b_n = p^m$, and $b_0 = p^m a_0$. By assumption, at least one b_i is not divisible by p. Thus

$$g(x) \equiv (b_n x^{n-k} + \cdots + b_k)x^k \pmod{p},$$

where k is the smallest index such that b_k is not divisible by p. By Exercise 10.2.7 (the polynomial form of Hensel's lemma) this lifts to a factorization in $\mathbb{Z}_p[x]$, which means that $g(x) = p^m f(x)$ is reducible, a contradiction, since $f(x)$ is the minimal polynomial of α. This completes the proof. □

Exercise 10.2.9 *Show that for $p \neq 2$, the only solution to $x^2 \equiv 1 \pmod{p^n}$ is $x = \pm 1$, for every $n \geq 1$.*

10.3 p-adic Interpolation

The notion of p-adic continuity is evident. We say that a function $f : \mathbb{Q}_p \to \mathbb{Q}_p$ is continuous if $f(x_n) \to f(x)$ whenever $x_n \to x$.

The problem of interpolation is this: Given a sequence $a_1, a_2, a_3, \ldots$ of elements in $\mathbb{Q}_p$, does there exist a continuous function $f : \mathbb{Z}_p \to \mathbb{Q}_p$ such that $f(n) = a_n$? Since the set of natural numbers is dense in $\mathbb{Z}_p$, there can exist at most one such function.

The classic example of interpolation is given by the Γ-function:

$$\Gamma(n+1) = \int_0^\infty e^{-x} x^n \, dx = n!.$$

Hence

$$\Gamma(s+1) = \int_0^\infty e^{-x} x^s \, dx$$

interpolates the sequence of factorials.

Exercise 10.3.1 *Show that there is no continuous function $f : \mathbb{Z}_p \to \mathbb{Q}_p$ such that $f(n) = n!$*

The difficulty in interpolation stems from $n!$ being highly divisible by p. Thus, a natural idea is to consider the sequence

$$\prod_{\substack{1 \leq j \leq n \\ (j,p)=1}} j$$

instead of the factorials and hope that this works.

A continuous function $f : \mathbb{Z}_p \to \mathbb{Q}_p$ is in fact determined by its restriction to natural numbers. Thus given a sequence of integers $\{a_k\}_{k=0}^{\infty}$, we need only verify that for any natural number m, there is an integer $N = N(m)$ such that

$$k \equiv k' \pmod{p^N} \Rightarrow a_k \equiv a_{k'} \pmod{p^m}. \tag{10.1}$$

That is, whenever k and k' are close p-adically, then a_k and $a_{k'}$ are close p-adically.

We first begin by showing that the sequence defined by

$$a_k = \prod_{\substack{j \leq k \\ (j,p)=1}} j$$

has almost the property (10.1). As we shall see, this is essentially Wilson's theorem of elementary number theory.

Exercise 10.3.2 *Let $p \neq 2$, be prime. Prove that for any natural numbers n, s we have*

$$\prod_{\substack{j=1 \\ (n+j,p)=1}}^{p^s-1} (n+j) \equiv -1 \pmod{p^s}.$$

Exercise 10.3.3 *Show that if $p \neq 2$,*

$$a_k = \prod_{\substack{j \leq k \\ (j,p)=1}} j,$$

then $a_{k+p^s} \equiv -a_k \pmod{p^s}$.

The previous exercise almost satisfies (10.1) apart from the sign. This motivates the definition of the p-adic gamma function:

$$\Gamma_p(n) := (-1)^n \prod_{\substack{j<n \\ (j,p)=1}} j.$$

Exercise 10.3.4 *Prove that for $p \neq 2$,*

$$\Gamma_p(k + p^s) \equiv \Gamma_p(k) \pmod{p^s}.$$

We now prove Mahler's interpolation theorem. As will be seen, the essential idea is combinatorial analysis based on a simplification due to Bojanic.

Exercise 10.3.5 *Let n, k be natural numbers and write*

$$n = a_0 + a_1 p + a_2 p^2 + \cdots ,$$

$$k = b_0 + b_1 p + b_2 p^2 + \cdots ,$$

for the p-adic expansions of n and k, respectively. Show that

$$\binom{n}{k} \equiv \binom{a_0}{b_0}\binom{a_1}{b_1}\binom{a_2}{b_2} \cdots \pmod p.$$

Exercise 10.3.6 *If p is prime, show that*

$$\binom{p^n}{k} \equiv 0 \pmod p$$

for $1 \leq k \leq p^n - 1$ and all n.

Exercise 10.3.7 (Binomial inversion formula) *Suppose for all n,*

$$b_n = \sum_{k=0}^{n} \binom{n}{k} a_k.$$

Show that

$$a_n = \sum_{k=0}^{n} \binom{n}{k} (-1)^{n-k} b_k,$$

and conversely.

Exercise 10.3.8 *Prove that*

$$\sum_{k=0}^{n} \binom{n}{k} (-1)^k \binom{k}{m} = \begin{cases} (-1)^m & \text{if} \quad n = m, \\ 0 & \text{otherwise.} \end{cases}$$

Exercise 10.3.7 suggests the following. If $f : \mathbb{Z}_p \to \mathbb{Q}_p$ is continuous, then let

$$a_n(f) = \sum_{k=0}^{n} \binom{n}{k} (-1)^{n-k} f(k),$$

so that

$$f(n) = \sum_{k=0}^{n} \binom{n}{k} a_k(f).$$

If we can show that the function

$$\sum_{k=0}^{\infty} \binom{x}{k} a_k(f)$$

is p-adically continuous, then this solves the interpolation problem. That is, if we can show that the series converges, we are done. This is the key idea of Mahler's theorem, namely, to show that $|a_k(f)|_p \to 0$ if the sequence $\{f(k)\}_{k=1}^{\infty}$ satisfies condition (10.1).

Exercise 10.3.9 *Define*

$$\Delta^n f(x) = \sum_{k=0}^{n} \binom{n}{k} (-1)^{n-k} f(x+k).$$

Show that

$$\Delta^n f(x) = \sum_{j=0}^{m} \binom{m}{j} \Delta^{n+j} f(x-m).$$

Exercise 10.3.10 *Prove that*

$$\sum_{j=0}^{m} \binom{m}{j} a_{n+j}(f) = \sum_{k=0}^{n} (-1)^{n-k} \binom{n}{k} f(k+m)$$

with $a_n(f)$ defined by

$$a_n(f) = \sum_{k=0}^{n} (-1)^{n-k} \binom{n}{k} f(k).$$

Exercise 10.3.11 *Show that the polynomial*

$$\binom{x}{n} = \begin{cases} \frac{x(x-1)\cdots(x-n+1)}{n!} & \text{if } n \geq 1, \\ 1 & \text{if } n = 0, \end{cases}$$

takes integer values for $x \in \mathbb{Z}$. Deduce that

$$\left| \binom{x}{n} \right|_p \leq 1$$

for all $x \in \mathbb{Z}_p$.

Theorem 10.3.12 (Mahler, 1961) *Suppose $f : \mathbb{Z}_p \to \mathbb{Q}_p$ is continuous. Let*

$$a_n(f) = \sum_{k=0}^{n} (-1)^{n-k} \binom{n}{k} f(k).$$

Then the series

$$\sum_{k=0}^{\infty} \binom{x}{k} a_k(f)$$

converges uniformly in $\mathbb{Z}_p$ and

$$f(x) = \sum_{k=0}^{\infty} \binom{x}{k} a_k(f).$$

Proof. We know that given any positive integer s, there exists a positive integer t such that for $x, y \in \mathbb{Z}_p$,

$$|x - y|_p \leq p^{-t} \Rightarrow |f(x) - f(y)|_p \leq p^{-s}.$$

In particular,

$$|f(k + p^t) - f(k)|_p \leq p^{-s}$$

for $k = 0, 1, 2, \ldots$.

Since f is continuous on $\mathbb{Z}_p$, it is bounded there (recall that $\mathbb{Z}_p$ is compact), and so we may suppose without loss of generality that $|f(x)|_p \leq 1$ for all $x \in \mathbb{Z}_p$. Hence,

$$|a_n(f)|_p \leq 1 \quad \text{for} \quad n = 0, 1, 2, \ldots.$$

Now by Exercise 10.3.10,

$$a_{n+p^t}(f) = -\sum_{j=1}^{p^t-1} \binom{p^t}{j} a_{n+j}(f) + \sum_{k=0}^{n} (-1)^{n-k} \binom{n}{k} \left\{ f(k+p^t) - f(k) \right\}.$$

By Exercise 10.3.6, $p | \binom{p^t}{j}$ for $1 \leq j \leq p^t - 1$, so that

$$|a_{n+p^t}(f)|_p \leq \max_{1 \leq j < p^t} \{p^{-1} |a_{n+j}(f)|_p, p^{-s}\}.$$

Since $|a_n(f)|_p \leq 1$, we obtain

$$|a_n(f)|_p \leq p^{-1} \quad \text{for} \quad n \geq p^t.$$

Replacing n by $n + p^t$ in the penultimate inequality and using the above inequality, we obtain

$$|a_n(f)|_p \leq p^{-2} \quad \text{for} \quad n \geq 2p^t.$$

Repeating the argument $(s-1)$ times gives

$$|a_n(f)|_p \leq p^{-s} \quad \text{for} \quad n \geq sp^t.$$

This proves $a_n(f) \to 0$ as $n \to \infty$. By Exercise 10.3.11, we have

$$\left|\binom{x}{n}\right|_p \leq 1$$

for $x \in \mathbb{Z}_p$. Therefore, the series

$$\sum_{k=0}^{\infty} \binom{x}{k} a_k(f)$$

converges uniformly on $\mathbb{Z}_p$ and thus defines a continuous function. Since this function agrees with $f(n)$ on the natural numbers and $\mathbb{N}$ is dense in $\mathbb{Z}_p$, we deduce the result. □

Exercise 10.3.13 *If $f(x) \in \mathbb{C}[x]$ is a polynomial taking integral values at integral arguments, show that*

$$f(x) = \sum_k c_k \binom{x}{k}$$

for certain integers c_k.

Exercise 10.3.14 *If $n \equiv 1 \pmod{p}$, prove that $n^{p^m} \equiv 1 \pmod{p^{m+1}}$. Deduce that the sequence $a_k = n^k$ can be p-adically interpolated.*

The previous exercise shows that if $n \equiv 1 \pmod{p}$, then $f(s) = n^s$ is a continuous function of a p-adic variables s. The next exercises show how this can be extended for other values of n.

Exercise 10.3.15 *Let $(n,p) = 1$. If $k \equiv k' \pmod{(p-1)p^N}$, then show that*

$$n^k \equiv n^{k'} \pmod{p^{N+1}}.$$

Exercise 10.3.16 *Fix $s_0 \in \{0, 1, 2, \ldots, p-2\}$ and let A_{s_0} be the set of integers congruent to $s_0 \pmod{p-1}$. Show that A_{s_0} is a dense subset of $\mathbb{Z}_p$.*

Exercise 10.3.17 *If $(n,p)=1$, show that $f(k)=n^k$ can be extended to a continuous function on A_{s_0}.*

Remark. By Exercise 10.3.16, we see that $f(s)=n^s$ is a continuous function $f_{s_0}:\mathbb{Z}_p\to\mathbb{Z}_p$ that interpolates n^s, for $s\equiv s_0(\mathrm{mod}\, p-1)$.

10.4 The p-adic Zeta-Function

We begin with a brief description of p-adic integration theory. For further details we refer the reader to Koblitz [K]. A **p-adic distribution** μ on $\mathbb{Z}_p$ is a $\mathbb{Q}_p$-valued additive map from the set of compact open subsets in $\mathbb{Z}_p$. It is called a **measure** if there is a constant $B\in\mathbb{R}$ such that

$$|\mu(U)|_p\le B$$

for all compact open $U\subseteq\mathbb{Z}_p$.

To define a distribution or measure on $\mathbb{Z}_p$, it suffices to define it on subsets of the form

$$I=\left\{a+p^N\mathbb{Z}_p,\quad 0\le a\le p^N-1,\quad N=1,2,\ldots\right\},$$

since any open subset of $\mathbb{Q}_p$ is a union of subsets of this type.

It is not difficult to verify that a map $\mu: I\to\mathbb{Q}_p$ satisfying

$$\mu(a+p^n\mathbb{Z}_p)=\sum_{b=0}^{p-1}\mu(a+bp^n+p^{n+1}\mathbb{Z}_p)$$

extends uniquely to a p-adic distribution on $\mathbb{Z}_p$.

We define the Bernoulli distributions. Let

$$b_0(x)=1,\quad b_1(x)=x-\frac{1}{2},\quad b_2(x)=x^2-x+\frac{1}{6},\quad \cdots$$

be the sequence of Bernoulli polynomials. Define

$$\mu_k(a+p^n\mathbb{Z}_p)=p^{n(k-1)}b_k\left(\frac{a}{p^n}\right).$$

Exercise 10.4.1 *Verify that μ_k extends to a distribution on $\mathbb{Z}_p$.*

If μ is a p-adic measure, one can define a good theory of integration:

Theorem 10.4.2 *Let μ be a p-adic measure on $\mathbb{Z}_p$, and let $f : \mathbb{Z}_p \to \mathbb{Q}_p$ be a continuous function. Then the "Riemann sums"*

$$S_N := \sum_{a \le a \le p^N - 1} f(x_{a,N})\mu(a + p^N \mathbb{Z}_p),$$

where $x_{a,N}$ is any element in the "interval" $a + p^n\mathbb{Z}$, converge to a limit in $\mathbb{Q}_p$ as $N \to \infty$, and this limit is independent of the choices $\{x_{a,N}\}$.

Proof. We first show that the sequence of S_N is Cauchy. By the continuity of f, we assume that N is large enough so that $|f(x) - f(y)| < \epsilon$ whenever $x \equiv y (\operatorname{mod} p^N)$. Now let $M > N$. By the additivity of μ, we can rewrite

$$S_N = \sum_{0 \le a \le p^M - 1} f(x_{\overline{a},N})\mu\left(a + p^M \mathbb{Z}_p\right),$$

where $\overline{a}$ denotes the least nonnegative residue of $a \left(\operatorname{mod} p^N\right)$. Since $x_{\overline{a},N} \equiv x_{a,M} \left(\operatorname{mod} p^N\right)$,

$$\begin{aligned} |S_N - S_M|_p &= \left| \sum_{0 \le a \le p^M - 1} (f(x_{\overline{a},N}) - f(x_{a,M}))\mu\left(a + p^M \mathbb{Z}_p\right) \right|_p \\ &\le B\epsilon, \end{aligned}$$

where $|\mu(U)|_p \le B$ for all compact open U. Since $\mathbb{Q}_p$ is complete, the sequence of S_N's converges to a limit. This limit is easily seen to be independent of the choice of the $x_{a,N}$'s. □

If μ is a measure on $\mathbb{Z}_p$ and $f : \mathbb{Z}_p \to \mathbb{Q}_p$ is a continuous function, we denote by $\int_{\mathbb{Z}_p} f(x) d\mu(x)$ the limit of the "Riemann sums" of Theorem 10.4.2.

We now introduce the Mazur measure. Let $\alpha \in \mathbb{Z}_p$. We let $(\alpha)_N$ be the rational integer between 0 and $p^N - 1$ that is congruent to $\alpha \,(\operatorname{mod} p^N)$. If μ is a distribution and $\alpha \in \mathbb{Q}_p$, it is clear that $\alpha\mu$ is again a distribution. If $\alpha \in \mathbb{Z}_p^*$, then μ' defined by $\mu'(U) = \mu(\alpha U)$ is again a distribution. Now let α be any rational integer coprime to p and unequal to 1. We define the "regularized" Bernoulli distribution by setting

$$\mu_{k,\alpha}(U) = \mu_k(U) - \alpha^{-k}\mu_k(\alpha U)$$

for any compact open set U. It can be shown that $\mu_{k,\alpha}$ is a measure.

Exercise 10.4.3 *Show that $\mu_{1,\alpha}$ is a measure.*

The measure $\mu_{1,\alpha}$ is called the Mazur measure. Its significance is disclosed by Theorem 10.4.7.

Exercise 10.4.4 *Let d_k be the least common multiple of the denominators of coefficients of $b_k(x)$. Show that*

$$d_k\mu_{k,\alpha}(a + p^N\mathbb{Z}_p) \equiv d_k k a^{k-1}\mu_{1,\alpha}(a + p^N\mathbb{Z}_p) \pmod{p^N}.$$

Exercise 10.4.5 *Show that*

$$\int_{\mathbb{Z}_p} d\mu_{k,\alpha} = k\int_{\mathbb{Z}_p} x^{k-1}d\mu_{1,\alpha}.$$

For any compact open set U and a continuous function $f : X \to \mathbb{Q}_p$, we define

$$\int_U f d\mu = \int_{\mathbb{Z}_p} f(x)\chi_U(x)d\mu.$$

Exercise 10.4.6 *If $\mathbb{Z}_p^*$ is the group of units of $\mathbb{Z}_p$, show that*

$$\mu_{k,\alpha}(\mathbb{Z}_p^*) = (1-\alpha^{-k})(1-p^{k-1})B_k,$$

where B_k is the kth Bernoulli number.

Putting these two exercises together gives the following important theorem:

Theorem 10.4.7 (Mazur, 1972)

$$-(1-p^{k-1})B_k/k = \frac{1}{\alpha^{-k}-1}\int_{\mathbb{Z}_p^*} x^{k-1}d\mu_{1,\alpha}.$$

By Exercise 8.2.12, we can interpret the left hand side of the equation in Theorem 10.4.7 as

$$\left(1-p^{k-1}\right)\zeta(1-k).$$

The theorem allows us to show that these values can be p-adically interpolated, provided that k lies in a fixed residue class $\pmod{p-1}$.

Exercise 10.4.8 (Kummer congruences) *If $(p-1)\nmid i$ and $i \equiv j \pmod{p^n}$, show that*

$$(1-p^{i-1})B_i/i \equiv (1-p^{j-1})B_j/j \pmod{p^{n+1}}.$$

Exercise 10.4.9 (Kummer) *If $(p-1) \nmid i$, show that $|B_i/i|_p \leq 1$.*

Exercise 10.4.10 (Clausen and von Staudt) *If $(p-1)|i$ and i is even, then*

$$pB_i \equiv -1 \pmod{p}.$$

Theorem 10.4.7 and the Kummer congruences motivate the definition of the p-adic ζ-function. If k is in a fixed residue class $s_0 \pmod{p-1}$, then the Kummer congruences imply that the numbers

$$\left(1 - p^{k-1}\right) \zeta(1-k)$$

can be p-adically interpolated. By Theorem 10.4.7 we see that this function must be

$$\frac{1}{\alpha^{-(s_0+(p-1)s)} - 1} \int_{\mathbb{Z}_p^*} x^{s_0+(p-1)s-1} d\mu_{1,\alpha},$$

and we designate it as $\zeta_{p,s_0}(s)$, and call it the p-adic zeta function. One can show that $\zeta_{p,s_0}(s)$ does not depend on the choice of α.

This observation of Kubota and Leopoldt in 1964 initiated a rich theory of p-adic zeta and L-functions. We refer the reader to Koblitz [K] and Washington [W] for further details.

10.5 Supplementary Problems

Exercise 10.5.1 *Let $1 \leq a \leq p-1$, and set $\phi(a) = (a^{p-1} - 1)/p$. Prove that $\phi(ab) \equiv \phi(a) + \phi(b) \pmod{p}$.*

Exercise 10.5.2 *With ϕ as in the previous exercise, show that*

$$\phi(a + pt) \equiv \phi(a) - \overline{a}t \pmod{p},$$

where $a\overline{a} \equiv 1 \pmod{p}$.

Exercise 10.5.3 *Let $[x]$ denote the greatest integer less than or equal to x. For $1 \leq a \leq p-1$, show that*

$$\frac{a^p - a}{p} \equiv \sum_{j=1}^{p-1} \frac{1}{j} \left[\frac{aj}{p}\right] \pmod{p}.$$

Exercise 10.5.4 *Prove the following generalization of Wilson's theorem:*

$$(p-k)!(k-1)! \equiv (-1)^k \pmod p$$

for $1 \leq k \leq p-1$.

Exercise 10.5.5 *Prove that for an odd prime* p,

$$\frac{2^{p-1}-1}{p} \equiv \sum_{j=1}^{p-1} \frac{(-1)^{j+1}}{2j} \pmod p.$$

Deduce that $2^{p-1} \equiv 1 \pmod{p^2}$ *if and only if the numerator of*

$$1 - \frac{1}{2} + \frac{1}{3} - \cdots - \frac{1}{p-1}$$

is divisible by p.

Exercise 10.5.6 *Let* p *be an odd prime. Show that for all* $x \in \mathbb{Z}_p$, $\Gamma_p(x+1) = h_p(x)\Gamma_p(x)$, *where*

$$h_p(x) = \begin{cases} -x & \text{if } |x|_p = 1, \\ \\ -1 & \text{if } |x|_p < 1. \end{cases}$$

Exercise 10.5.7 *For* $s \geq 2$, *show that the only solutions of* $x^2 \equiv 1 \pmod{2^s}$ *are* $x \equiv 1, -1, 2^{s-1}-1$, *and* $2^{s-1}+1$.

Exercise 10.5.8 (The 2-adic Γ-function) *Show that the sequence defined by*

$$\Gamma_2(n) = (-1)^n \prod_{\substack{1 \leq j < n \\ (j,2)=1}} j$$

can be extended to a continuous function on $\mathbb{Z}_2$.

Exercise 10.5.9 *Prove that for all natural numbers* n,

$$\Gamma_p(-n)\Gamma_p(n+1) = (-1)^{[n/p]+n+1}.$$

Exercise 10.5.10 *If* p *is an odd prime, prove that for* $x \in \mathbb{Z}_p$,

$$\Gamma_p(x)\Gamma_p(1-x) = (-1)^{\ell(x)},$$

where $\ell(x)$ *is defined as the element of* $\{1, 2, \ldots, p\}$ *satisfying* $\ell(x) \equiv x \pmod p$. (This is the p-adic analogue of Exercise 6.3.4.)

Exercise 10.5.11 *Show that*

$$\Gamma_p(1/2)^2 = \begin{cases} 1 & \text{if} \quad p \equiv 3 \pmod{4}, \\ -1 & \text{if} \quad p \equiv 1 \pmod{4}. \end{cases}$$

11 Equidistribution

11.1 Uniform distribution modulo 1

The theory of uniform distribution of sequences is vast and varied. A good reference book is [KN]. Here, we give the most basic introduction to this important chapter of analytic number theory.

A sequence of real numbers $\{x_n\}_{n=1}^{\infty}$ is said to be **uniformly distributed modulo 1** (abbreviated u.d.) if for every pair of real numbers a, b with $0 \leq a < b \leq 1$, we have

$$\lim_{N \to \infty} \frac{\#\{n \leq N : (x_n) \in [a, b]\}}{N} = b - a,$$

where $(x_n) := x_n - [x_n]$ denotes the fractional part of x_n.

Usually, it is convenient to take a sequence $\{x_n\}_{n=1}^{\infty}$ satisfying $0 \leq x_n < 1$ in discussing uniform distribution and we assume this is the case in the discussion below. It is clear from the definition that if a sequence $\{x_n\}_{n=1}^{\infty}$ is u.d. then it is also dense in the unit interval.

Exercise 11.1.1 *Let us write the sequence of non-zero rational numbers in* $[0, 1]$ *as follows:*

$$1, \frac{1}{2}, \frac{1}{3}, \frac{2}{3}, \frac{1}{4}, \frac{3}{4}, \frac{1}{5}, \frac{2}{5}, \frac{3}{5}, \frac{4}{5}, \frac{1}{6}, \frac{5}{6}, \ldots$$

where we successively write all the fractions with denominator b for $b = 1, 2, 3, \dots$. Show that this sequence is u.d. mod 1.

Exercise 11.1.2 *If a sequence of real numbers $\{x_n\}_{n=1}^{\infty}$ is u.d., show that for any a with $0 \le a < 1$, we have*

$$\#\{n \le N : x_n = a\} = o(N).$$

Exercise 11.1.3 *If the sequence $\{x_n\}_{n=1}^{\infty}$ is u.d. and $f : [0,1] \to \mathbb{C}$ is a continuous function, show that*

$$\lim_{N\to\infty} \frac{1}{N} \sum_{n=1}^{N} f(x_n) \to \int_0^1 f(x)dx,$$

and conversely.

Exercise 11.1.4 *If $\{x_n\}_{n=1}^{\infty}$ is u.d. then*

$$\lim_{N\to\infty} \frac{1}{N} \sum_{n=1}^{N} f(x_n) = \int_0^1 f(x)dx,$$

for any piecewise C^1-function $f : [0,1] \to \mathbb{C}$.

In particular, if $\{x_n\}_{n=1}^{\infty}$ is u.d. then for the functions $f_m(x) = e^{2\pi i m x}$, we have

$$\lim_{N\to\infty} \frac{1}{N} \sum_{n\le N} e^{2\pi i m x_n} = 0,$$

for all non-zero integers m. Weyl's criterion (to be proved below) is that the converse is true.

Theorem 11.1.5 [Weyl, 1916] *A sequence $\{x_n\}_{n=1}^{\infty}$ is u.d. if and only if*

$$\sum_{n=1}^{N} e^{2\pi i m x_n} = o(N), \qquad m = \pm 1, \pm 2, \dots \tag{11.1}$$

Proof. As observed earlier, the necessity is clear. For sufficiency, let $\epsilon > 0$ and f a continuous function $f : [0,1] \to \mathbb{C}$. By the Weierstrass approximation theorem, there is a trigonometric polynomial $\phi(x)$ such that $\deg \phi \le M$, with M depending on ϵ such that

$$\sup_{0\le x\le 1} |f(x) - \phi(x)| \le \epsilon. \tag{11.2}$$

Then,

$$\left| \int_0^1 f(x)dx - \frac{1}{N}\sum_{n=1}^{N} f(x_n) \right|$$

$$\leq \left| \int_0^1 (f(x) - \phi(x)dx \right| + \left| \int_0^1 \phi(x)dx - \frac{1}{N}\sum_{n=1}^{N} f(x_n) \right|.$$

By (11.2), the first term is $\leq \epsilon$. The second term is

$$\leq \left| \int_0^1 \phi(x)dx - \frac{1}{N}\sum_{n=1}^{N} \phi(x_n) \right| + \left| \frac{1}{N}\sum_{n=1}^{N} (\phi(x_n) - f(x_n)) \right|.$$

Again by (11.2), the last term is $\leq \epsilon$. Writing

$$\phi(x) = \sum_{|m| \leq M} a_m e^{2\pi i m x},$$

we see that

$$\int_0^1 \phi(x)dx = a_0,$$

and

$$\frac{1}{N}\sum_{n=1}^{N} \phi(x_n) = a_0 + \sum_{1 \leq |m| \leq M} a_m \left(\sum_{n=1}^{N} e^{2\pi i m x_n} \right),$$

so that

$$\left| \int_0^1 \phi(x)dx - \frac{1}{N}\sum_{n=1}^{N} \phi(x_n) \right| \leq \sum_{1 \leq |m| \leq M} |a_m| \left| \frac{1}{N}\sum_{n=1}^{N} e^{2\pi i m x_n} \right|.$$

Let $T = \sum_{1 \leq |m| \leq M} |a_m|$. We may choose N (which depends on M) sufficiently large so that all of the inner terms above are $\leq \epsilon/T$ by virtue of (11.1). Thus, this term is also $\leq \epsilon$. Hence,

$$\lim_{N \to \infty} \frac{1}{N}\sum_{n=1}^{N} f(x_n) = \int_0^1 f(x)dx.$$

This completes the proof. □

Exercise 11.1.6 *Show that Weyl's criterion need only be checked for positive integers m.*

Exercise 11.1.7 *Show that the sequence $\{x_n\}_{n=1}^{\infty}$ is u.d. mod 1 if and only if*

$$\lim_{N\to\infty} \frac{1}{N}\sum_{n=1}^{N} f(x_n) = \int_0^1 f(x)dx,$$

for any family of functions f which is dense in $C[0,1]$. Here, $C[0,1]$ is the metric space of continuous functions on $[0,1]$ with the sup norm.

Exercise 11.1.8 *Let θ be an irrational number. Show that the sequence $x_n = n\theta$ is u.d.*

Exercise 11.1.9 *If θ is rational, show that the sequence $x_n = n\theta$ is not u.d.*

Exercise 11.1.10 *Show that the sequence $x_n = \log n$ is not u.d. but is dense mod 1.*

Exercise 11.1.11 *Let $0 \leq x_n < 1$. Show that the sequence $\{x_n\}_{n=1}^{\infty}$ is u.d. mod 1 if and only if*

$$\lim_{N\to\infty} \frac{1}{N}\sum_{n=1}^{N} x_n^r = \frac{1}{r+1},$$

for every natural number r.

Exercise 11.1.12 *If $\{x_n\}_{n=1}^{\infty}$ is u.d. mod 1, then show that $\{mx_n\}_{n=1}^{\infty}$ is u.d. mod 1 for m a non-zero integer.*

Exercise 11.1.13 *If $\{x_n\}_{n=1}^{\infty}$ is u.d. mod 1, and c is a constant, show that $\{x_n + c\}_{n=1}^{\infty}$ is u.d. mod 1.*

Exercise 11.1.14 *If $\{x_n\}_{n=1}^{\infty}$ is u.d. mod 1 and $y_n \to c$ as $n \to \infty$, show that $\{x_n + y_n\}_{n=1}^{\infty}$ is u.d. mod 1.*

Exercise 11.1.15 *Let F_n denote the nth Fibonacci number defined by the recursion $F_0 = 1, F_1 = 1, F_{n+1} = F_n + F_{n-1}$. Show that $\log F_n$ is u.d. mod 1.*

To study the equidistribution of various sequences, an important technique was introduced by Weyl and van der Corput. The technique is based on the following simple inequality.

Theorem 11.1.16 (van der Corput, 1931) *Let $y_1, ..., y_N$ be complex numbers. Let H be an integer with $1 \le H \le N$. Then*

$$\left|\sum_{n=1}^{N} y_n\right|^2 \le$$

$$\frac{N+H}{H+1}\sum_{n=1}^{N}|y_n|^2 + \frac{2(N+H)}{H+1}\sum_{r=1}^{H}\left(1-\frac{r}{H+1}\right)\left|\sum_{n=1}^{N-r} y_{n+r}\overline{y}_n\right|.$$

Proof. It is convenient to set $y_n = 0$ for $n \le 0$ and $n > N$. Clearly,

$$(H+1)^2\left|\sum_{n} y_n\right|^2 = \left|\sum_{h=0}^{H}\sum_{n} y_{n+h}\right|^2 = \left|\sum_{n}\sum_{h=0}^{H} y_{n+h}\right|^2.$$

We note that the inner sum is zero if $n \ge N+1$ or $n \le -H$. Thus, in the outer sum, n is restricted to the interval $[-H+1, N]$. Applying the Cauchy-Schwarz inequality, we get that this is

$$\le (N+H)\sum_{n}\left|\sum_{h=0}^{H} y_{n+h}\right|^2.$$

Expanding the sum, we obtain

$$\sum_{n}\sum_{h=0}^{H}\sum_{k=0}^{H} y_{n+h}\overline{y}_{n+k} = (H+1)\sum_{n}|y_n|^2 + \sum_{n}\sum_{h\neq k} y_{n+h}\overline{y}_{n+k}.$$

In the second sum, we combine the terms corresponding to (h, k) and (k, h) to get that it is

$$2\,\mathrm{Re}\left(\sum_{n}\sum_{h=0}^{H}\sum_{k<h} y_{n+h}\overline{y}_{n+k}\right).$$

We write $m = n + k$ and re-write this as

$$2\,\mathrm{Re}\left(\sum_{m}\sum_{h=0}^{H}\sum_{k<h} y_{m-k+h}\overline{y}_m\right) = 2\,\mathrm{Re}\left(\sum_{m}\sum_{r=1}^{H} y_{m+r}\overline{y}_m \sum_{k<h;h-k=r} 1\right).$$

The innermost sum is easily seen to be $H+1-r$. Therefore,

$$\left|\sum_{n=1}^{N} y_n\right|^2 \le$$

$$\frac{N+H}{H+1}\sum_{n=1}^{N}|y_n|^2 + \frac{2(N+H)}{H+1}\sum_{r=1}^{H}\left(1-\frac{r}{H+1}\right)\left|\sum_{n=1}^{N-r} y_{n+r}\overline{y}_n\right|.$$

This completes the proof. □

Corollary 11.1.17 (van der Corput, 1931) *If for each positive integer r, the sequence $x_{n+r} - x_n$ is u.d. mod 1, then the sequence x_n is u.d. mod 1.*

Proof. We apply Theorem 11.1.16 with $y_n = e^{2\pi i m x_n}$ to get

$$\left|\frac{1}{N}\sum_{n=1}^{N} e^{2\pi i m x_n}\right|^2$$

$$\leq \frac{1+H/N}{H+1} + \frac{2(N+H)}{N^2(H+1)}\sum_{r=1}^{H}\left(1-\frac{r}{H+1}\right)\left|\sum_{n=1}^{N-r} e^{2\pi i m(x_{n+r}-x_n)}\right|.$$

Taking the limit as $N \to \infty$ and using the fact that $x_{n+r} - x_n$ is u.d. mod 1 for every $r \geq 1$, we see that

$$\lim_{N\to\infty}\left|\frac{1}{N}\sum_{n=1}^{N} e^{2\pi i m x_n}\right|^2 \ll \frac{1}{H},$$

for any H. Choosing H arbitrarily large gives the result. □

Exercise 11.1.18 *Let $y_1, ..., y_N$ be complex numbers. Let $\mathcal{H}$ be a subset of $[0, H]$ with $1 \leq H \leq N$. Show that*

$$\left|\sum_{n=1}^{N} y_n\right|^2 \leq \frac{N+H}{|\mathcal{H}|}\sum_{n=1}^{N}|y_n|^2 + \frac{2(N+H)}{|\mathcal{H}|^2}\sum_{r=1}^{H} N_r \left|\sum_{n=1}^{N-r} y_{n+r}\overline{y_n}\right|,$$

where N_r is the number of solutions of $h-k = r$ with $h > k$ and $h, k \in \mathcal{H}$.

Exercise 11.1.19 *Let θ be an irrational number. Show that the sequence $\{n^2\theta\}_{n=1}^{\infty}$ is u.d. mod 1.*

Exercise 11.1.20 *Show that the sequence $\{an^2 + bn\}_{n=1}^{\infty}$ is u.d. provided that one of a or b is irrational.*

Exercise 11.1.21 *Let $P(n) = a_d n^d + a_{d-1}n^{d-1} + \cdots + a_1 + a_0$ be a polynomial with real coefficients with at least one coefficient a_i with $i \geq 1$ irrational. Show that the sequence of fractional parts of $P(n)$ is u.d. mod 1.*

11.2 Normal numbers

Let x be a real number and $b \geq 2$ a positive integer. Then, x has a b-adic expansion

$$x = [x] + \sum_{n=1}^{\infty} \frac{a_n}{b^n},$$

with $0 \leq a_n < b$. This expansion is essentially unique. x is said to be **simply normal to the base** b if for each $0 \leq a < b$,

$$\lim_{N\to\infty} \frac{\#\{n \leq N : a_n = a\}}{N} = \frac{1}{b}.$$

In other words, each digit occurs with equal frequency in the b-adic expansion of x. More generally, we may consider a block of digits of length k and inquire how frequently this block appears in the b-adic expansion. To be precise, let B_k be a natural number whose b-adic expansion is of the form $b_1 b_2 \cdots b_k$. A number x is called **normal to the base** b if

$$\lim_{N\to\infty} \frac{1}{N}\#\{n \leq N - k + 1 : a_{n+j-1} = b_j \text{ for } 1 \leq j \leq k\} = \frac{1}{b^k}.$$

For instance, the number

$$0.010101\cdots = \sum_{n=1}^{\infty} \frac{1}{2^{2n}},$$

is simply normal to the base 2 but not normal to the base 2 since the block 11 does not occur at all in the expression.

Exercise 11.2.1 *Show that a normal number is irrational.*

Theorem 11.2.2 *The number x is normal to the base b if and only if the sequence (xb^n) is u.d. mod 1.*

Proof. Let $B_k = b_1 b_2 \cdots b_k$ be a block of k digits. The block

$$a_m a_{m+1} \cdots a_{m+k-1}$$

in the b-adic expansion of x is identical with B_k if and only if

$$\frac{B_k}{b^k} \leq (xb^{m-1}) < \frac{B_k + 1}{b^k}.$$

Let I_k denote this interval of length $1/b^k$. If the sequence $\{xb^n\}_{n=1}^{\infty}$ is u.d. then

$$\#\{m \leq N-k+1 : xb^{m-1} \in I_k\} \sim N/b^k$$

as N tends to infinity. Thus, x is normal to the base b. Conversely, if x is normal to the base b, then for any rational number of the form $y = a/b^k$, we have

$$\#\{n \leq N : (xb^n) < a/b^k\} = \#\{m \leq N-k+1 : (xb^{m-1}) < a/b^k\} + O(k).$$

This is easily seen to be equal to

$$\sum_{B_k < a} \#\{m \leq N-k+1 : (xb^{m-1}) \in I_k\} + O(k)$$

$$= \sum_{B_k < a} \left(N/b^k + o(N)\right) + O(k)$$

which is $aN/b^k + o(N)$ since x is normal to the base b. Since the numbers of the form a/b^k are dense in $[0,1]$, the asymptotic above extends to all y with $0 \leq y < 1$. This completes the proof. □

Exercise 11.2.3 *If x is normal to the base b, show that mx is normal to the base b for any non-zero integer m.*

We will now show that almost all numbers are normal (in the sense of Lebesgue measure).

Exercise 11.2.4 *Let $\{v_n\}_{n=1}^{\infty}$ be a sequence of distinct integers and set for a non-zero integer h,*

$$S(N,x) = \frac{1}{N}\sum_{n=1}^{N} e^{2\pi i v_n x h}.$$

Show that

$$\int_0^1 |S(N,x)|^2 dx = \frac{1}{N},$$

and

$$\sum_{N=1}^{\infty} \int_0^1 |S(N^2,x)|^2 dx < \infty.$$

From the previous exercise and Fatou's lemma, we deduce

$$\int_0^1 \sum_{N=1}^{\infty} |S(N^2, x)|^2 dx < \infty,$$

so that

$$\sum_{N=1}^{\infty} |S(N^2, x)|^2 < \infty,$$

for almost all x. Therefore,

$$\lim_{N\to\infty} S(N^2, x) = 0$$

for almost all x. Now, given any $N \geq 1$, we can find an m such that

$$m^2 \leq N < (m+1)^2.$$

Thus,

$$|S(N, x)| \leq |S(m^2, x)| + \frac{2m}{N} \leq |S(m^2, x)| + \frac{2}{\sqrt{N}}.$$

Thus,

$$\lim_{N\to\infty} S(N, x) = 0,$$

for all $x \notin V_h$, with V_h a set of measure zero. Since a countable union of sets of measure zero is still measure zero, we have proved:

Theorem 11.2.5 *Let v_n be a distinct sequence of natural numbers. For almost all x, the sequence $\{v_n x\}_{n=1}^{\infty}$ is u.d. mod 1.*

Applying the above theorem with $v_n = b^n$ and using Theorem 11.2.2, we deduce that almost all numbers are normal to every base b.

Exercise 11.2.6 *Show that the sequence $n!e$ is not u.d. mod 1.*

The determination of which numbers are normal is not an easy one. For instance, it is known that the number

$$0.12345678910111213...$$

called Champerowne's number, obtained by writing all the numbers in sequence is a normal number to the base 10. In 1946, Copeland and Erdös showed that

$$0.23571113171923...$$

obtained writing the sequence of prime numbers is normal to the base 10. It is unknown at present if numbers such as $\sqrt{2}, \log 2, e$ or π are normal numbers to any base b. In fact, there are no concrete examples of numbers which are normal to any base b, although almost all numbers are normal to any base b.

Exercise 11.2.7 *If x is normal to the base b, show that it is simply normal to the base b^m for every natural number m.*

11.3 Asymptotic distribution functions mod 1

Let $\{x_n\}_{n=1}^{\infty}$ be a sequence of real numbers and let $S(x; N) = \#\{n \leq N : 0 \leq (x_n) \leq x\}$. A sequence $\{x_n\}_{n=1}^{\infty}$ is said to have the **asymptotic distribution function** (abbreviated a.d.f. mod 1 or simply a.d.f.) $g(x)$ if

$$\lim_{N\to\infty} \frac{S(x; N)}{N} = g(x)$$

for all $0 \leq x \leq 1$. Clearly, g is non-decreasing and we have $g(0) = 0$ and $g(1) = 1$. A sequence which is u.d. mod 1 has asymptotic distribution function $g(x) = x$. Thus, this is a generalization of the concept discussed in the first section. As stated earlier, we assume we have a sequence $\{x_n\}_{n=1}^{\infty}$ with $0 \leq x_n < 1$.

Exercise 11.3.1 *A sequence $\{x_n\}_{n=1}^{\infty}$ has a.d.f. $g(x)$ if and only if for every piecewise continuous function f on $[0, 1]$, we have*

$$\lim_{N\to\infty} \frac{1}{N} \sum_{n=1}^{N} f(x_n) = \int_0^1 f(x) dg(x).$$

Exercise 11.3.2 *A sequence $\{x_n\}_{n=1}^{\infty}$ has a.d.f. $g(x)$ if and only if*

$$\lim_{N\to\infty} \frac{1}{N} \sum_{n=1}^{N} e^{2\pi i m x_n} = \int_0^1 e^{2\pi i m x} dg(x),$$

for all integers m.

Theorem 11.3.3 [Wiener - Schoenberg, 1928] *The sequence* $\{x_n\}_{n=1}^{\infty}$ *has a continuous a.d.f. if and only if for every integer* m, *the limit*

$$a_m := \lim_{N\to\infty} \frac{1}{N}\sum_{n=1}^{N} e^{2\pi i m x_n}$$

exists and

$$\sum_{m=1}^{N} |a_m|^2 = o(N). \tag{11.3}$$

Proof. Suppose the sequence has a continuous a.d.f. $g(x)$. The existence of the limits is clear. Now, by Exercise 11.3.2, we have

$$a_m = \int_0^1 e^{2\pi i m x} dg(x).$$

Thus,

$$\lim_{N\to\infty} \frac{1}{N}\sum_{m=1}^{N} |a_m|^2 = \lim_{N\to\infty} \frac{1}{N}\sum_{m=1}^{N} \int_0^1 \int_0^1 e^{2\pi i m(x-y)} dg(x)dg(y).$$

This is equal to

$$\lim_{N\to\infty} \int_0^1 \int_0^1 \left(\frac{1}{N}\sum_{m=1}^{N} e^{2\pi i m(x-y)} \right) dg(x)dg(y).$$

By the Lebesgue dominated convergence theorem this is equal to

$$\int_0^1 \int_0^1 \left(\lim_{n\to\infty} \frac{1}{N}\sum_{m=1}^{N} e^{2\pi i m(x-y)} \right) dg(x)dg(y).$$

The integrand is zero unless $x - y \in \mathbb{Z}$, in which case it is 1. The set of such $(x, y) \in [0, 1]^2$ is a set of measure zero. Therefore the limit is zero. Conversely, suppose that the limit is zero. By the Riesz representation theorem, there is a measurable function $g(x)$ such that

$$a_m = \int_0^1 e^{2\pi i m x} dg(x).$$

Consequently,

$$\int_0^1 \int_0^1 f(x-y)dg(x)dg(y) = 0,$$

where $f(x-y) = 0$ unless $x - y \in \mathbb{Z}$, in which case it is 1. We want to show that this implies that g is continuous. Indeed, if g has a jump discontinuity at c (say), the double integral is at least $[g(c+) - g(c-)]^2 > 0$. This completes the proof. □

Exercise 11.3.4 *Suppose that $\{x_n\}_{n=1}^{\infty}$ is a sequence such that for all integers m, the limits*

$$a_m := \lim_{N\to\infty} \frac{1}{N} \sum_{n=1}^{N} e^{2\pi i m x_n},$$

exist and

$$\sum_{m=-\infty}^{\infty} |a_m|^2 < \infty.$$

Put

$$g_1(x) = \sum_{m=-\infty}^{\infty} a_m e^{2\pi i m x}.$$

Show that

$$\lim_{N\to\infty} \frac{\#\{n \leq N : x_n \in [\alpha, \beta]\}}{N} = \int_\alpha^\beta g_1(x)dx,$$

for any interval $[\alpha, \beta]$ contained in $[0, 1]$.

11.4 Discrepancy

Given a sequence $\{x_n\}_{n=1}^{\infty}$, we define the sequence D_N by setting

$$D_N := \sup_{0\leq a<b\leq 1} \left| \frac{\#\{n \leq N : a \leq (x_n) \leq b\}}{N} - (b-a) \right|$$

and call this the **discrepancy** of the sequence.

Exercise 11.4.1 *Show that the sequence $\{x_n\}_{n=1}^{\infty}$ is u.d. mod 1 if and only if $D_N \to 0$ as $N \to \infty$.*

Exercise 11.4.2 *Show that*

$$\left(\frac{\sin \pi z}{\pi}\right)^2 \sum_{n=-\infty}^{\infty} \frac{1}{(z-n)^2} = 1, \quad z \notin \mathbb{Z}$$

The proof of Weyl's criterion relied on the existence of finite trigonometric polynomials which approximate the characteristic function of an interval. It will be useful to have a quantitative version of these approximations. There are several ways of obtaining such a version. A most expedient route was discovered by Montgomery [Mo] using functions that had earlier been discovered by Beurling and Selberg and utilised by the latter to obtain sharp constants in the large sieve inequality. This is the route we shall follow.

For $z \in \mathbb{C}$, we define $\operatorname{sgn} z = 1$ if $\operatorname{Re}(z) \geq 0$ and $\operatorname{sgn} z = -1$ if $\operatorname{Re}(z) < 0$.

Theorem 11.4.3 (Beurling, 1938) *Let*

$$B(z) = \left(\frac{\sin \pi z}{\pi}\right)^2 \left(\sum_{n=0}^{\infty} \frac{1}{(z-n)^2} - \sum_{n=1}^{\infty} \frac{1}{(z+n)^2} + \frac{2}{z}\right).$$

Then

1. $B(z)$ *is entire;*

2. $B(x) \geq \operatorname{sgn} x$ *for real* x*;*

3. $B(z) = \operatorname{sgn} z + O(e^{2\pi|\operatorname{Im} z|}/|z|)$ *;*

4. $$\int_{-\infty}^{\infty} (B(x) - \operatorname{sgn} x)dx = 1.$$

Proof. The first assertion is clear since $\sin \pi z$ has simple zeros for $z \in \mathbb{Z}$. To prove the second assertion, we observe that

$$\left(\frac{\sin \pi z}{\pi}\right)^2 \sum_{n=-\infty}^{\infty} \frac{1}{(z-n)^2} = 1. \tag{11.4}$$

which is the content of Exercise 11.4.2. For $x > 0$, we also have

$$\sum_{n=1}^{\infty} \frac{1}{(x+n)^2}$$

$$\leq \sum_{n=1}^{\infty} \int_{x+n-1}^{x+n} \frac{du}{u^2} = \int_{x}^{\infty} \frac{du}{u^2} = \frac{1}{x} = \sum_{n=0}^{\infty} \int_{x+n}^{x+n+1} \frac{du}{u^2} \leq \sum_{n=0}^{\infty} \frac{1}{(x+n)^2}, \tag{11.5}$$

by the method used to prove the integral test. From (11.4), we have

$$B(z) - \operatorname{sgn} z = \left(\frac{\sin \pi z}{\pi}\right)^2 \left(\frac{2}{z} - 2\sum_{n=1}^{\infty} \frac{1}{(z+n)^2}\right)$$

for $\operatorname{Re}(z) > 0$ and

$$B(z) - \operatorname{sgn} z = \left(\frac{\sin \pi z}{\pi}\right)^2 \left(\frac{2}{z} + 2\sum_{n=0}^{\infty} \frac{1}{(z-n)^2}\right)$$

for $\operatorname{Re}(z) < 0$. The second assertion follows immediately from these identities and (11.5). For the third assertion, we note that

$$\sin^2 \pi z = O(e^{2\pi|\operatorname{Im}(z)|}).$$

In addition, for $x, y > 0$, we have

$$\begin{aligned} \sum_{n=0}^{\infty} \frac{1}{(x+n)^2 + y^2} &\leq \frac{1}{x^2+y^2} + \min\left(\int_0^{\infty} \frac{dt}{(x+t)^2}, \int_0^{\infty} \frac{dt}{t^2+y^2}\right) \\ &= \frac{1}{x^2+y^2} + \min\left(\frac{1}{x}, \frac{\pi}{2y}\right). \end{aligned}$$

Therefore,

$$\sum_{n=1}^{\infty} \frac{1}{|z+n|^2} = O(1/|z|), \quad \text{for } \operatorname{Re}(z) \geq 0,$$

and

$$\sum_{n=0}^{\infty} \frac{1}{|z-n|^2} = O(1/|z|), \quad \text{for } \operatorname{Re}(z) < 0.$$

The third assertion is now immediate from these observations. Finally, for the last assertion, we note from the second assertion that the integrand is non-negative. Also,

$$\int_{-A}^{A} (B(x) - \operatorname{sgn} x)\, dx = \int_0^A (B(x) + B(-x))\, dx = \int_0^A \left(\frac{\sin \pi x}{\pi}\right)^2 \frac{2}{x^2}\, dx,$$

after a short computation. The last integral tends to 1 as A tends to infinity. This completes the proof. □

Following Selberg, we now use this theorem to majorize and minorize the characteristic function of an interval via finite trigonometric polynomials.

Theorem 11.4.4 (Selberg, 1970) *Let $I = [a, b]$ be an interval and χ_I its characteristic function. Then, there are continuous functions $S_+(x)$ and $S_-(x)$ in $L^1(\mathbb{R})$ such that*

$$S_-(x) \leq \chi_I(x) \leq S_+(x),$$

with

$$\hat{S}_\pm(t) = 0, \quad \text{for} \quad |t| \geq 1.$$

In addition,

$$\int_{-\infty}^{\infty} (\chi_I(x) - S_-(x))dx = 1$$

and

$$\int_{-\infty}^{\infty} (S^+(x) - \chi_I(x))dx = 1.$$

Proof. With B as in Theorem 11.4.3, let

$$S_+(x) = \frac{1}{2}(B(x-a) + B(b-x)).$$

Then,

$$S_+(x) \geq \frac{1}{2}(\text{sgn}\,(x-a) + \text{sgn}\,(b-x)) = \chi_I(x),$$

and

$$\int_{-\infty}^{\infty} (S_+(x) - \chi_I(x))dx = 1$$

by the last assertion of Theorem 11.4.3. Hence, $S_+ \in L^1(\mathbb{R})$. Moreover, the function S_+ is continuous, being the restriction of an entire function. Now we will show that for $t > 1$,

$$\hat{S}_+(t) = \int_{-\infty}^{\infty} S_+(x)e(-tx)dx = 0, \quad e(u) = e^{2\pi i u}.$$

To this end, we begin by showing

$$J(A, B) = \int_{-A}^{B} S_+(x)e(-tx)dx = O\left(\frac{1}{A} + \frac{1}{B}\right)$$

as A, B tend to infinity. By contour integration, $J(A, B)$ can be written as the sum of three line integrals, two of them being along the vertical line segments $[-A, -A - iT]$ and $[B - iT, B]$ and one being along the horizontal segment $[-A - iT, B - iT]$. This last integral is easily estimated using Theorem 11.4.3. It is bounded by

$$\int_{-A'}^{B'} |B(x - iT)| e^{-2\pi tT} dx \ll \int_{-A'}^{B'} e^{2\pi T} e^{-2\pi tT} dx,$$

where $A' = A + \max(|a|, |b|)$ and $B' = B + \max(|a|, |b|)$. This integral tends to zero as T tends to infinity. The other two integrals are similarly estimated. For $z = -A + iy$, we have

$$B(z - a) = -1 + O\left(\frac{e^{-2\pi y}}{A}\right), \quad \text{for } A > |a|,$$

$$B(b - z) = 1 + O\left(\frac{e^{-2\pi y}}{A}\right), \quad \text{for } A > |b|,$$

so that

$$S_+(z) \ll \frac{e^{-2\pi y}}{A}$$

and the integral over the left vertical line is

$$\ll \frac{1}{A} \int_{-\infty}^{0} e^{-2\pi y} e^{2\pi t y} dy \ll \frac{1}{A}.$$

The other vertical line integral is similarly estimated so that letting A, B tend to infinity, we deduce that $\hat{S}_+(t) = 0$ for $t > 1$. For $t < -1$, we use $\hat{S}_+(-t) = \overline{\hat{S}_+(t)}$ and deduce the desired result from this. For $t = \pm 1$, the result follows by continuity of S_+. Finally, we set

$$S_-(x) = -\frac{1}{2}(B(x - a) + B(x - b)),$$

and proceed analogously to complete the proof of the theorem. □

Exercise 11.4.5 *For any $\delta > 0$, and any interval $I = [a, b]$, show that there are continuous functions $H_+(x), H_-(x) \in L_1(\mathbb{R})$ such that*

$$H_-(x) \leq \chi_I(x) \leq H_+(x),$$

with $\hat{H}_\pm(t) = 0$ for $|t| \geq \delta$ and

$$\int_{-\infty}^{\infty} (\chi_I(x) - H_-(x)) dx = \int_{-\infty}^{\infty} (H_+(x) - \chi_I(x)) dx = \frac{1}{\delta}.$$

Exercise 11.4.6 *Let $f \in L^1(\mathbb{R})$. Show that the series*

$$F(x) = \sum_{n \in \mathbb{Z}} f(n + x),$$

is absolutely convergent for almost all x, has period 1 and satisfies $\hat{F}(k) = \hat{f}(k)$.

Theorem 11.4.7 *Let M be a natural number. For any interval $I = [a, b]$ with length $b - a < 1$, write*

$$\Xi_I(x) = \sum_{n \in \mathbb{Z}} \chi_I(n + x).$$

Then, there are trigonometric polynomials

$$S_M^{\pm}(x) = \sum_{|m| \le M} \hat{S}_M^{\pm}(m) e(mx),$$

such that for all x

$$S_M^{-}(x) \le \Xi_I(x) \le S_M^{+}(x),$$

and

$$\hat{S}_M^{-}(0) = b - a - \frac{1}{M+1}, \quad \hat{S}_M^{+}(0) = b - a + \frac{1}{M+1}.$$

Proof. Take $\delta = M + 1$ in Exercise 11.4.5 and let $H_{\pm}$ be the functions obtained by that exercise. Put

$$V_{\pm}(x) = \sum_{n \in \mathbb{Z}} H_{\pm}(n + x).$$

By Exercise 11.4.6, $V_{\pm}(x) \in L^1(0, 1)$ and $\hat{V}_{\pm}(t) = 0$ for $|t| \ge M + 1$. Thus,

$$V_{\pm}(x) = \sum_{|m| \le M} \hat{V}_{\pm}(m) e(mx),$$

almost everywhere. Now set

$$S_M^{\pm} = \sum_{|m| \le M} \hat{V}_{\pm}(m) e(mx).$$

Since $\chi_I(x) \ge H_{-}(x)$, we get

$$\Xi(x) = \sum_{n \in \mathbb{Z}} \chi_I(n + x) \ge \sum_{n \in \mathbb{Z}} H_{-}(n + x) = V_{-}(x)$$

for almost all x. By continuity, we deduce that $\Xi_I(x) \geq S_M^-(x)$ for all x. Similary, we deduce $\Xi_I(x) \leq S_M^+(x)$ for all x. In addition, we have

$$\hat{S}_M^{\pm}(0) = b - a \pm \frac{1}{M+1}.$$

□

We now prove the following theorem due to Erdös and Turán. The proof given below is due to Montgomery [Mo].

Theorem 11.4.8 (Erdös-Turán, 1948) *For any integer $M \geq 1$,*

$$D_N \leq \frac{1}{M+1} + 3\sum_{m=1}^{M} \frac{1}{Nm}\left|\sum_{n=1}^{N} e^{2\pi i m x_n}\right|.$$

Proof. Let χ_I be the characteristic function of the interval $I = [a, b]$. Using Theorem 11.4.7, we have

$$\sum_{n=1}^{N} \Xi_I(x_n) \leq \sum_{n=1}^{N} S_M^+(x_n)$$

$$\leq N(b-a) + \frac{N}{M+1} + \sum_{0<|m|\leq M} |\hat{S}_M^+(m)| \left|\sum_{n=1}^{N} e^{2\pi i m x_n}\right|.$$

To estimate $\hat{S}_M^+(m)$, we use

$$\hat{S}_M^+(m) = \int_0^1 \Xi_I(t)e(-mt)dt + \int_0^1 (S_M^+(t) - \Xi_I(t))e(-mt)dt,$$

to deduce

$$|\hat{S}_M^+(m)| \leq \int_0^1 (S_M^+(x) - \Xi_I(x))dx + |\hat{\Xi}_I(m)|.$$

The integral is $1/(M+1)$ and

$$\hat{\Xi}_I(m) = e(-\frac{1}{2}m(a+b))\frac{\sin \pi(b-a)m}{\pi m},$$

from which we get

$$|\hat{S}_M^+(m)| \leq \frac{1}{M+1} + \left|\frac{\sin \pi(b-a)m}{\pi m}\right| \leq \frac{3}{2|m|}.$$

Thus,

$$\sum_{n=1}^{N} \Xi_I(x_n) \leq N(b-a) + \frac{N}{M+1} + 3\sum_{m=1}^{M} \frac{1}{m} \left| \sum_{n=1}^{N} e^{2\pi i m x_n} \right|.$$

Similarly, we obtain

$$\sum_{n=1}^{N} \Xi_I(x_n) \geq N(b-a) - \frac{N}{M+1} - 3\sum_{m=1}^{M} \frac{1}{m} \left| \sum_{n=1}^{N} e^{2\pi i m x_n} \right|,$$

from which the theorem follows. □

Exercise 11.4.9 *Let $x_1, ..., x_N$ be N points in $[0,1]$. For $0 \leq x \leq 1$, let*

$$R_N(x) = \#\{m \leq N : 0 \leq x_m \leq x\} - Nx.$$

Show that

$$\int_0^1 R_N^2(x)dx = \left(\sum_{n=1}^{N} (x_n - 1/2) \right)^2 + \frac{1}{2\pi^2} \sum_{h=1}^{\infty} \frac{1}{h^2} \left| \sum_{n=1}^{N} e^{2\pi i h x_n} \right|^2.$$

Exercise 11.4.10 *Let α be irrational. Let $||x||$ denote the distance of x from the nearest integer. Show that the discrepancy D_N of the sequence $n\alpha$ satisfies*

$$D_N \ll \frac{1}{M} + \frac{1}{N} \sum_{m=1}^{M} \frac{1}{m||m\alpha||},$$

for any natural number M.

11.5 Equidistribution and L-functions

We will discuss a general formalism to study equidistribution due to Serre [Se]. Let G be a compact group with Haar measure μ normalized so that $\mu(G) = 1$. The space X of conjugacy classes of G inherits a natural topology from that of G as well as the measure. Let K be a number field and for each place v of K, let Nv denote its norm and let us suppose we have a map

$$v \mapsto x_v \in X.$$

For each representation,

$$\rho : G \to GL(V),$$

we set

$$L(s, \rho) = \prod_v \det(1 - \rho(x_v) Nv^{-s})^{-1}.$$

Exercise 11.5.1 *Show that* $L(s, \rho)$ *defines an analytic function in the region* $\text{Re}(s) > 1$.

The sequence x_v is said to be μ-equidistributed in X if for any continuous function f on X we have

$$\lim_{x \to \infty} \frac{1}{\pi_K(x)} \sum_{Nv \leq x} f(x_v) = \int_G f(x) d\mu(x),$$

where $\pi_K(x)$ denotes the number of places v with $Nv \leq x$.

By the celebrated Peter-Weyl theorem, every continuous function f can be approximated by a finite linear combination of irreducible characters χ. Thus, it suffices to verify the existence of the limit when f is restricted to an irreducible character. The orthogonality relations now give us:

Theorem 11.5.2 (Weyl criterion for compact groups) *Let G be a compact group with normalized Haar measure μ. Let X be the space of conjugacy classes of G as above. A sequence x_v is is μ-equidistributed if and only if for every irreducible character $\chi \neq 1$ of G, we have*

$$\lim_{x \to \infty} \frac{1}{\pi_K(x)} \sum_{Nv \leq x} \chi(x_v) = 0.$$

Exercise 11.5.3 (Serre) *Suppose that for each irreducible representation $\rho \neq 1$, we have that $L(s, \rho)$ extends to an analytic function for* $\text{Re}(s) \geq 1$ *and does not vanish there. Prove that the sequence x_v is μ-equidistributed in the space of conjugacy classes, with respect to the image of the normalized Haar measure μ of G.*

This formalism includes many of the classical prime number theorems. Indeed, if $G = (\mathbb{Z}/m\mathbb{Z})^*$ is the group of coprime residue classes mod m, and K is the rational number field, then we can associate to each prime p coprime with m, the residue class it belongs to mod m. The associated L-functions are the Dirichlet L-functions. Their analytic continuation and non-vanishing on $\mathrm{Re}(s) = 1$ is the content of Chapter 3. In this way, we deduce Dirichlet's theorem concerning the distribution of primes in a given arithmetic progression. More generally, if $K/\mathbb{Q}$ is a Galois extension with group G and we associate to each unramified prime p of $\mathbb{Q}$ the conjugacy class of the Frobenius automorphism of the prime ideal $\mathfrak{p}$ lying above p, then, the corresponding equidistribution theorem is the Chebotarev density theorem.

A conjectural example is given by the Ramanujan τ function. Recall that this function is defined by the product expansion

$$\sum_{n=1}^{\infty} \tau(n) q^n = q \prod_{n=1}^{\infty} (1 - q^n)^{24}.$$

In 1916, Ramanujan conjectured that $\tau(n)$ is a multiplicative function and this was proved a year later by Mordell. He also conjectured that for every prime p, we have

$$|\tau(p)| \leq 2p^{11/2}.$$

Thus, we may write

$$\tau(p) = 2p^{11/2} \cos \theta_p,$$

for some unique $\theta_p \in [0, \pi]$. Inspired by the Sato-Tate conjecture in the theory of elliptic curves, Serre [Se] made the following conjecture. Let $G = SU(2)$ be the special unitary group of 2×2 matrices over the complex numbers. The conjugacy classes of G are parametrized by elements of the interval $[0, \pi]$. More precisely, for each $\theta \in [0, \pi]$ the corresponding conjugacy class $X(\theta)$ has the element

$$\begin{pmatrix} e^{i\theta} & 0 \\ 0 & e^{-i\theta} \end{pmatrix}.$$

In Serre's formalism, we can construct a family of L-functions attached to the irreducible representations of $SU(2)$ via the mapping

$$p \mapsto X(\theta_p).$$

$SU(2)$ has a standard 2-dimensional representation given by the natural map ρ into $GL(2)$. It is known that all the irreducible representations of G are the m-th symmetric powers $\mathrm{Sym}^m(\rho)$. The Sato-Tate conjecture (as formulated by Serre in this context) is the assertion that the elements $X(\theta_p)$ are equidistributed in the space of conjugacy classes of $SU(2)$ with respect to the Haar measure, which one can show is

$$\frac{2}{\pi}\sin^2\theta d\theta.$$

To prove this conjecture, it suffices to show that each of the L-series attached to these representations extends to $\mathrm{Re}(s) \geq 1$ and does not vanish there. This conjecture fits neatly into a larger package of conjectures in the Langlands program. In fact, the example given above with the Ramanujan τ function is a special case of a larger family of conjectures, one for each Hecke eigenform, and more generally, for automorphic representations on $GL(2)$. At present, it is known that for $m \leq 9$, the m-th symmetric power L-series has the predicted analytic continuation and non-vanishing property. Recently, Taylor has announced a proof of the Sato-Tate conjecture for elliptic curves (which is the original context in which such equidistribution conjectures were made).

Exercise 11.5.4 *Let G be the additive group of residue classes mod k. Show that a sequence of natural numbers $\{x_n\}_{n=1}^{\infty}$ is equidistributed in G if and only if*

$$\sum_{n=1}^{N} e^{2\pi i a x_n/k} = o(N),$$

for $a = 1, 2, ..., k-1$.

Exercise 11.5.5 *Let p_n denote the n-th prime. Show that the sequence $\{\log p_n\}_{n=1}^{\infty}$ is not u.d. mod 1.*

Exercise 11.5.6 *Let $v_1, v_2, ...$ be a sequence of vectors in $\mathbb{R}^k/\mathbb{Z}^k$. Show that the sequence is equidistributed in $\mathbb{R}^k/\mathbb{Z}^k$ if and only if*

$$\sum_{n=1}^{N} e^{2\pi i b\cdot v_n} = o(N),$$

for every $b \in \mathbb{Z}^k$ with b unequal to the zero vector.

Exercise 11.5.7 *Let $1, \alpha_1, \alpha_2, ..., \alpha_k$ be linearly independent over $\mathbb{Q}$. Show that the vectors $v_n = (n\alpha_1, ..., n\alpha_k)$ are equidistributed in $\mathbb{R}^k/\mathbb{Z}^k$.*

Exercise 11.5.8 *Let a be a squarefree number and for primes p coprime to a, consider the map*

$$p \mapsto x_p := \left(\frac{a}{p}\right),$$

where (a/p) denotes the Legendre symbol. Show that the sequence of x_p's is equidistributed in the group of order 2 consisting of $\{\pm 1\}$.

11.6 Supplementary Problems

Exercise 11.6.1 *Show that Exercise 1.1.2 cannot be extended to Lebesgue integrable functions f.*

Exercise 11.6.2 (Féjer) *Let f be a real valued differentiable function, with $f'(x) > 0$ and monotonic. If $f(x) = o(x)$ and $xf'(x) \to \infty$ when $x \to \infty$, show that the sequence $\{f(n)\}_{n=1}^{\infty}$ is u.d. mod 1.*

Exercise 11.6.3 *For any $c \in (0,1)$, and $\alpha \neq 0$, show that the sequence αn^c is u.d. mod 1.*

Exercise 11.6.4 *For any $c > 1$, show that the sequence $(\log n)^c$ is u.d. mod 1.*

Exercise 11.6.5 *Let f be real valued and have a monotone derivative f' in $[a,b]$ with $f'(x) \geq \lambda > 0$. Show that*

$$\left| \int_a^b e^{2\pi i f(x)} dx \right| \leq \frac{2}{\pi \lambda}.$$

Exercise 11.6.6 *Let f be as in the previous exercise but now assume that $f'(x) \leq -\lambda < 0$. Show that the integral estimate is still valid.*

Exercise 11.6.7 *Let f be real-valued and twice differentiable on $[a,b]$ with $f''(x) \geq \delta > 0$. Prove that*

$$\left| \int_a^b e^{2\pi i f(x)} dx \right| \leq \frac{4}{\sqrt{\delta}}.$$

Exercise 11.6.8 *Let $b - a \geq 1$. Let $f(x)$ be a real-valued function on $[a,b]$ with $f''(x) \geq \delta > 0$ on $[a,b]$. Show that*

$$\left| \sum_{a<n<b} e^{2\pi i f(n)} \right| \ll \frac{f'(b) - f'(a) + 1}{\sqrt{\delta}}.$$

Exercise 11.6.9 *Show that the estimate in the previous exercise is still valid if $f''(x) \leq -\delta < 0$.*

Exercise 11.6.10 *Show that the sequence* $\{\log n!\}_{n=1}^{\infty}$ *is u.d mod 1.*

Exercise 11.6.11 *Let* $\zeta(s)$ *denote the Riemann zeta function and assume the Riemann hypothesis. Let* $1/2 + i\gamma_1, 1/2 + i\gamma_2, \ldots$ *denote the zeros of* $\zeta(s)$ *with positive imaginary part, arranged so that* $\gamma_1 \le \gamma_2 \le \gamma_3 \cdots$. *Show that the sequence* $\{\gamma_n\}$ *is uniformly distributed mod 1.*

Exercise 11.6.12 *Let* A_n *be a sequence of sets of real numbers with* $\#A_n \to \infty$. *We will say that this sequence is* **set equidistributed** *mod 1 (s.e.d. for short) if for any* $[a, b] \subseteq [0, 1]$ *we have*

$$\lim_{n\to\infty} \frac{\#\{t \in A_n : a \le (t) \le b\}}{\#A_n} = b - a.$$

The usual notion of u.d. mod 1 is obtained as a special case of this by taking $A_n = \{x_1, \ldots, x_n\}$. *Show that the sequence of sets* A_n *is s.e.d. mod 1 if and only if for any continuous function* $f : [0, 1] \to \mathbb{C}$, *we have*

$$\lim_{n\to\infty} \frac{1}{\#A_n} \sum_{t\in A_n} f(t) = \int_0^1 f(x)dx.$$

Exercise 11.6.13 *Show that the sequence of sets* A_n *is s.e.d mod 1 if and only if for every non-zero integer* m, *we have*

$$\lim_{n\to\infty} \frac{1}{\#A_n} \sum_{t\in A_n} e^{2\pi imt} = 0.$$

Exercise 11.6.14 *Let* A_n *be the finite set of rational numbers with denominator* n. *Show the sequence* A_n *is set equidistributed mod 1.*

Exercise 11.6.15 *A sequence of sets* A_n *with* $A_n \subseteq [0, 1]$ *and* $\#A_n \to \infty$ *is said to have* **set asymptotic distribution function** *(s.a.d.f. for short)* $g(x)$ *if*

$$\lim_{n\to\infty} \frac{\#\{t \in A_n : 0 \le t \le x\}}{\#A_n} = g(x).$$

Show that the sequence has s.a.d.f. $g(x)$ *if and only if for every continuous function* f, *we have*

$$\lim_{n\to\infty} \frac{1}{\#A_n} \sum_{t\in A_n} f(t) = \int_0^1 f(x)dg(x).$$

Exercise 11.6.16 (Generalized Wiener-Schoenberg criterion) *Show that the sequence of sets* $\{A_n\}_{n=1}^{\infty}$ *with* $A_n \subseteq [0,1]$ *and* $\#A_n \to \infty$ *has a continuous s.a.d.f. if and only if for all* $m \in \mathbb{Z}$ *the limit*

$$a_m := \lim_{n\to\infty} \frac{1}{\#A_n} \sum_{t\in A_n} e^{2\pi i m t}$$

exists and

$$\sum_{m=1}^{N} |a_m|^2 = o(N).$$

Part II

Solutions

1

Arithmetic Functions

1.1 The Möbius Inversion Formula and Applications

1.1.1 *Prove that*

$$\sum_{d|n} \mu(d) = \begin{cases} 1 & \textit{if} \quad n = 1, \\ 0 & \textit{otherwise.} \end{cases}$$

Let $n = p_1^{\alpha_1} \cdots p_k^{\alpha_k}$ be the unique factorization of n as a product of powers of primes. Let $N = p_1 \cdots p_k$. Then

$$\sum_{d|n} \mu(d) = \sum_{d|N} \mu(d),$$

since the Möbius function vanishes on numbers that are not squarefree. Any divisor of N corresponds to a subset of $\{p_1, \ldots, p_k\}$. Thus, for $n > 1$,

$$\sum_{d|n} \mu(d) = \sum_{r=0}^{k} \binom{k}{r} (-1)^r = (1-1)^k = 0.$$

The result is clear if $n = 1$. $\square$

1.1.2 (The Möbius inversion formula)*Show that*

$$f(n) = \sum_{d|n} g(d) \qquad \forall n \in \mathbb{N}$$

if and only if

$$g(n) = \sum_{d|n} \mu(d) f(n/d) \qquad \forall n \in \mathbb{N}.$$

We have

$$\begin{aligned}
\sum_{d|n} \mu(d) f\left(\frac{n}{d}\right) &= \sum_{d|n} \mu(d) \sum_{e|\frac{n}{d}} g(e) \\
&= \sum_{des=n} \mu(d) g(e) \\
&= \sum_{e|n} g(e) \sum_{d|\frac{n}{e}} \mu(d) \\
&= g(n),
\end{aligned}$$

since the inner sum in the penultimate step is zero unless $n/e = 1$. The converse is also easily established as follows. Suppose

$$g(n) = \sum_{d|n} \mu(d) f(n/d).$$

Then

$$\begin{aligned}
\sum_{d|n} g(d) &= \sum_{d|n} \sum_{e|d} \mu(e) f(d/e) \\
&= \sum_{est=n} \mu(e) f(s) \\
&= \sum_{s|n} f(s) \sum_{e|\frac{n}{s}} \mu(e) \\
&= f(n),
\end{aligned}$$

since the inner sum is again by (1.1.1) equal to zero unless $n/s = 1$. □

1.1.3 *Show that*

$$\sum_{d|n} \varphi(d) = n.$$

We shall count the residue classes (mod n) in two different ways. On the one hand, there are n residue classes. Each residue class representative u can be written as dn_0, where $d = (u, n)$. Thus $(n_0, n/d) = 1$. Thus, we can partition the residue classes $u (\mathrm{mod}\, n)$ according to the value of $\gcd(u, n)$. The number of classes corresponding to a given $d|n$ is precisely $\varphi(n/d)$. Thus

$$n = \sum_{d|n} \varphi(n/d) = \sum_{d|n} \varphi(d),$$

as desired. □

1.1.4 *Show that*

$$\frac{\varphi(n)}{n} = \sum_{d|n} \frac{\mu(d)}{d}.$$

This is immediate from the Möbius inversion formula and Exercise 1.1.3.

1.1.5 *Let f be multiplicative. Suppose that*

$$n = \prod_{p^\alpha || n} p^\alpha$$

is the unique factorization of n into powers of distinct primes. Show that

$$\sum_{d|n} f(d) = \prod_{p^\alpha || n} (1 + f(p) + f(p^2) + \cdots + f(p^\alpha)).$$

Deduce that the function $g(n) = \sum_{d|n} f(d)$ is also multiplicative. The notation $p^\alpha || n$ means that p^α is the exact power dividing n.

A typical divisor d of n is of the form $d = \prod_{p|n} p^{\beta(p)}$, where $\beta(p) \leq \alpha$ and $p^\alpha || n$. Thus $f(d) = \prod_{p|n} f\left(p^{\beta(p)}\right)$, which is a typical term appearing in the expansion of the product on the right-hand side. Clearly, if n_1 and n_2 are coprime, then

$$\begin{aligned} g(n_1 n_2) &= \prod_{p^\alpha || n_1 n_2} (1 + f(p) + \cdots + f(p^\alpha)) \\ &= g(n_1) g(n_2), \end{aligned}$$

since we can decompose the product into two parts, namely those primes dividing n_1 and those dividing n_2. (This result can be used to give an alternative solution of Exercise 1.1.4.) □

1.1.6 *Show that*

$$\sum_{d|n} \Lambda(d) = \log n.$$

Deduce that

$$\Lambda(n) = -\sum_{d|n} \mu(d) \log d.$$

This is immediate from the unique factorization theorem:

$$n = p_1^{\alpha_1} \cdots p_k^{\alpha_k},$$

where the p_i are distinct primes. Then

$$\log n = \sum_{i=1}^{k} \alpha_i \log p_i = \sum_{d|n} \Lambda(d).$$

The equality

$$\Lambda(n) = \sum_{d|n} \mu(d) \log \frac{n}{d}$$

follows from Möbius inversion. Therefore,

$$\Lambda(n) = -\sum_{d|n} \mu(d) \log d,$$

since $\sum_{d|n} \mu(d) = 0$ unless $n = 1$ (by Exercise 1.1.1).

1.1.7 *Show that*

$$\sum_{d^2|n} \mu(d) = \begin{cases} 1 & \text{if } n \text{ is square-free} \\ 0 & \text{otherwise.} \end{cases}$$

Clearly, the sum on the left-hand side is a multiplicative function. It therefore suffices to evaluate it when n is a prime power. If $n = p^\alpha$, we see that

$$\sum_{d^2|p^\alpha} \mu(d) = \begin{cases} 1 & \text{if} \quad \alpha \le 1, \\ 0 & \text{otherwise.} \end{cases}$$

The result is now clear from this fact. □

1.1.8 *Show that for any natural number k,*

$$\sum_{d^k|n} \mu(d) = \begin{cases} 1 & \textit{if } n \textit{ is } k\textit{th power-free}, \\ 0 & \textit{otherwise}. \end{cases}$$

Since the left-hand side is a multiplicative function of n, it suffices to evaluate it when n is a prime power. Thus

$$\sum_{d^k|p^\alpha} \mu(d) = \begin{cases} 1 & \text{if} \quad \alpha \le k-1, \\ 0 & \text{otherwise}, \end{cases}$$

from which the result follows. □

1.1.9 *If for all positive x,*

$$G(x) = \sum_{n \le x} F\Big(\frac{x}{n}\Big),$$

show that

$$F(x) = \sum_{n \le x} \mu(n) G\Big(\frac{x}{n}\Big)$$

and conversely.

We have

$$\begin{aligned} \sum_{n \le x} \mu(n) G\Big(\frac{x}{n}\Big) &= \sum_{n \le x} \mu(n) \sum_{m \le \frac{x}{n}} F\Big(\frac{x}{mn}\Big) \\ &= \sum_{mn \le x} \mu(n) F\Big(\frac{x}{mn}\Big) \\ &= \sum_{r \le x} F\Big(\frac{x}{r}\Big) \sum_{n|r} \mu(n) \\ &= F(x) \end{aligned}$$

by an application of Exercise 1.1.1. For the converse,

$$\begin{aligned}
\sum_{n\le x} F\Big(\frac{x}{n}\Big) &= \sum_{n\le x}\sum_{m\le x/n} \mu(m)G\Big(\frac{x}{mn}\Big)\\
&= \sum_{r\le x} G\Big(\frac{x}{r}\Big)\sum_{m|r}\mu(m)\\
&= G(x),
\end{aligned}$$

as required. □

1.1.10 *Suppose that*

$$\sum_{k=1}^{\infty} d_3(k)|f(kx)| < \infty,$$

where $d_3(k)$ denotes the number of factorizations of k as a product of three numbers. Show that if

$$g(x) = \sum_{m=1}^{\infty} f(mx),$$

then

$$f(x) = \sum_{n=1}^{\infty} \mu(n)g(nx)$$

and conversely. We have, by absolute convergence of the series involved,

$$\begin{aligned}
\sum_{n=1}^{\infty} \mu(n)g(nx) &= \sum_{n=1}^{\infty}\mu(n)\sum_{m=1}^{\infty} f(mnx)\\
&= \sum_{r=1}^{\infty} f(rx)\sum_{n|r}\mu(n)\\
&= f(x)
\end{aligned}$$

by Exercise 1.1.1. For the converse,

$$\begin{aligned}\sum_{m=1}^{\infty} f(mx) &= \sum_{m=1}^{\infty}\sum_{n=1}^{\infty} \mu(n)g(mnx)\\ &= \sum_{r=1}^{\infty} g(rx)\sum_{n|r}\mu(n)\\ &= g(x),\end{aligned}$$

as required. In the first case, the rearrangement of the series is justified by the absolute convergence of

$$\sum_{m,n} f(mnx) = \sum_{k=1}^{\infty} d(k)f(kx),$$

where $d(k)$ is the number of divisors of k. In the second case, the absolute convergence of

$$\sum_{m,n} g(mnx)$$

follows from the convergence of

$$\sum_{k} d_3(k)|f(kx)|.$$

□

1.1.11 *Let $\lambda(n)$ denote Liouville's function given by $\lambda(n) = (-1)^{\Omega(n)}$, where $\Omega(n)$ is the total number (counting multiplicity) of prime factors of n. Show that*

$$\sum_{d|n} \lambda(d) = \begin{cases} 1 & \textit{if } n \textit{ is a square,} \\ 0 & \textit{otherwise.} \end{cases}$$

The left-hand side is multiplicative and therefore it suffices to compute it for prime powers. We have

$$\sum_{d|p^\alpha} \lambda(d) = \begin{cases} 1 & \text{if } \alpha \text{ is even,} \\ 0 & \text{if } \alpha \text{ is odd,} \end{cases}$$

from which the result follows immediately. □

1.1.12 *The Ramanujan sum $c_n(m)$ is defined as*

$$c_n(m) = \sum_{\substack{1 \le h \le n \\ (h,n)=1}} e\left(\frac{hm}{n}\right),$$

where $e(t) = e^{2\pi i t}$. Show that

$$c_n(m) = \sum_{d|(m,n)} d\mu(n/d).$$

Let

$$g(n) = \sum_{1 \le h \le n} e\left(\frac{hm}{n}\right).$$

Since this is the sum of a geometric progression, we find that

$$g(n) = \begin{cases} n & \text{if} \quad n|m, \\ 0 & \text{otherwise.} \end{cases}$$

But we can write

$$\begin{aligned} g(n) &= \sum_{d|n} \sum_{\substack{1 \le h \le n \\ (h,n)=d}} e\left(\frac{hm}{n}\right) \\ &= \sum_{d|n} \sum_{\substack{1 \le h_1 \le n_1 \\ (h_1,n_1)=1}} e\left(\frac{h_1 m}{n_1}\right), \end{aligned}$$

where we have written $h = dh_1$, $n = dn_1$ with $(h_1, n_1) = 1$ in the last sum. Thus,

$$g(n) = \sum_{d|n} c_{n/d}(m),$$

which by Möbius inversion (Exercise 1.1.2) gives

$$c_n(m) = \sum_{d|n} \mu(d) g(n/d).$$

But $g(d) = d$ if $d|m$ and vanishes otherwise. Therefore,

$$c_n(m) = \sum_{d|(n,m)} d\mu(n/d)$$

as required. □

1.1.13 *Show that*

$$\mu(n) = \sum_{\substack{1\le h\le n\\(h,n)=1}} e\left(\frac{h}{n}\right).$$

Set $m = 1$ in the previous exercise. □

1.1.14 *Let* $\delta = (n, m)$. *Show that*

$$c_n(m) = \mu(n/\delta)\varphi(n)/\varphi(n/\delta).$$

We have (by Exercise 1.1.12)

$$\begin{aligned} c_n(m) &= \sum_{d|\delta} d\mu(n/d) \\ &= \sum_{de=\delta} d\mu(ne/\delta) \\ &= \sum_{de=\delta} d\mu(n_1 e), \end{aligned}$$

where $n = \delta n_1$. Now, $\mu(n_1 e) = \mu(n_1)\mu(e)$ if $(n_1, e) = 1$ and 0 otherwise. Thus,

$$\begin{aligned} c_n(m) &= \sum_{\substack{de=\delta\\(n_1,e)=1}} d\mu(n_1)\mu(e) \\ &= \mu(n_1)\delta \sum_{\substack{e|\delta\\(n_1,e)=1}} \frac{\mu(e)}{e} \\ &= \mu(n_1)\delta \prod_{\substack{p|\delta\\p\nmid n_1}} \left(1 - \frac{1}{p}\right). \end{aligned}$$

By Exercise 1.1.4,

$$\frac{\varphi(n)}{\varphi(n/\delta)} = \frac{n}{n/\delta} \prod_{\substack{p|n \\ p\nmid n_1}} \left(1 - \frac{1}{p}\right) = \delta \prod_{\substack{p|\delta \\ p\nmid n_1}} \left(1 - \frac{1}{p}\right),$$

from which the result follows. □

1.2 Formal Dirichlet Series

1.2.1 *Let f be a multiplicative function. Show that*

$$D(f, s) = \prod_p \left(\sum_{\nu=0}^{\infty} f(p^\nu) p^{-\nu s} \right).$$

This is more or less an extension of Exercise 1.1.5 and is immediate upon expansion of the infinite product on the right-hand side and the unique factorization theorem. □

1.2.2 *If*

$$\zeta(s) = D(1, s) = \sum_{n=1}^{\infty} \frac{1}{n^s},$$

show that

$$D(\mu, s) = 1/\zeta(s).$$

By Exercise 1.2.1,

$$\zeta(s) = \prod_p \left(1 + \frac{1}{p^s} + \frac{1}{p^{2s}} + \cdots\right) = \prod_p \left(1 - \frac{1}{p^s}\right)^{-1}.$$

Again by Exercise 1.2.1,

$$D(\mu, s) = \prod_p (1 - \frac{1}{p^s}).$$

The result is now immediate. □

1.2.3 *Show that*

$$D(\Lambda, s) = \sum_{n=1}^{\infty} \frac{\Lambda(n)}{n^s} = -\frac{\zeta'}{\zeta}(s),$$

where $-\zeta'(s) = \sum_{n=1}^{\infty} (\log n) n^{-s}$. Since

$$-\zeta'(s) = \sum_{n=1}^{\infty} (\log n) n^{-s}$$

and

$$\frac{1}{\zeta(s)} = \sum_{n=1}^{\infty} \mu(n) n^{-s}$$

by the previous exercise, we obtain upon multiplying the two series,

$$D(\mu * (-\log), s),$$

which by Exercise 1.1.6 is the formal series attached to Λ. □

1.2.4 *Suppose that*

$$f(n) = \sum_{d|n} g(d).$$

Show that $D(f, s) = D(g, s)\zeta(s)$.

This is immediate from the formula for the multiplication of formal series. □

1.2.5 *Let* $\lambda(n)$ *be the Liouville function defined by* $\lambda(n) = (-1)^{\Omega(n)}$, *where* $\Omega(n)$ *is the total number of prime factors of* n. *Show that*

$$D(\lambda, s) = \frac{\zeta(2s)}{\zeta(s)}.$$

Since λ is multiplicative, by Exercise 1.2.1 we have

$$\begin{aligned} D(\lambda, s) &= \prod_p \left(1 - \frac{1}{p^s} + \frac{1}{p^{2s}} - \frac{1}{p^{3s}} + \cdots\right) \\ &= \prod_p \left(1 + \frac{1}{p^s}\right)^{-1} = \prod_p \left(1 - \frac{1}{p^{2s}}\right)^{-1} \left(1 - \frac{1}{p^s}\right) \\ &= \frac{\zeta(2s)}{\zeta(s)} \end{aligned}$$

by an application of Exercise 1.2.2.

1.2.6 *Prove that*

$$\sum_{n=1}^{\infty} \frac{2^{\nu(n)}}{n^s} = \frac{\zeta^2(s)}{\zeta(2s)}.$$

Since $2^{\nu(n)}$ is multiplicative,

$$\begin{aligned}
\sum_{n=1}^{\infty} \frac{2^{\nu(n)}}{n^s} &= \prod_p \left(1 + \frac{2}{p^s} + \frac{2}{p^{2s}} + \cdots\right) \\
&= \prod_p \left(1 + \frac{2}{p^s}\left(1 - \frac{1}{p^s}\right)^{-1}\right) \\
&= \prod_p \left(1 - \frac{1}{p^s}\right)^{-1} \left(\frac{2}{p^s} + \left(1 - \frac{1}{p^s}\right)\right) \\
&= \prod_p \left(1 - \frac{1}{p^s}\right)^{-1} \left(1 + \frac{1}{p^s}\right) \\
&= \zeta(s) \prod_p \left(1 + \frac{1}{p^s}\right).
\end{aligned}$$

The latter product is $\zeta(s)/\zeta(2s)$ by Exercise 1.2.5, so that the result is now immediate. □

1.2.7 *Show that*

$$\sum_{n=1}^{\infty} \frac{|\mu(n)|}{n^s} = \frac{\zeta(s)}{\zeta(2s)}.$$

Since $|\mu|$ is a multiplicative function, we obtain

$$\sum_{n=1}^{\infty} \frac{|\mu(n)|}{n^s} = \prod_p \left(1 + \frac{1}{p^s}\right) = \frac{\zeta(s)}{\zeta(2s)}.$$

by Exercise 1.2.5. □

1.2.8 *Let $d(n)$ denote the number of divisors of n. Prove that*

$$\sum_{n=1}^{\infty} \frac{d^2(n)}{n^s} = \frac{\zeta^4(s)}{\zeta(2s)}$$

(This example is due to Ramanujan.)

We observe the following identity due to Ramanujan:

$$\sum_{n=0}^{\infty}(\frac{\alpha^{n+1}-\beta^{n+1}}{\alpha-\beta})(\frac{\gamma^{n+1}-\delta^{n+1}}{\gamma-\delta})T^n$$

$$=\frac{1-\alpha\beta\gamma\delta T^2}{(1-\alpha\gamma T)(1-\alpha\delta T)(1-\beta\gamma T)(1-\beta\delta T)},$$

which is proved easily using the formula for the sum of a geometric series. This identity is useful in other contexts, and so we record it here for future use.

If we write

$$\frac{\alpha^{n+1}-\beta^{n+1}}{\alpha-\beta}=\alpha^n+\alpha^{n-1}\beta+\cdots+\alpha\beta^{n-1}+\beta^n,$$

we see that the special case $\alpha=\beta=\gamma=\delta=1$ gives the identity

$$\sum_{n=1}^{\infty}(n+1)^2T^n=\frac{1-T^2}{(1-T)^4}.$$

Thus,

$$\begin{aligned}\sum_{n=1}^{\infty}\frac{d^2(n)}{n^s} &= \prod_p\left(\sum_{\alpha=0}^{\infty}\frac{(\alpha+1)^2}{p^{\alpha s}}\right)\\ &= \prod_p\left(1-\frac{1}{p^{2s}}\right)\left(1-\frac{1}{p^s}\right)^{-4}\\ &= \frac{\zeta^4(s)}{\zeta(2s)},\end{aligned}$$

as desired. □

1.2.9 *For any complex numbers a, b, show that*

$$\sum_{n=1}^{\infty}\frac{\sigma_a(n)\sigma_b(n)}{n^s}=\frac{\zeta(s)\zeta(s-a)\zeta(s-b)\zeta(s-a-b)}{\zeta(2s-a-b)}.$$

We have

$$\sum_{n=1}^{\infty}\frac{\sigma_a(n)\sigma_b(n)}{n^s}=\prod_p\left(\sum_{\alpha=0}^{\infty}\frac{\sigma_a(p^\alpha)\sigma_b(p^\alpha)}{p^{\alpha s}}\right).$$

Now,

$$\sigma_a(p^\alpha) = 1 + p^a + p^{2a} + \cdots + p^{\alpha a} = \frac{p^{a(\alpha+1)} - 1}{p^a - 1}.$$

We apply Ramanujan's identity (see Exercise 1.2.8) to deduce

$$\sum_{\alpha=0}^{\infty} \left(\frac{p^{a(\alpha+1)} - 1}{p^a - 1}\right)\left(\frac{p^{b(\alpha+1)} - 1}{p^b - 1}\right) T^\alpha$$

$$= \frac{1 - p^{a+b}T^2}{(1 - p^{a+b}T)(1 - p^a T)(1 - p^b T)(1 - T)}.$$

Putting $T = p^{-s}$ in this identity, we deduce the stated result. □

1.2.10 *Let $q_k(n)$ be 1 if n is kth power-free and 0 otherwise. Show that*

$$\sum_{n=1}^{\infty} \frac{q_k(n)}{n^s} = \frac{\zeta(s)}{\zeta(ks)}.$$

If we multiply out the series on the right-hand side, we obtain

$$\sum_{d,e} \frac{\mu(d)}{d^{ks} e^s} = \sum_{n=1}^{\infty} \frac{1}{n^s} \Big(\sum_{d^k e = n} \mu(d) \Big).$$

The inner sum is $q_k(n)$ by Exercise 1.1.8. □

1.3 Orders of Some Arithmetical Functions

1.3.1 *Show that $d(n) \le 2\sqrt{n}$, where $d(n)$ is the number of divisors of n.*

Each divisor α of n corresponds to a factorization $\alpha\beta = n$. One of α or β must be less than or equal to $\sqrt{n}$. Thus, the number of divisors of n is less than or equal to $2\sqrt{n}$. □

1.3.2 *For any $\epsilon > 0$, there is a constant $C(\epsilon)$ such that $d(n) \le C(\epsilon) n^\epsilon$.*

Observe that

$$\frac{d(n)}{n^\epsilon} = \prod_{p^\alpha || n} \frac{\alpha + 1}{p^{\alpha\epsilon}}.$$

We decompose the product into two parts: those $p < 2^{1/\epsilon}$ and those $p \geq 2^{1/\epsilon}$.

In the second part, $p^\epsilon \geq 2$, so that $p^{\alpha\epsilon} \geq 2^\alpha$ and

$$\frac{\alpha+1}{p^{\alpha\epsilon}} \leq \frac{\alpha+1}{2^\alpha} \leq 1.$$

Thus, we must estimate the first part. Notice that

$$\frac{\alpha+1}{p^{\alpha\epsilon}} \leq 1 + \frac{\alpha}{p^{\alpha\epsilon}} \leq 1 + \frac{1}{\epsilon \log 2},$$

since

$$\alpha\epsilon \log 2 \leq e^{\alpha\epsilon \log 2} = 2^{\alpha\epsilon} \leq p^{\alpha\epsilon}.$$

Hence

$$\prod_{p<2^{1/\epsilon}} \left(1 + \frac{1}{\epsilon \log 2}\right) = C(\epsilon)$$

is the desired constant. □

1.3.3 *For any $\eta > 0$, show that*

$$d(n) < 2^{(1+\eta)\log n/\log\log n}$$

for all n sufficiently large.

We refine the argument of Exercise 1.3.2, where we now set

$$\epsilon = \frac{(1+\frac{\eta}{2})\log 2}{\log\log n}$$

in the proof. The estimate for the second part of the product remains valid. We must estimate (by applying $1 + x \leq e^x$)

$$\prod_{p<2^{1/\epsilon}} \left(1 + \frac{1}{\epsilon \log 2}\right) \leq \exp\left\{\frac{1}{\epsilon \log 2} 2^{1/\epsilon}\right\}.$$

Now,

$$2^{1/\epsilon} = (\log n)^{1/(1+\frac{\eta}{2})},$$

so that

$$\begin{aligned} C(\epsilon) &\leq \exp\left\{\frac{\log\log n}{(1+\frac{\eta}{2})\log^2 2}(\log n)^{1/(1+\frac{\eta}{2})}\right\} \\ &\leq \exp\left\{\frac{\eta}{2}\frac{(\log 2)\log n}{\log\log n}\right\} \end{aligned}$$

for $n \geq n_0(\eta)$. □

1.3.4 *Prove that* $\sigma_1(n) \leq n(\log n + 1)$.

We have

$$\sigma_1(n) = \sum_{d|n} d = \sum_{d|n} \frac{n}{d} \leq n \sum_{d \leq n} \frac{1}{d}.$$

Now,

$$\sum_{2 \leq d \leq n} \frac{1}{d} \leq \int_1^n \frac{dt}{t} = \log n.$$

1.3.5 *Prove that*

$$c_1 n^2 \leq \phi(n)\sigma_1(n) \leq c_2 n^2$$

for certain positive constants c_1 *and* c_2.

We have

$$\frac{\phi(n)}{n} = \prod_{p|n} \left(1 - \frac{1}{p}\right)$$

and

$$\sigma_1(n) = \prod_{p^\alpha || n} \frac{p^{\alpha+1} - 1}{p - 1}.$$

Now,

$$\frac{\sigma_1(n)}{n} = \prod_{p^\alpha || n} \left(1 + \frac{1}{p} + \frac{1}{p^2} + \cdots + \frac{1}{p^\alpha},\right)$$

so that

$$\frac{\phi(n)\sigma_1(n)}{n^2} = \prod_{p^\alpha || n} \left(1 - \frac{1}{p^{\alpha+1}}\right).$$

Since each factor in the product is less than or equal to 1, we have

$$\frac{\phi(n)\sigma_1(n)}{n^2} \leq 1.$$

Also,

$$\begin{aligned} \prod_{p^\alpha || n} \left(1 - \frac{1}{p^{\alpha+1}}\right) &\geq \prod_{p|n} \left(1 - \frac{1}{p^2}\right) \\ &\geq \prod_{i \leq \nu(n)} \left(1 - \frac{1}{p_i^2},\right) \end{aligned}$$

where $\nu(n)$ denotes the number of distinct prime factors of n, and p_i is the ith prime.

Recall that an infinite product

$$\prod_{n=1}^{\infty}(1+a_n)$$

converges if and only if $\sum_{n=1}^{\infty}|a_n| < \infty$. Therefore

$$\frac{1}{\zeta(2)} = \prod_{p}\left(1-\frac{1}{p^2}\right) < \infty.$$

In addition, $\zeta(2) \neq 0$. Since the product converges to a nonzero limit, it is clear that there is a $c_1 > 0$ such that

$$\prod_{i\leq\nu(n)}\left(1-\frac{1}{p_i^2}\right) \geq c_1.$$

□

1.3.6 *Let $\nu(n)$ denote the number of distinct prime factors of n. Show that*

$$\nu(n) \leq \frac{\log n}{\log 2}.$$

Writing $n = p_1^{\alpha} \cdots p_k^{\alpha_k}$, where the p_i are distinct primes, we obtain

$$\sum_i \alpha_i \log p_i \leq \log n.$$

Since each $p_i \geq 2$, we deduce the stronger result

$$(\log 2)\Omega(n) \leq \log n,$$

where $\Omega(n) = \sum_{i=1}^{k} \alpha_i$. □

1.4 Average Orders of Arithmetical Functions

1.4.1 *Show that the average order of $d(n)$ is $\log n$.*

We have

$$\sum_{n\leq x} d(n) = \sum_{ab\leq x} 1 = \sum_{a\leq x}\left[\frac{x}{a}\right].$$

Now,

$$\left[\frac{x}{a}\right] = \frac{x}{a} + O(1).$$

Thus,

$$\sum_{a \le x} \left[\frac{x}{a}\right] = x \sum_{a \le x} \frac{1}{a} + O(x).$$

We can compare

$$\sum_{a \le x} \frac{1}{a}$$

with the integral

$$\int_1^x \frac{dt}{t} = \log x,$$

and we easily obtain

$$\sum_{a \le x} \frac{1}{a} = \log x + O(1).$$

Thus,

$$\sum_{n \le x} d(n) = x \log x + O(x).$$

$\square$

1.4.2 *Show that the average order of $\phi(n)$ is cn for some constant c.*

By Exercise 1.1.4, we obtain

$$\begin{aligned} \sum_{n \le x} \varphi(n) &= \sum_{ab \le x} \mu(a) b \\ &= \sum_{a \le x} \mu(a) \sum_{b \le \frac{x}{a}} b. \end{aligned}$$

The inner sum is

$$\frac{1}{2}\left[\frac{x}{a}\right]\left(\left[\frac{x}{a}\right] + 1\right),$$

which is equal to

$$\frac{1}{2}\left(\frac{x}{a} + O(1)\right)^2 = \frac{x^2}{2a^2} + O\left(\frac{x}{a}\right).$$

Inserting this into the penultimate sum, we obtain

$$\begin{aligned}\sum_{n\leq x}\varphi(n) &= \sum_{a\leq x}\mu(a)\left(\frac{x^2}{2a^2}+O(\frac{x}{a})\right)\\ &= \frac{x^2}{2}\sum_{a\leq x}\frac{\mu(a)}{a^2}+O(x\log x).\end{aligned}$$

Now,

$$\sum_{a\leq x}\frac{\mu(a)}{a^2}=\sum_{a=1}^{\infty}\frac{\mu(a)}{a^2}+O\left(\frac{1}{x}\right)$$

by an easy application of the integral test.

The series

$$\sum_{a=1}^{\infty}\frac{\mu(a)}{a^2}$$

converges by the comparison test. This completes the proof. (Later, we shall see that the value of the series is $6/\pi^2$.) □

1.4.3 *Show that the average order of $\sigma_1(n)$ is $c_1 n$ for some constant c_1.*

We have

$$\sum_{n\leq x}\sigma_1(n)=\sum_{n\leq x}\sum_{d|n}d=\sum_{de\leq x}d.$$

Now,

$$\begin{aligned}\sum_{de\leq x}d &= \sum_{e\leq x}\sum_{d\leq x/e}d\\ &= \sum_{e\leq x}\frac{1}{2}\left[\frac{x}{e}\right]\left(\left[\frac{x}{e}\right]+1\right)\\ &= \frac{1}{2}\sum_{e\leq x}\left[\frac{x}{e}\right]\left(\frac{x}{e}+O(1)\right)\\ &= \frac{x}{2}\sum_{e\leq x}\left[\frac{x}{e}\right]\frac{1}{e}+O(x\log x).\end{aligned}$$

Also,

$$\sum_{e\leq x}\left[\frac{x}{e}\right]\frac{1}{e}=\sum_{e\leq x}\frac{x}{e^2}+O(\log x),$$

so that

$$\sum_{n \le x} \sigma_1(n) = \frac{x^2}{2} \sum_{e \le x} \frac{1}{e^2} + O(x \log x).$$

Since $\sum 1/e^2 < \infty$, we deduce

$$\sum_{n \le x} \sigma_1(n) \sim c_1 x^2$$

for some constant c_1. □

1.4.4 *Let $q_k(n) = 1$ if n is kth power-free and zero otherwise. Show that*

$$\sum_{n \le x} q_k(n) = c_k x + O\left(x^{1/k}\right),$$

where

$$c_k = \sum_{n=1}^{\infty} \frac{\mu(n)}{n^k}.$$

By Exercise 1.1.8,

$$q_k(n) = \sum_{d^k | n} \mu(d),$$

so that

$$\begin{aligned} \sum_{n \le x} q_k(n) &= \sum_{d^k e \le x} \mu(d) \\ &= \sum_{d^k \le x} \mu(d) \left[\frac{x}{d^k}\right] \\ &= \sum_{d^k \le x} \mu(d) \frac{x}{d^k} + O(x^{1/k}). \end{aligned}$$

By the integral test,

$$\sum_{d > x^{1/k}} \frac{\mu(d)}{d^k} \ll \int_{x^{1/k}}^{\infty} \frac{dt}{t^k} \ll x^{-1+\frac{1}{k}},$$

so that the desired result follows immediately. □

1.5.1 *Prove that*

$$\sum_{\substack{n\le x\\(n,k)=1}} \frac{1}{n} \sim \frac{\phi(k)}{k}\log x$$

as $x \to \infty$.

By Exercise 1.1.1, the left-hand side can be written as

$$\begin{aligned}\sum_{n\le x}\frac{1}{n}\sum_{d|(n,k)}\mu(d) &= \sum_{d|k}\mu(d)\sum_{\substack{n\le x\\ d|n}}\frac{1}{n}\\ &= \sum_{d|k}\frac{\mu(d)}{d}\sum_{t\le x/d}\frac{1}{t}\\ &= \sum_{d|k}\frac{\mu(d)}{d}\left(\log\frac{x}{d}+O(1)\right)\end{aligned}$$

by the solution of Exercise 1.4.1. Therefore,

$$\sum_{\substack{n\le x\\(n,k)=1}} \frac{1}{n} = \Big(\sum_{d|k}\frac{\mu(d)}{d}\Big)\log x + O(1),$$

where the O-constant now may depend on k. But, by Exercise 1.1.4,

$$\sum_{d|k}\frac{\mu(d)}{d} = \frac{\phi(k)}{k},$$

which completes the proof. □

1.5.2 *Let* $J_r(n)$ *be the number of* r*-tuples* $(a_1, a_2, \dots, a_r)$ *satisfying* $a_i \le n$ *and* $\gcd(a_1,\dots,a_r,n) = 1$. *Show that*

$$J_r(n) = n^r \prod_{p|n}\left(1-\frac{1}{p^r}\right).$$

($J_r(n)$ is called Jordan's totient function. For $r = 1$, this is, of course, Euler's ϕ-function.)

We partition the total number of r-tuples $(a_1, a_2, \dots, a_r)$ according to $d = \gcd(a_1,\dots,a_r,n)$. Thus, $1 = \gcd(a_1/d,\dots,a_r/d,n/d)$ and each $a_i \le n$, so that we have

$$n^r = \sum_{d|n} J_r(n/d).$$

By Möbius inversion, the result is now immediate. □

1.5 Supplementary Problems

1.5.3 *For $r \geq 2$, show that there are positive constant c_1 and c_2 such that*

$$c_1 n^r \leq J_r(n) \leq c_2 n^r.$$

Since each factor of

$$\prod_{p|n}\left(1-\frac{1}{p^r}\right)$$

is less than 1, we can take $c_2 = 1$. For the lower bound, we have

$$\prod_{p|d}\left(1-\frac{1}{p^r}\right) \geq \prod_{i\leq\nu(n)}\left(1-\frac{1}{p_i^r}\right),$$

which converges to a nonzero limit as $\nu(n) \to \infty$. Thus, there is a constant c_1 such that

$$J_r(n) \geq c_1 n^r.$$

□

1.5.4 *Show that the average order of $J_r(n)$ is cn^r for some constant $c > 0$.* We have (by Exercise 1.5.2)

$$\begin{aligned}\sum_{n\leq x} J_r(n) &= \sum_{n\leq x} n^r \sum_{d|n} \frac{\mu(d)}{d^r}\\ &= \sum_{d\leq x} \frac{\mu(d)}{d^r} \sum_{\substack{n\leq x\\ d|n}} n^r\\ &= \sum_{d\leq x} \mu(d) \sum_{t\leq x/d} t^r,\end{aligned}$$

where we have written $n = dt$ in the inner sum of the penultimate step. Now,

$$\sum_{k=1}^{N}\int_{k-1}^{k} v^r\,dr \le \sum_{1\le t\le N} t^r \le \sum_{k=1}^{N}\int_{k}^{k+1} v^r\,dv$$

by a comparison of areas. Thus

$$\begin{aligned}\sum_{1\le t\le N} t^r &= \int_1^N v^r\,dv + O(N^r)\\ &= \frac{N^{r+1}}{r+1} + O(N^r).\end{aligned}$$

Thus,

$$\sum_{n\le x} J_r(n) = \sum_{d\le x}\mu(d)\left\{\frac{x^{r+1}}{d^{r+1}} + O\Big(\Big(\frac{x}{d}\Big)^r\Big)\right\}$$

from which we deduce

$$\sum_{n\le x} J_r(n) = c_r x^{r+1} + O(x^r),$$

where

$$c_r = \frac{1}{r+1}\sum_{d=1}^{\infty}\frac{\mu(d)}{d^{r+1}} \ne 0,$$

since it can be written as

$$\frac{1}{r+1}\prod_p\left(1-\frac{1}{p^{r+1}}\right). \qquad \square$$

1.5.5 *Let $d_k(n)$ be the number of ways of writing n as a product of k positive numbers. Show that*

$$\sum_{n=1}^{\infty}\frac{d_k(n)}{n^s} = \zeta^k(s).$$

Clearly,

$$d_k(n) = \sum_{\delta|n} d_{k-1}(\delta),$$

since for each factorization $n = \delta e$ we can count the number of ways of writing δ as a product of $k-1$ numbers to enumerate $d_k(n)$. This shows that

$$\sum_{n=1}^{\infty} \frac{d_k(n)}{n^s} = \zeta(s) \sum_{n=1}^{\infty} \frac{d_{k-1}(n)}{n^s}.$$

Since $d_2(n) = d(n)$ satisfies

$$\sum_{n=1}^{\infty} \frac{d(n)}{n^s} = \zeta^2(s),$$

the desired result follows by induction. □

1.5.6 *If $d_k^*(n)$ denotes the number of factorizations of n as a product of k positive numbers each greater than 1, show that*

$$\sum_{n=1}^{\infty} \frac{d_k^*(n)}{n^s} = (\zeta(s) - 1)^k.$$

Expanding the right-hand side as a Dirichlet series and collecting terms we get the desired result. □

1.5.7*Let $\Delta(n)$ be the number of nontrivial factorization of n. Show that*

$$\sum_{n=1}^{\infty} \frac{\Delta(n)}{n^s} = (2 - \zeta(s))^{-1},$$

as a formal Dirichlet series. We can write

$$\Delta(n) = 1 + \sum_{k=2}^{\infty} d_k^*(n),$$

so that

$$\sum_{n=1}^{\infty} \frac{\Delta(n)}{n^s} = 1 + (\zeta(s) - 1) + \sum_{k=2}^{\infty} (\zeta(s) - 1)^k,$$

which is equal to

$$\frac{1}{2 - \zeta(s)}$$

as required. □

1.5.8 *Show that*

$$\sum_{\substack{n\le x\\(n,k)=1}} n = \frac{\phi(k)}{2k}x^2 + O(d(k)x),$$

where $d(k)$ denotes the number of divisors of k.

We have

$$\begin{aligned}\sum_{\substack{n\le x\\(n,k)=1}} n &= \sum_{n\le x} n \sum_{\substack{d|n\\d|k}} \mu(d)\\ &= \sum_{d|k}\mu(d)\sum_{\substack{n\le x\\d|n}} n\\ &= \sum_{d|k}\mu(d)d\sum_{t\le x/d} t\\ &= \sum_{d|k}\mu(d)d\left\{\frac{1}{2}\left[\frac{x}{d}\right]\left(\left[\frac{x}{d}\right]+1\right)\right\}\\ &= \sum_{d|k}\mu(d)d\left\{\frac{x^2}{2d^2}+O\left(\frac{x}{d}\right)\right\}\end{aligned}$$

which is equal to

$$\frac{x^2\phi(k)}{2k} + O(xd(k)),$$

as required. □

1.5.9 *Prove that*

$$\sum_{\substack{d|n\\\nu(d)\le r}} \mu(d) = (-1)^r\binom{\nu(n)-1}{r},$$

where $\nu(n)$ denotes the number of distinct prime factors of n.

By comparing the coefficient of x^r on both sides of the identity,

$$(1-x)^{-1}(1-x)^{\nu} = (1-x)^{\nu-1},$$

we deduce

$$\sum_{k\le r}(-1)^k\binom{\nu}{k}=(-1)^r\binom{\nu-1}{r}.$$

Now, if N is the product of the distinct prime divisors of n, then

$$\sum_{\substack{d|n\\ \nu(d)\le r}}\mu(d)=\sum_{\substack{d|N\\ \nu(d)\le r}}\mu(d),$$

and the latter sum is

$$\sum_{k\le r}(-1)^k\binom{\nu(n)}{k}=(-1)^r\binom{\nu(n)-1}{r}$$

by our initial observation. □

1.5.10 *Let $\pi(x,z)$ denote the number of $n\le x$ coprime to all the prime numbers $p\le z$. Show that*

$$\pi(x,z)=x\prod_{p\le z}\left(1-\frac{1}{p}\right)+O(2^z).$$

Let P_z denote the product of the primes less than or equal to z. Then

$$\begin{aligned}\pi(x,z) &= \sum_{\substack{n\le x\\ d|(n,P_z)}}\mu(d)\\ &= \sum_{d|P_z}\mu(d)]\left[\frac{x}{d}\right]\\ &= x\sum_{d|P_z}\frac{\mu(d)}{d}+O(2^z)\\ &= x\prod_{p\le z}\left(1-\frac{1}{p}\right)+O(2^z)\end{aligned}$$

by Exercise 1.1.4, as required. □

1.5.11 *Prove that*

$$\sum_{p\le x}\frac{1}{p}\ge \log\log x + c$$

for some constant c.

Since every natural number can be written as a product of prime numbers, we have

$$\sum_{n\le x}\frac{1}{n}\le \prod_{p\le x}\left(1-\frac{1}{p}\right)^{-1}.$$

Taking logarithms and using the fact that

$$\sum_{n\le x}\frac{1}{n}=\log x + O(1),$$

we deduce

$$-\sum_{p\le x}\log\left(1-\frac{1}{p}\right)\ge \log\log x + O(1).$$

Now,

$$-\log\left(1-\frac{1}{p}\right)=\frac{1}{p}+O\left(\frac{1}{p^2}\right),$$

so that

$$\sum_{p\le x}\frac{1}{p}\ge \log\log x + O(1),$$

since $\sum_p 1/p^2 < \infty$. This completes the proof. □

1.5.12 *Let* $\pi(x)$ *be the number of primes less than or equal to* x. *Choosing* $z=\log x$ *in Exercise 1.5.10, deduce that*

$$\pi(x)=O\left(\frac{x}{\log\log x}\right).$$

Clearly,

$$\pi(x)\le \pi(x,z)+z.$$

Now,

$$\pi(x,z)=x\prod_{p\le z}\left(1-\frac{1}{p}\right)+O(2^z)$$

by Exercise 1.5.10. Choosing $z = \log x$ and observing that

$$-\sum_{p\le z}\log\left(1-\frac{1}{p}\right)=\sum_{p\le z}\frac{1}{p}+O(1),$$

we deduce

$$\begin{aligned}\pi(x,z) &= x\exp\Big(-\sum_{p\le z}\frac{1}{p}+O(1)\Big)+O(x^{\log 2})\\ &= O\Big(\frac{x}{\log\log x}\Big)\end{aligned}$$

by the previous exercise. This completes the proof. □

1.5.13 *Let $M(x) = \sum_{n\le x}\mu(n)$. Show that*

$$\sum_{n\le x} M\Big(\frac{x}{n}\Big) = 1.$$

We have

$$\sum_{n\le x} M\Big(\frac{x}{n}\Big) = \sum_{n\le x}\sum_{d\le x/n}\mu(d) = \sum_{dn\le x}\mu(d) = \sum_{r\le x}\Big(\sum_{d|r}\mu(d)\Big).$$

The inner sum is 1 if $r = 1$, and 0 otherwise by Exercise 1.1.1. The result is now immediate. □

1.5.14 *Let $\mathbb{F}_p[x]$ denote the polynomial ring over the finite field of p elements. Let N_d be the number of monic irreducible polynomials of degree d in $\mathbb{F}_p[x]$. Using the fact that every monic polynomial in $\mathbb{F}_p[x]$ can be factored uniquely as a product of monic irreducible polynomials, show that*

$$p^n = \sum_{d|n} dN_d.$$

Consider the formal power series

$$\sum_f T^{\deg f},$$

where the summation is over monic polynomials f in $\mathbb{F}_p[x]$. Since every f can be written uniquely as a product of monic irreducible polynomials and $\deg f_1 f_2 = \deg f_1 + \deg f_2$, we obtain

$$\sum_f T^{\deg f} = \prod_v \left(1 + T^{\deg v} + T^{2\deg v} + \cdots\right),$$

where the product is over monic irreducible polynomials v of $\mathbb{F}_p[x]$. Thus,

$$\begin{aligned}\sum_f T^{\deg f} &= \prod_v \left(1 - T^{\deg v}\right)^{-1} \\ &= \prod_{d=1}^{\infty} \left(1 - T^d\right)^{-N_d}.\end{aligned}$$

But the left-hand side is

$$\sum_{n=1}^{\infty} p^n T^n = (1 - pT)^{-1},$$

since the number of monic polynomials of degree n in p^n. Therefore,

$$-\log(1 - pT) = -\sum_{d=1}^{\infty} N_d \log\left(1 - T^d\right)$$

so that

$$\begin{aligned}\sum_{n=1}^{\infty} \frac{p^n T^n}{n} &= \sum_{d=1}^{\infty} N_d \sum_{e=1}^{\infty} \frac{T^{de}}{e} \\ &= \sum_{n=1}^{\infty} \frac{T^n}{n} \Big(\sum_{de=n} dN_d \Big).\end{aligned}$$

Comparing coefficients of T^n gives us the result. □

1.5.15 *With the notation as in the previous exercise, show that*

$$N_n = \frac{1}{n} \sum_{d|n} \mu(d) p^{n/d}$$

and that $N_n \geq 1$. Deduce that there is always an irreducible polynomial of degree n in $\mathbb{F}_p[x]$.

The formula for N_n is immediate upon Möbius inversion of the result derived in the previous exercise. Notice that

$$nN_n = \sum_{d|n} \mu(d) p^{n/d}.$$

The right hand side can be viewed as the difference of two numbers in base p with the larger number having $(n+1)$ digits and the smaller one at most $n/2+1$ digits. Thus, the righthand side is not zero, so that $nN_n \geq 1$, which implies $N_n \geq 1/n$. Since N_n is an integer, we get $N_n \geq 1$. (This fact is used to establish the existence of finite fields $\mathbb{F}_{p^n}$ for every n.)

1.5.16 *Suppose $f(d) = \sum_{d|n} g(n)$, where the summation is over all multiples of d. Show that*

$$g(d) = \sum_{d|n} \mu\left(\frac{n}{d}\right) f(n)$$

and conversely (assuming that all the series are absolutely convergent).

We have

$$\begin{aligned}
\sum_{d|n} \mu\left(\frac{n}{d}\right) f(n) &= \sum_{t} \mu(t) f(dt) \\
&= \sum_{t} \mu(t) \sum_{r} g(dtr) \\
&= \sum_{m} g(dm) \Big(\sum_{tr=m} \mu(t) \Big) = g(m),
\end{aligned}$$

since the inner sum is 1 if $m = 1$, and zero otherwise. Similarly, for the converse,

$$\begin{aligned}\sum_{d|n} g(n) &= \sum_t g(dt) \\ &= \sum_t \sum_r \mu(r) f(dtr) \\ &= \sum_m f(dm) \Big(\sum_{tr=m} \mu(r) \Big) = f(d),\end{aligned}$$

since the inner sum is again 1 if $m = 1$, and zero otherwise. □

1.5.17 *Prove that*

$$\sum_{n \le x} \frac{\varphi(n)}{n} = cx + O(\log x)$$

for some constant $c > 0$. We have

$$\frac{\varphi(n)}{n} = \sum_{d|n} \frac{\mu(d)}{d},$$

so that

$$\sum_{n \le x} \frac{\varphi(n)}{n} = \sum_{d \le x} \frac{\mu(d)}{d} \left[\frac{x}{d} \right].$$

Hence,

$$\sum_{d \le x} \frac{\mu(d)}{d} \left[\frac{x}{d} \right] = x \sum_{d \le x} \frac{\mu(d)}{d^2} + O(\log x).$$

Now,

$$\sum_{d \le x} \frac{\mu(d)}{d^2} = \sum_{d=1}^{\infty} \frac{\mu(d)}{d^2} - \sum_{d > x} \frac{\mu(d)}{d^2},$$

and the latter sum is $O(1/x)$. Thus,

$$\sum_{n \le x} \frac{\varphi(n)}{n} = cx + O(\log x)$$

with $c = \sum_{d=1}^{\infty} \mu(d)/d^2 = \prod_p (1 - 1/p^2) \neq 0$. □

1.5.18 *For* $\mathrm{Re}(s) > 2$, *prove that*

$$\sum_{n=1}^{\infty} \frac{\varphi(n)}{n^s} = \frac{\zeta(s-1)}{\zeta(s)}.$$

Since

$$\varphi(n) = \sum_{d|n} \mu(d)(n/d),$$

we have

$$\begin{aligned} \sum_{n=1}^{\infty} \frac{\varphi(n)}{n^s} &= \Big(\sum_{n=1}^{\infty} \frac{\mu(n)}{n^s}\Big)\Big(\sum_{n=1}^{\infty} \frac{n}{n^s}\Big) \\ &= \frac{\zeta(s-1)}{\zeta(s)}, \end{aligned}$$

as required. □

1.5.19 *Let* k *be a fixed natural number. Show that if*

$$f(n) = \sum_{d^k|n} g(n/d^k),$$

then

$$g(n) = \sum_{d^k|n} \mu(d) f(n/d^k)$$

and conversely.

We have

$$\begin{aligned} \sum_{d^k e=n} \mu(d) f(e) &= \sum_{d^k e=n} \mu(d) \sum_{\delta^k|e} g(e/\delta^k) \\ &= \sum_{d^k \delta^k t=n} \mu(d) g(t) \\ &= \sum_{r^k|n} g(n/r^k)\Big(\sum_{d\delta=r} \mu(d)\Big), \end{aligned}$$

and the inner sum is 1 if $r = 1$, and 0 otherwise. Therefore,

$$g(n) = \sum_{d^k|n} \mu(d) f(n/d^k).$$

For the converse,

$$\sum_{d^k e=n} g(n/d^k) = \sum_{d^k e=n}\sum_{\delta^k|e} \mu(\delta) f(e/\delta^k)$$

$$= \sum_{d^k \delta^k t=n} \mu(\delta) f(t)$$

$$= \sum_{r^k|n} f(n/r^k)\Big(\sum_{d\delta=r} \mu(\delta)\Big) = f(n),$$

as required. □

1.5.20 *The mth cyclotomic polynomial is defined as*

$$\phi_m(x) = \prod_{\substack{1\le i\le m\\(i,m)=1}} (x - \zeta_m^i),$$

where ζ_m denotes a primitive mth root of unity. Show that

$$x^m - 1 = \prod_{d|m} \phi_d(x).$$

We have

$$x^m - 1 = \prod_{1\le i\le m} (x - \zeta_m^i).$$

We can partition the right-hand side according to $d = \gcd(i, m)$. Then, $(i/d, m/d) = 1$, and

$$\zeta_m^i = \zeta_{m/d}^{i/d}$$

is a primitive (m/d)th root of unity. Also, every primitive (m/d)th root of unity is a root of $x^m - 1$. Thus,

$$x^m - 1 = \prod_{d|m} \phi_{m/d}(x) = \prod_{d|m} \phi_d(x),$$

as required. □

1.5.21 *With the notation as in the previous exercise, show that the coefficient of*

$$x^{\varphi(m)-1}$$

in $\phi_m(x)$ *is* $-\mu(m)$.

The coefficient of $x^{\phi(m)-1}$ in $\phi_m(x)$ is clearly

$$-\sum_{\substack{1 \le i \le m \\ (i,m)=1}} \zeta_m^i,$$

which is the Ramanujan sum $-c_m(1) = -\mu(m)$ by Exercise 1.1.13. □

1.5.22 *Prove that*

$$\phi_m(x) = \prod_{d|m} (x^d - 1)^{\mu(m/d)}.$$

By Exercise 1.5.20,

$$x^m - 1 = \prod_{d|m} \phi_d(x),$$

so that

$$\log(x^m - 1) = \sum_{d|m} \log \phi_d(x),$$

as formal series. By Möbius inversion,

$$\log \phi_m(x) = \sum_{d|m} \mu(d) \log(x^{m/d} - 1).$$

Hence

$$\phi_m(x) = \prod_{d|m} (x^d - 1)^{\mu(m/d)},$$

as required. □

1.5.23 *If* $\phi_m(x)$ *is the* m*th cyclotomic polynomial, prove that*

$$\phi_m(1) = \begin{cases} p & \text{if} \quad m = p^\alpha, \\ 1 & \text{otherwise,} \end{cases}$$

where p *is a prime number.*

We have

$$\frac{x^m - 1}{x - 1} = \prod_{\substack{d|m \\ d \ne 1}} \phi_d(x).$$

The left-hand side is $1 + x + x^2 + \cdots + x^{m-1}$. Evaluating both sides of the equation at $x = 1$ gives

$$\log m = \sum_{\substack{d|m \\ d\neq 1}} \log \phi_d(1).$$

Set $g(d) = \log \phi_d(1)$, if $d \neq 1$ and $g(1) = 0$ otherwise. Thus,

$$\log m = \sum_{d|m} g(d),$$

and by Möbius inversion, we have

$$\begin{aligned} g(m) &= \sum_{d|m} \mu(d) \log m/d \\ &= -\sum_{d|m} \mu(d) \log d \end{aligned}$$

for $m \neq 1$. By Exercise 1.1.6, $g(m) = \Lambda(m)$ as required. □

1.5.24 *Prove that $\phi_m(x)$ has integer coefficients.* We proceed by induction on m. For $m = 1$, this is clear. Writing

$$\begin{aligned} x^m - 1 &= \prod_{d|m} \phi_d(x) = \phi_m(x)\Big(\prod_{\substack{d|m \\ d<m}} \phi_d(x)\Big) \\ &= \phi_m(x)v(x) \quad \text{(say)}, \end{aligned}$$

we find that $v(x)$ has integer coefficients by induction. Also note that $v(x)$ is monic. Thus, by long division, we can write

$$x^m - 1 = q(x)v(x) + r(x),$$

where $q(x), r(x)$ have integer coefficients and either $r = 0$ or degree of $r <$ degree of v. For every complex root α of $v(x)$, we have $\alpha^m - 1 = 0$ so that $r(\alpha) = 0$. This forces $r = 0$ for otherwise it will have more complex roots than its degree. Hence $q(x) = \phi_m(x)$ has integer coefficients. □

1.5.25 *Let q be a prime number. Show that any prime divisor p of $a^q - 1$ satisfies $p \equiv 1 \pmod{q}$ or $p|(a-1)$.* We have

$$a^q \equiv 1 (\text{mod}\, p).$$

Thus, the order of $a \pmod p$ divides q. Since q is not a prime, it must be either 1 or q. If it is 1, then $a \equiv 1 \pmod p$, so that $p|(a-1)$. If the order is q, then $q|p-1$, since the group of coprime residue classes $\pmod p$ has order $p-1$. □

1.5.26 *Let q be a prime number. Show that any prime divisor p of $1 + a + a^2 + \cdots + a^{q-1}$ satisfies $p \equiv 1 \pmod q$ or $p = q$. Deduce that there are infinitely many primes $p \equiv 1 \pmod q$.*

Notice that

$$1 + a + a^2 + \cdots + a^{q-1} = \frac{a^q - 1}{a - 1}$$

if $a \neq 1$. Hence if

$$1 + a + a^2 + \cdots + a^{q-1} \equiv 0 \pmod p,$$

then either $a^q \equiv 1 \pmod p$ and $a \not\equiv 1 \pmod p$ or $a \equiv 1 \pmod p$. In the former case $q|p-1$, since a has order q. Notice that any prime divisor of $2^q - 1$ is congruent to $1 \pmod q$, by the previous exercise. Thus, there is at least one prime congruent to $1 \pmod q$. If there are only finitely many such primes, let us list them as

$$p_1,\ p_2,\ p_3, \ldots,\ p_k.$$

Then, putting $a = qp_1p_2\cdots p_k$, we find that any prime divisor p of

$$1 + a + a^2 + \cdots a^{q-1}$$

is first, coprime to $a = qp_1 \cdots p_k$, and second, must be congruent to $1 \pmod q$ or equal to q, which is a contradiction. □

1.5.27 *Let q be a prime number. Show that any prime divisor p of*

$$1 + b + b^2 + \cdots + b^{q-1}$$

with $b = a^{q^{k-1}}$ satisfies $p \equiv 1 \pmod{q^k}$ or $p = q$.

If $b \not\equiv 1 \pmod p$, then

$$1 + b + \cdots + b^{q-1} = \frac{b^q - 1}{b - 1} \equiv 0 \pmod p$$

implies that a has order q^k, so $p \equiv 1 \pmod{q^k}$. If $b \equiv 1 \pmod p$, then $p = q$, as required. □

1.5.28 *Using the previous exercise, deduce that there are infinitely many primes $p \equiv 1 \pmod{q^k}$, for any positive integer k.* In the previous exercise, we set $b = a^{q^{k-1}}$ to deduce that

$$1 + b + b^2 + \cdots + b^{q-1}$$

has a prime divisor congruent to $1 \pmod{q^k}$. Thus, there is at least one prime congruent to $1 \pmod{q^k}$. Now suppose there are only finitely many such primes, $p_1, p_2, \ldots, p_r$ (say). Then, with

$$b = (qp_1 \cdots p_r)^{q^{k-1}}$$

we deduce $1 + b + \cdots + b^{q-1}$ has a prime divisor congruent to $1 \pmod{q^k}$ different from $p_1, \ldots, p_r$, a contradiction. □

1.5.29 *Let p be a prime not dividing m. Show that $p|\phi_m(a)$ if and only if the order of $a \bmod p$ is m. (Here $\phi_m(x)$ is the mth cyclotomic polynomial.)*

Since

$$x^m - 1 = \prod_{d|m} \phi_d(x),$$

we deduce $a^m \equiv 1 \pmod{p}$. If k is the order of $a \pmod{p}$, then

$$a^k - 1 = \prod_{d|k} \phi_d(a) \equiv 0 \pmod{p},$$

so that $\phi_d(a) \equiv 0 \pmod{p}$ for some $d|k$. If $k < m$, then

$$a^m - 1 = \phi_m(a)\phi_d(a) \text{ (other factors)} \equiv 0 \pmod{p^2}.$$

Since $\phi_m(a+p) \equiv \phi_m(a) (\bmod\, p)$ we deduce

$$(a+p)^m \equiv 1 \pmod{p^2},$$

on the one hand, and

$$(a+p)^m \equiv a^m + ma^{m-1}p \pmod{p^2},$$

on the other. Thus, $ma^{m-1}p \equiv 0 \pmod{p^2}$, so that $p|m$ (because $(a,p) = 1$). This is a contradiction. Hence $k = m$. For the converse, if a has order m, then $a^m \equiv 1 \pmod{p}$. From

$$a^m - 1 = \prod_{d|m} \phi_d(a)$$

we deduce $\phi_d(a) \equiv 0 \pmod{p}$ for some $d \leq m$. If $d < m$, then

$$a^d - 1 = \prod_{\delta | d} \phi_\delta(a)$$

is divisible by p, implying $a^d \equiv 1 \pmod{p}$. This contradicts the fact that a has order m. □

1.5.30 *Using the previous exercise, deduce the infinitude of primes $p \equiv 1 \pmod{m}$.*

Observe that from

$$x^m - 1 = \prod_{d | m} \phi_d(x)$$

we deduce that $\phi_d(0) = \pm 1$ for any d. Thus, $\phi_m(m^r)$ is coprime to m. As r varies over positive integers, only a finite number of them can be equal to ± 1 since $\phi_m(x)$ has degree $\phi(m)$. Thus, for some r,

$$|\phi_m(m^r)| > 1,$$

and so there is a prime divisor p of $\phi_m(m^r)$. The order of $m^r \pmod{p}$ is m. Hence, there is a prime $p \equiv 1 \pmod{m}$. If there are only finitely many such primes $p_1, p_2, \ldots, p_t$ (say), then

$$\phi_m(mp_1p_2 \cdots p_t)$$

must have a prime divisor $p \equiv 1 \pmod{p}$ different from $p_1, \ldots, p_t$. This is a contradiction. □

2

Primes in Arithmetic Progressions

2.1 Characters mod q

2.1.2 *Show that*

$$\sum_{n \leq x} \log n = x \log x - x + O(\log x).$$

Put $f(t) = \log t$, $a_n = 1$ in Theorem 2.1.1. We obtain

$$\begin{aligned} \sum_{n \leq x} \log n &= [x] \log x - \int_1^x [t] \frac{dt}{t} \\ &= x \log x - x + O(\log x) \end{aligned}$$

upon writing $[t] = t - \{t\}$, where $\{t\}$ denotes the fractional part of t, in the integral. $\square$

2.1.3 *Show that*

$$\sum_{n \leq x} \frac{1}{n} = \log x + O(1).$$

In fact, show that

$$\lim_{x \to \infty} \left(\sum_{n \leq x} \frac{1}{n} - \log x \right)$$

exists.

Put $a_n = 1$, $f(t) = 1/t$ in Theorem 2.1.1. Notice that for x a positive integer, we have

$$\begin{aligned}\sum_{2\leq n\leq x} \frac{1}{n} - \log x &= \sum_{2\leq n\leq x} \frac{1}{n} - \int_1^x \frac{dt}{t} \\ &= \sum_{2\leq n\leq x} \left(\frac{1}{n} - \int_{n-1}^n \frac{dt}{t}\right) \\ &= \sum_{2\leq n\leq x} \left(\frac{1}{n} + \log\left(1 - \frac{1}{n}\right)\right).\end{aligned}$$

Since

$$-\log\left(1 - \frac{1}{n}\right) = \frac{1}{n} + \frac{1}{2}\left(\frac{1}{n}\right)^2 + \cdots = \frac{1}{n} + O\left(\frac{1}{n^2}\right),$$

we deduce that

$$\sum_{2\leq n\leq x} \frac{1}{n} - \log x$$

converges to a limit as $x \to \infty$.

2.1.4 *Let $d(n)$ denote the number of divisors of a natural number n. Show that*

$$\sum_{n\leq x} d(n) = x\log x + O(x).$$

Since $d(n) = \sum_{\delta|n} 1$, we have

$$\sum_{n\leq x} d(n) = \sum_{\delta\leq x} \left[\frac{x}{\delta}\right] = x\sum_{\delta\leq x} \frac{1}{\delta} + O(x),$$

and by Exercise 2.1.3, we are done. □

2.1.5 *Suppose $A(x) = O(x^\delta)$. Show that for $s > \delta$,*

$$\sum_{n=1}^{\infty} \frac{a_n}{n^s} = s\int_1^\infty \frac{A(t)}{t^{s+1}}dt.$$

Hence the Dirichlet series converges for $s > \delta$.

By Theorem 2.1.1, with $f(n) = n^{-s}$,

$$\sum_{n \le x} a_n n^{-s} = \frac{A(x)}{x^s} + s \int_1^x \frac{A(t)}{t^{s+1}} dt.$$

For s fixed, $s > \delta$, we know that $A(x) = O(x^\delta)$, so that

$$\lim_{x \to \infty} \frac{A(x)}{x^s} = 0.$$

Thus

$$\sum_{n=1}^{\infty} \frac{a_n}{n^s} = s \int_1^{\infty} \frac{A(t)}{t^{s+1}} dt,$$

for any $s > \delta$.

2.1.6 *Show that for* $s > 1$,

$$\zeta(s) = \frac{s}{s-1} - s \int_1^{\infty} \frac{\{x\}}{x^{s+1}} dx,$$

where $\{x\} = x - [x]$. *Deduce that* $\lim_{s \to 1^+} (s-1)\zeta(s) = 1$.

By Exercise 2.1.5, we get

$$\begin{aligned} \zeta(s) &= s \int_1^{\infty} \frac{[x]}{x^{s+1}} dx \\ &= s \int_1^{\infty} \frac{x - \{x\}}{x^{s+1}} dx \\ &= \frac{s}{s-1} - s \int_1^{\infty} \frac{\{x\}}{x^{s+1}} dx. \end{aligned}$$

Also,

$$(s-1)\zeta(s) = s - s(s-1) \int_1^{\infty} \frac{\{x\}}{x^{s+1}} dx,$$

so that

$$\lim_{s \to 1^+} (s-1)\zeta(s) = 1,$$

since the integral converges for $s > 0$. □

2.1.7 *Prove that*

$$F(x,t) = \sum_{r=0}^{\infty} b_r(x)\frac{t^r}{r!} = \frac{te^{xt}}{e^x - 1}.$$

By the recursion for $b_r(x)$, we have

$$\begin{aligned}\frac{d}{dx}F(x,t) &= \sum_{r=1}^{\infty} b_r'(x)\frac{t^r}{r!}\\ &= \sum_{r=1}^{\infty} rb_{r-1}(x)\frac{t^r}{r!}\\ &= t\cdot F(x,t).\end{aligned}$$

Thus,

$$\log F(x,t) = tx + c(t).$$

Exponentiating, we get

$$F(x,t) = e^{tx+c(t)}.$$

On the other hand,

$$\int_0^1 F(x,t)dx = \int_0^1 \Big(\sum_{r=0}^{\infty} b_r(x)\frac{t^r}{r!}\Big)dx = 1.$$

Thus,

$$\begin{aligned}1 &= \textstyle\int_0^1 (e^{tx+c(t)})dx = e^{c(t)}\cdot\left[\frac{e^{tx}}{t}\right]_{x=0}^{x=1}\\ &= e^{c(t)}\frac{e^t-1}{t},\end{aligned}$$

so that

$$F(x,t) = \frac{te^{xt}}{e^t - 1},$$

as desired. □

2.1.8 *Show that* $B_{2r+1} = 0$ *for* $r \geq 1$.

Since

$$\frac{t}{2} + \sum_{r=0}^{\infty} b_r(0)\frac{t^r}{r!} = \frac{t}{e^t-1} + \frac{t}{2} = \frac{t(e^t+1)}{2(e^t-1)}$$

and the right-hand side is an even function, it follows that $b_r(0) = 0$ for r odd, $r \geq 3$. $\square$

2.1.11 *Show that for some constant B,*

$$\sum_{n \leq x} \frac{1}{\sqrt{n}} = 2\sqrt{x} + B + O\left(\frac{1}{\sqrt{x}}\right).$$

Suppose first that x is a natural number. Put $f(t) = 1/\sqrt{t}$ in Theorem 2.1.9, $a = 1$ and $b = x$. Take $k = 0$. Then,

$$\sum_{1<n \leq x} \frac{1}{\sqrt{n}} = 2\left(\sqrt{x} - 1\right) + \frac{1}{2}\left(\frac{1}{\sqrt{x}} - 1\right) + \frac{1}{2}\int_1^t \left(-\frac{1}{2}\right) t^{-3/2} B_2(t) dt.$$

The integral

$$\int_1^\infty \frac{B_2(t)}{t^{3/2}} dt$$

converges, and we may write for some constant B',

$$\int_1^x \frac{B_2(t)}{t^{3/2}} dt = B' - \int_x^\infty \frac{B_2(t)}{t^{3/2}} dt.$$

The latter integral is $O(1/\sqrt{x})$, whence

$$\sum_{n \leq x} \frac{1}{\sqrt{n}} = 2\sqrt{x} + B + O\left(\frac{1}{\sqrt{x}}\right)$$

for some constant B. If x is not a natural number, notice that

$$\left(\sqrt{x} - \sqrt{[x]}\right)\left(\sqrt{x} + \sqrt{[x]}\right) = x - [x] \leq 1.$$

From this inequality, the result is clear for all x. $\square$

2.1.12 *For $z \in \mathbb{C}$, and $|\arg z| \leq \pi - \delta$, where $\delta > 0$, show that*

$$\sum_{j=0}^{n} \log(z+j) = \left(z + n + \frac{1}{2}\right)\log(z+n)$$

$$-n - \left(z - \frac{1}{2}\right)\log z + \int_0^n \frac{B_1(x)dx}{z+x}.$$

We apply Theorem 2.1.9 for $k = 1$:

$$\sum_{a<j\le b} f(j) = \int_a^b f(t)dt + \frac{1}{2}(f(b) - f(a))$$

$$- \int_a^b B_1(t)f'(t)dt.$$

Now set $f(j) = \log(z+j)$ which is analytic in $|\arg z| \le \pi - \delta$. The result is now immediate.

2.2.1 *Prove that χ is a completely multiplicative function.*

We must show that $\chi(mn) = \chi(m)\chi(n)$ for all natural numbers m, n. If m or n is not coprime to q, then both sides of the equation are zero, and the result is clear. If m and n are coprime to q, then since χ is a homomorphism, the result is immediate. □

2.2.2 *Prove that for* $\mathrm{Re}(s) > 1$,

$$L(s,\chi) = \prod_p \left(1 - \frac{\chi(p)}{p^s}\right)^{-1},$$

where the product is over prime numbers p.

Since χ is multiplicative, so is $\chi(n)/n^s$, and so

$$L(s,\chi) = \prod_p \left(1 + \frac{\chi(p)}{p^s} + \frac{\chi(p^2)}{p^{2s}} + \cdots\right).$$

Now, $\chi(p^m) = \chi(p)^m$ so that

$$\sum_{m=0}^{\infty} \frac{\chi(p^m)}{p^{ms}} = \left(1 - \frac{\chi(p)}{p^s}\right)^{-1},$$

and the result is now clear. □

2.2.3 *Show that $(\mathbb{Z}/p\mathbb{Z})^*$ is cyclic if p is a prime.*

We first list all the possible orders of elements of $(\mathbb{Z}/p\mathbb{Z})^*$:

$$d_1, d_2, \ldots, d_r \qquad \text{(say)}.$$

Let e be the least common multiple of $d_1, d_2, \ldots, d_r$ and factor

$$e = p_1^{a_1} p_2^{a_2} \cdots p_k^{a_k}$$

as a product of distinct prime powers. For each $p_i^{a_i}$ there is some d_j divisible by it. Thus

$$d_j = p_i^{a_i} t$$

for some t coprime to p_i. Since the d_j's are orders of elements of $(\mathbb{Z}/p\mathbb{Z})^*$, there is an element x_i whose order is $p_i^{a_i} t$. Therefore, the element $y_i = x_i^t$ has order $p_i^{a_i}$. Hence, the element $y_1 y_2 \cdots y_k$ has order e. Thus, we have found an element of order e. Therefore, $e|p-1$. But the polynomial

$$x^e - 1$$

has $(p-1)$ roots (mod p), since every nonzero element of $\mathbb{Z}/p\mathbb{Z}$ is a root. Since $\mathbb{Z}/p\mathbb{Z}$ is a field, any polynomial of degree e cannot have more than e roots. Thus, $(p-1) \le e$. Since $e|p-1$, we deduce $e = p-1$. Thus, we have found an element of order $p-1$. □

2.2.4 *Let p be an odd prime. Show that $(\mathbb{Z}/p^a\mathbb{Z})^*$ is cyclic for any $a \ge 1$.*

For $a = 1$, we are done by Exercise 2.2.3. Let g be a primitive root $(\text{mod } p)$. We first find a t such that

$$(g + pt)^{p-1} \not\equiv 1 \pmod{p^2}.$$

Indeed, if $g^{p-1} \not\equiv 1 \pmod{p^2}$, then we can take $t = 0$. Otherwise,

$$\begin{aligned}(g+pt)^{p-1} &\equiv g^{p-1} + p(p-1)tg^{p-2} \pmod{p^2} \\ &\equiv 1 + p(p-1)tg^{p-2} \pmod{p^2},\end{aligned}$$

so that $t = 1$ works. Let $g + pt$ have order d $(\text{mod}\, p^a)$. Then $d|\varphi(p^a)$ by Euler's theorem. Thus, $d|p^{a-1}(p-1)$. Since g is a primitive root mod p, $(p-1)|d$, and so $d = p^{r-1}(p-1)$ for some $r \le a$. We also know that

$$(g + pt)^{p-1} = 1 + pu_1,$$

where u_1 is not divisible by p. Thus

$$\begin{aligned}(g+pt)^{p(p-1)} &= (1 + pu_1)^p \\ &= 1 + p(pu_1) + \binom{p}{2}(pu_1)^2 + \cdots .\end{aligned}$$

Because p is odd, $\binom{p}{2} = \frac{p(p-1)}{2} \equiv 0 \pmod{p}$. Thus

$$(g+pt)^{p(p-1)} \equiv 1 + p^2 u_1 \pmod{p^3}.$$

By induction,

$$(g+pt)^{p^{b-1}(p-1)} \equiv 1 + p^b u_1 \pmod{p^{b+1}}.$$

Now, $g + pt$ has order $d = p^{r-1}(p-1) \pmod{p^a}$ implies

$$(g+pt)^{p^{r-1}(p-1)} \equiv 1 \pmod{p^a}.$$

But then $1 + p^r u_1 \equiv 1 \pmod{p^{r+1}}$ if $r \leq a-1$, which implies $p|u_1$, a contradiction. Thus, $r = a$, and we are done. □

2.2.5 *Let $a \geq 3$. Show that $5 \pmod{2^a}$ has order 2^{a-2}.*

We will prove by induction that

$$5^{2^{n-3}} \equiv 1 + 2^{n-1} \pmod{2^n}$$

for $n \geq 3$. For $n = 3$, this is clear, since $5 \equiv 1 + 4 \pmod{8}$. Suppose we know

$$5^{2^{n-3}} \equiv 1 + 2^{n-1} + 2^n u.$$

Then squaring both sides, we obtain

$$\begin{aligned} 5^{2^{n-2}} &= (1 + 2^{n-1} + 2^n u)^2 \\ &= 1 + 2^{2n-2} + 2^{2n}u^2 + 2^n + 2^{n+1}u + 2^{2n}u \\ &= 1 + 2^n + 2^{n+1}\left\{u + 2^{n-1}u + 2^{n-1}u^2 + 2^{n-3}\right\}, \end{aligned}$$

from which the result is immediate.

It is also clear that

$$\begin{aligned} 5^{2^{n-2}} &\equiv (1+2^{n-1})^2 \pmod{2^n} \\ &\equiv 1 \pmod{2^n}. \end{aligned}$$

Thus, 5 has order $2^{n-2} \pmod{2^n}$. □

2.2.6 *Show that $(\mathbb{Z}/2^a\mathbb{Z})^*$ is isomorphic to $(\mathbb{Z}/2\mathbb{Z}) \times (\mathbb{Z}/2^{a-2}\mathbb{Z})$, for $a \geq 3$.*

By Exercise 2.2.5, we see that 5 has order $2^{a-2} \pmod{2^a}$. Observe that if $5^j \equiv -1 \pmod{2^a}$, then $1 \equiv -1 \pmod 4$, a contradiction. Thus -1 is not in the subgroup generated by 5 $\pmod{2^a}$. Hence, every coprime residue class can be written as $\pm 5^j$. $\square$

2.2.7 *Show that the group of characters* $\pmod q$ *has order* $\varphi(q)$.

Since

$$(\mathbb{Z}/q\mathbb{Z})^* \simeq (\mathbb{Z}/p_1^{a_1}\mathbb{Z})^* \times \cdots \times (\mathbb{Z}/p_k^{a_k}\mathbb{Z})^*,$$

where $q = p_1^{a_1} \cdots p_k^{a_k}$ is the unique factorization of q into prime powers, we see that any character $\chi \pmod q$ decomposes uniquely as

$$\chi_1 \chi_2 \cdots \chi_k,$$

where χ_i is a character of $(\mathbb{Z}/p_i^{a_i}\mathbb{Z})^*$. If p_i is odd, the latter group is cyclic of order $\varphi(p_i^{a_i})$, so that the number of choices for χ_i is $\varphi(p_i^{a_i})$. If $p_i = 2$, then χ_i is a character of $\mathbb{Z}/2\mathbb{Z} \times \mathbb{Z}/2^{a_i-2}\mathbb{Z}$, and again the number of such characters is $\varphi(2^{a_i})$. Thus, the total number of characters is $\varphi(p_1^{a_1}) \cdots \varphi(p_k^{a_k}) = \varphi(q)$. $\square$

2.2.8 *If* $\chi \neq \chi_0$, *show that*

$$\sum_{a \pmod q} \chi(a) = 0.$$

Since $\chi \neq \chi_0$, there is a $b \pmod q$ such that $(b, q) = 1$ and $\chi(b) \neq 1$. Then

$$s = \sum_{a \pmod q} \chi(a) = \sum_{a \pmod q} \chi(ab) = \chi(b)s,$$

since ab runs through coprime residue classes as a does. Hence

$$(1 - \chi(b))s = 0.$$

Therefore, $s = 0$, since $\chi(b) \neq 1$. $\square$

2.2.9 *Show that*

$$\sum_{\chi \pmod q} \chi(n) = \begin{cases} \varphi(q) & \text{if } n \equiv 1 \pmod q \\ 0 & \text{otherwise.} \end{cases}$$

If $n \equiv 1 \pmod{q}$, the result is clear. If $n \not\equiv 1 \pmod{q}$ and $(n, q) = 1$, then there is a character ψ such that $\psi(n) \neq 1$. Thus

$$T = \sum_{\chi \pmod{q}} \chi(n) = \sum_{\chi \pmod{q}} (\psi\chi)(n) = \psi(n) \sum_{\chi \pmod{q}} \chi(n)$$

because $\psi\chi$ ranges over all the characters $(\bmod\, q)$ as χ does. But then

$$(1 - \psi(n)) \sum_{\chi \pmod{q}} \chi(n) = 0,$$

so that $\sum_{\chi \pmod{q}} \chi(n) = 0$, since $\psi \neq 1$. □

2.2 Dirichlet's Theorem

2.3.1 *Let* $\chi = \chi_0$ *be the trivial character* $\pmod{q}$. *Show that*

$$\lim_{s \to 1+} \log L(s, \chi_0) = +\infty.$$

Since $L(s, \chi_0) = \zeta(s) \prod_{p|q}(1 - \frac{1}{p^s})$, the result is clear. □

2.3.2 *Show that for* $s > 1$,

$$\sum_{\chi \pmod{q}} \log L(s, \chi) = \varphi(q) \sum_{n \geq 1} \sum_{p^n \equiv 1 \pmod{q}} \frac{1}{np^{ns}}.$$

Since $L(s, \chi) = \prod_p \left(1 - \frac{\chi(p)}{p^s}\right)^{-1}$, we have

$$\begin{aligned} \sum_{\chi \pmod{q}} \log L(s, \chi) &= \sum_{\chi \pmod{q}} \sum_{p} \left(\sum_{n \geq 1} \frac{\chi(p^n)}{np^{ns}} \right) \\ &= \sum_{p,n} \frac{1}{np^{ns}} \Big(\sum_{\chi \pmod{q}} \chi(p^n) \Big), \end{aligned}$$

the interchange of summation being justified because the series converge absolutely for $s > 1$. By Exercise 2.2.9, we find that the inner sum is 0 unless $p^n \equiv 1 \bmod q$ in which case it is $\varphi(q)$. The result is now immediate. □

2.3.3 *Show that for $s > 1$ the Dirichlet series*

$$\sum_{n=1}^{\infty} \frac{a_n}{n^s} := \prod_{\chi \pmod q} L(s, \chi)$$

has the property that $a_1 = 1$ and $a_n \geq 0$ for $n \geq 2$.

If we exponentiate the identity of Exercise 2.3.2 and use the series

$$e^x = 1 + x + \frac{x^2}{2!} + \cdots,$$

the result is clear. □

2.3.4 *For $\chi \neq \chi_0$, a Dirichlet character $\pmod q$, show that*

$$\left| \sum_{n \leq x} \chi(n) \right| \leq q.$$

Deduce that

$$L(s, \chi) = \sum_{n=1}^{\infty} \frac{\chi(n)}{n^s}$$

converges for $s > 0$.

By Exercise 2.1.5

$$L(s, \chi) = s \int_1^{\infty} \frac{S(t)}{t^{s+1}} dt,$$

where $S(t)=\sum_{n \leq t} \chi(n)$. By Exercise 2.3.8, we know that $\sum_{n \leq q} \chi(n)= 0$. Since χ is periodic with period q, $\sum_{n \leq kq} \chi(n) = 0$ for any k. Let k satisfy $kq \leq t \leq (k+1)q$. Then

$$S(t) = \sum_{n \leq kq} \chi(n) + \sum_{kq < n \leq t} \chi(n).$$

The first sum is zero, and the latter sum cannot exceed q. Thus, the series converges for $s > 0$. □

2.3.5 *If $L(1, \chi) \neq 0$, show that $L(1, \overline{\chi}) \neq 0$, for any character $\chi \neq \chi_0$* $\pmod q$.

We know that

$$L(1,\chi) = \lim_{x\to\infty} \sum_{n\le x} \frac{\chi(n)}{n},$$

since the series converges by Exercise 2.3.4. Now,

$$L(1,\overline{\chi}) = \lim_{x\to\infty} \sum_{n\le x} \overline{\frac{\chi(n)}{n}} = \overline{L(1,\chi)},$$

from which the result follows. □

2.3.6 *Show that*

$$\lim_{s\to 1^+} (s-1)L(s,\chi_0) = \varphi(q)/q.$$

Since

$$L(s,\chi_0) = \zeta(s) \prod_{p|q} \left(1 - \frac{1}{p^s}\right),$$

we obtain

$$\lim_{s\to 1^+} (s-1)L(s,\chi_0) = \lim_{s\to 1^+} [(s-1)\zeta(s)] \prod_{p|q} \left(1 - \frac{1}{p^s}\right) = \frac{\varphi(q)}{q}$$

by Exercise 2.1.6. □

2.3.7 *If* $L(1,\chi) \neq 0$ *for every* $\chi \neq \chi_0$, *deduce that*

$$\lim_{s\to 1^+} (s-1) \prod_{\chi (\mathrm{mod}\ q)} L(s,\chi) \neq 0$$

and hence

$$\sum_{p\equiv 1 (\mathrm{mod}\ q)} \frac{1}{p} = +\infty.$$

(That is, there are infinitely many primes congruent to $1 \pmod q$.)

We have

$$\begin{aligned} \lim_{s\to 1^+} (s-1) \prod_{\chi (\mathrm{mod}\ q)} L(s,\chi) &= \lim_{s\to 1^+} (s-1)L(s,\chi_0) \prod_{\chi\neq\chi_0} L(s,\chi) \\ &= \frac{\varphi(q)}{q} \prod_{\chi\neq\chi_0} L(1,\chi) \neq 0. \end{aligned}$$

On the other hand, by Exercise 2.3.2,

$$\prod_{\chi (\text{mod } q)} L(s,\chi) = \exp\Big(\varphi(q) \sum_{\substack{n,p \\ p^n \equiv 1 (\text{mod } q)}} \frac{1}{np^{ns}}\Big).$$

Observe that we can write the exponential as

$$\varphi(q)\Big(\sum_{\substack{p \\ p \equiv 1 (\text{mod } q)}} \frac{1}{p^s} + \sum_{\substack{p, n \geq 2 \\ p^n \equiv 1 (\text{mod } q)}} \frac{1}{np^{ns}}\Big),$$

and we clearly see that

$$\begin{aligned} \lim_{s \to 1^+} \sum_{p, n \geq 2} \frac{1}{np^{ns}} &\leq \sum_{p,n \geq 2} \frac{1}{np^n} \\ &\leq \sum_p \frac{1}{p(p-1)} < \infty. \end{aligned}$$

Thus,

$$\lim_{s \to 1^+} (s-1) \exp\Big(\varphi(q) \sum_{p \equiv 1 (\text{mod } q)} \frac{1}{p^s}\Big) \neq 0.$$

It is now immediate that $\sum_{p \equiv 1 (\text{mod } q)} \frac{1}{p} = +\infty$. □

2.3.8 *Fix* $(a,q) = 1$. *Show that*

$$\sum_{\chi (\text{mod } q)} \overline{\chi(a)} \chi(n) = \begin{cases} \varphi(q) & \text{if } n \equiv a \ (\text{mod } q) \\ 0 & \text{otherwise.} \end{cases}$$

Note that $\chi(a)\overline{\chi(a)} = 1$. Also, $\chi(a)\chi(a^{-1}) = 1$. Hence $\overline{\chi(a)} = \chi(a^{-1})$, where a^{-1} is the inverse of a in $(\mathbb{Z}/q\mathbb{Z})^*$. Therefore,

$$\sum_{\chi (\text{mod } q)} \overline{\chi(a)} \chi(n) = \sum_{\chi (\text{mod } q)} \chi(a^{-1} n),$$

which by Exercise 2.2.9 is $\varphi(q)$ if $a^{-1} n \equiv 1 \pmod q$, and 0 otherwise. Thus,

$$\sum_{\chi (\text{mod } q)} \overline{\chi(a)} \chi(n) = \begin{cases} \varphi(q) & \text{if } n \equiv a \ (\text{mod } q), \\ 0 & \text{otherwise.} \end{cases}$$

2.3.9 *Fix* $(a, q) = 1$. *If* $L(1, \chi) \neq 0$, *show that*

$$\lim_{s \to 1^+} (s-1) \prod_{\chi (\bmod q)} L(s, \chi)^{\overline{\chi(a)}} \neq 0.$$

Deduce that

$$\sum_{p \equiv a \ (\bmod q)} \frac{1}{p} = +\infty.$$

We see that for $s > 1$,

$$\begin{aligned}\sum_{\chi (\bmod q)} \overline{\chi(a)} \log L(s, \chi) &= \sum_{\chi (\bmod q)} \overline{\chi(a)} \sum \frac{\chi(p^n)}{np^{ns}} \\ &= \sum_{p,n} \frac{1}{np^{ns}} \Big(\sum_{\chi (\bmod q)} \overline{\chi(a)} \chi(p^n) \Big)\end{aligned}$$

as in Exercise 2.3.2. The inner sum, by Exercise 2.3.8, is $\varphi(q)$ if $p^n \equiv a \pmod q$ and zero otherwise. Thus

$$\prod_{\chi (\bmod q)} L(s, \chi)^{\overline{\chi(a)}} = \exp\Big(\varphi(q) \sum_{\substack{n,p \\ p^n \equiv a (\bmod q)}} \frac{1}{np^{ns}} \Big).$$

As before

$$\begin{aligned}&\lim_{s \to 1^+} (s-1) \prod_{\chi (\bmod q)} L(s, \chi)^{\overline{\chi(a)}} \\ &\qquad = \lim_{s \to 1^+} (s-1) L(s, \chi_0) \prod_{\chi \neq \chi_0} L(s, \chi)^{\overline{\chi(a)}} \neq 0,\end{aligned}$$

since $L(1, \chi) \neq 0$. The result now follows as in Exercise 2.3.7. □

2.3.10 *Suppose* $\chi_1 \neq \overline{\chi}_1$ *(that is,* χ_1 *is not real-valued). Show that* $L(1, \chi_1) \neq 0$ *by considering* $F(s)$.

By Exercise 2.3.4, $L(s, \chi)$ converges for $s > 0$. If $L(1, \chi_1) = 0$ then set

$$L(s, \chi_1) = (s-1) g(s, \chi_1) \quad \text{(say)},$$

where $g(s, \chi)$ is continuous for $s > 0$, $s \neq 1$. Observe also that since

$$L(s, \chi) = s \int_1^\infty \frac{S(t)}{t^{s+1}} dt,$$

where $|S(t)| \leq q$, the integral is absolutely convergent for $s > 0$. We also see that $L(s, \chi)$ is differentiable. Thus if we set $g(1, \chi) = L'(1, \chi_1)$ above, then $g(s, \chi)$ is continuous for all $s > 0$. By Exercise 2.3.5, $L(1, \overline{\chi}_1) \neq 0$, and we can also write $L(s, \overline{\chi}_1) = (s-1)g(s, \overline{\chi}_1)$. Therefore,

$$\prod_{\chi} L(s, \chi) = L(s, \chi_0)(s-1)^2 g(s, \chi_1) g(s, \overline{\chi}_1) \prod_{\chi \neq \chi_1, \overline{\chi}_1, \chi_0} L(s, \chi)$$

and we see that

$$\lim_{s \to 1^+} \prod_{\chi} L(s, \chi)$$

$$= \lim_{s \to 1^+} (s-1) L(s, \chi_0)(s-1) g(s, \chi_1) g(s, \overline{\chi_1}) \prod_{\chi \neq \chi_1, \overline{\chi}_1, \chi_0} L(s, \chi)$$

$$= \frac{\varphi(q)}{q} \lim_{s \to 1^+} (s-1) g(s, \chi_1) g(s, \overline{\chi}_1) \prod_{\chi \neq \chi_1, \overline{\chi}_1, \chi_0} L(s, \chi) = 0.$$

However, writing

$$\sum_{n=1}^{\infty} \frac{a_n}{n^s} = \prod_{\chi (\bmod q)} L(s, \chi) = 1 + \sum_{n=2}^{\infty} \frac{a_n}{n^s},$$

we proved $a_n \geq 0$ in Exercise 2.3.3, so that

$$\lim_{s \to 1^+} \prod_{\chi (\bmod q)} L(s, \chi) \geq 1.$$

This contradiction implies $L(1, \chi_1) \neq 0$. □

2.3 Dirichlet's Hyperbola Method

2.4.2 *Prove that*

$$\sum_{n \leq x} \sigma_0(n) = x \log x + (2\gamma - 1)x + O\left(\sqrt{x}\right).$$

We have

$$\sigma_0(n) = \sum_{d|n} 1.$$

We can apply Theorem 2.4.2 with $f(n) = \sigma_0(n)$, $g = h = 1$, and $y = \sqrt{x}$.

$$\begin{aligned}\sum_{n \le x} \sigma_0(n) &= 2 \sum_{d \le \sqrt{x}} \left[\frac{x}{d}\right] - [\sqrt{x}]^2 \\ &= 2 \sum_{d \le \sqrt{x}} \frac{x}{d} - [\sqrt{x}]^2 + O(\sqrt{x}).\end{aligned}$$

By Exercise 2.1.10, we have

$$\sum_{d \le \sqrt{x}} \frac{1}{d} = \frac{1}{2} \log x + \gamma + O\left(\frac{1}{\sqrt{x}}\right),$$

so inserting this above leads to

$$\sum_{n \le x} \sigma_0(n) = x \log x + 2\gamma x - [\sqrt{x}]^2 + O(\sqrt{x}).$$

Now,

$$[\sqrt{x}]^2 = (\sqrt{x} - \{\sqrt{x}\})^2 = x + O(\sqrt{x})$$

from which we deduce the final result. □

2.4.3 *Let* χ *be a real character* (mod q). *Define*

$$f(n) = \sum_{d|n} \chi(d).$$

Show that $f(1) = 1$ *and* $f(n) \ge 0$. *In addition, show that* $f(n) \ge 1$ *whenever* n *is a perfect square.*

Since χ is multiplicative, so is f. If we write

$$n = p_1^{\alpha_1} \cdots p_k^{\alpha_k}$$

as the unique factorization of n as a product of prime powers, then

$$\begin{aligned} f(n) &= f(p_1^{\alpha_1}) \cdots f(p_k^{\alpha_k}) \\ &= \prod_{p^\alpha || n} \left(1 + \chi(p) + \chi(p^2) + \cdots + \chi(p^\alpha)\right).\end{aligned}$$

Since χ is real, $\chi(p) = \pm 1$ whenever p is coprime to q. If $\chi(p) = 1$, then

$$1 + \chi(p) + \cdots + \chi(p^\alpha) = \alpha + 1 \geq 0.$$

If $\chi(p) = -1$, the sum is either 0 or 1 according as α is odd or even. If $p|q$, the sum is 1. In every case we have $f(n) \geq 0$. Clearly, $f(1) = 1$ and when n is a perfect square, each α_i is even. Thus, each sum in the product is greater than or equal to 1. Hence $f(n) \geq 1$ whenever n is a perfect square. □

2.4.4 *Using Dirichlet's hyperbola method, show that*

$$\sum_{n \leq x} \frac{f(n)}{\sqrt{n}} = 2L(1, \chi)\sqrt{x} + O(1),$$

where $f(n) = \sum_{d|n} \chi(d)$ *and* $\chi \neq \chi_0$.

We let $g(d) = \chi(d)/\sqrt{d}$, $h(e) = 1/\sqrt{e}$ in Theorem 2.4.1. We choose $y = \sqrt{x}$. Therefore,

$$\sum_{n \leq x} \frac{f(n)}{\sqrt{n}} = \sum_{d \leq \sqrt{x}} \frac{\chi(d)}{\sqrt{d}} H\left(\frac{x}{d}\right) + \sum_{d \leq \sqrt{x}} \frac{1}{\sqrt{d}} G\left(\frac{x}{d}\right) - G(\sqrt{x})H(\sqrt{x})$$

with notation as in Theorem 2.4.1. Now,

$$H(x) = \sum_{n \leq x} \frac{1}{\sqrt{n}} = 2\sqrt{x} + B + O\left(\frac{1}{\sqrt{x}}\right)$$

by Exercise 2.1.11. Also,

$$G(x) = \sum_{n \leq x} \frac{\chi(n)}{\sqrt{n}} = L\left(\frac{1}{2}, \chi\right) + O\left(\frac{1}{\sqrt{x}}\right),$$

since

$$\sum_{n=1}^{\infty} \frac{\chi(n)}{\sqrt{n}} = L\left(\frac{1}{2}, \chi\right),$$

and by partial summation,

$$\sum_{n \geq x} \frac{\chi(n)}{\sqrt{n}} \ll \int_x^\infty \frac{S(t)}{t^{3/2}} dt \ll \frac{1}{\sqrt{x}},$$

where $S(t) = \sum_{n\le t}\chi(n)$. Therefore,

$$G(x) = L\left(\frac{1}{2},\chi\right) + O\left(\frac{1}{\sqrt{x}}\right).$$

We will write

$$\sum_{n\le x}\frac{f(n)}{\sqrt{n}} = \sum_{d\le\sqrt{x}}\frac{\chi(d)}{\sqrt{d}}H\Big(\frac{x}{d}\Big) + \sum_{d\le\sqrt{x}}\frac{1}{\sqrt{d}}\Big(G\Big(\frac{x}{d}\Big) - G(\sqrt{x})\Big),$$

so that

$$G\Big(\frac{x}{d}\Big) - G(\sqrt{x}) = O\left(\frac{\sqrt{d}}{\sqrt{x}}\right) + O\Big(x^{-1/4}\Big).$$

Observe that

$$\sum_{d\le\sqrt{x}}\frac{1}{\sqrt{d}}\Big(G\Big(\frac{x}{d}\Big) - G(\sqrt{x})\Big) = O(1).$$

Hence,

$$\begin{aligned}\sum_{n\le x}\frac{f(n)}{\sqrt{n}} &= \sum_{d\le x}\frac{\chi(d)}{\sqrt{d}}\left\{2\sqrt{\frac{x}{d}} + B + O\left(\frac{\sqrt{d}}{\sqrt{x}}\right)\right\} + O(1)\\ &= 2\sqrt{x}L(1,\chi) + O(1),\end{aligned}$$

where we have used

$$\sum_{n\le x}\frac{\chi(n)}{n} = L(1,\chi) + O\Big(\frac{1}{x}\Big),$$

which is easily deduced by partial summation. This completes the proof. □

2.4.5 *If $\chi \neq \chi_0$ is a real character, deduce from the previous exercise that $L(1,\chi) \neq 0$.*

Suppose $L(1,\chi) = 0$. Then

$$\sum_{n\le x}\frac{f(n)}{\sqrt{n}} = O(1).$$

On the other hand, by Exercise 2.4.3, $f(n) \geq 0$, and $f(n) \geq 1$ when n is a perfect square, so that

$$\sum_{n \leq x} \frac{f(n)}{\sqrt{n}} \geq \sum_{m \leq \sqrt{x}} \frac{1}{m} \gg \log x,$$

a contradiction. □

2.4.6 *Prove that*

$$\sum_{n>x} \frac{\chi(n)}{n} = O\left(\frac{1}{x}\right)$$

whenever χ is a nontrivial character (mod q).

By partial summation, we have

$$\sum_{n>x} \frac{\chi(n)}{n} \ll \int_x^\infty \frac{s(t)dt}{t^2},$$

where $s(t) = \sum_{n \leq t} \chi(n)$. But $|s(t)| \leq q$, so that the estimate is now immediate. □

2.4.7 *Let*

$$a_n = \sum_{d|n} \chi(d),$$

where χ is a nonprincipal character (mod q). *Show that*

$$\sum_{n \leq x} a_n = xL(1, \chi) + O(\sqrt{x}).$$

We apply Dirichlet's hyperbola method:

$$\sum_{n \leq x} a_n = \sum_{d \leq y} \chi(d) \left[\frac{x}{d}\right] + \sum_{d \leq \frac{x}{y}} s\left(\frac{x}{d}\right) - s(y) \left[\frac{x}{y}\right],$$

where $s(y) = \sum_{n \leq y} \chi(n)$. Since $|s(y)| \leq q$, we get

$$\sum_{n \leq x} a_n = x \sum_{d \leq y} \frac{\chi(d)}{d} + O(y) + O\left(\frac{x}{y}\right).$$

Choosing $y = \sqrt{x}$, we obtain

$$\sum_{n \leq x} a_n = x \sum_{d \leq \sqrt{x}} \frac{\chi(d)}{d} + O(\sqrt{x}).$$

Finally, by the previous exercise,

$$\begin{aligned}\sum_{d\leq\sqrt{x}}\frac{\chi(d)}{d} &= L(1,\chi)-\sum_{d>\sqrt{x}}\frac{\chi(d)}{d}\\ &= L(1,\chi)+O\Big(\frac{1}{\sqrt{x}}\Big),\end{aligned}$$

which implies the required result. □

2.4.8 *Deduce from the previous exercise that $L(1,\chi)\neq 0$ for χ real.*

Consider the Dirichlet series

$$\sum_{n=1}^{\infty}\frac{a_n}{n^s}$$

with $a_n=\sum_{d|n}\chi(d)\geq 0$, as in Exercise 2.4.3. Then, if $L(1,\chi)=0$, by Exercise 2.4.7

$$F(x)=\sum_{n\leq x}a_n=O\left(\sqrt{x}\right).$$

A summation by parts, as in Exercise 2.1.5 gives

$$\sum_{n=1}^{\infty}\frac{a_n}{n^s}=s\int_1^{\infty}\frac{F(t)}{t^{s+1}}\,dt$$

for $s>1/2$ and the Dirichlet series converges for $\operatorname{Re}s>1/2$. By Exercise 1.2.4

$$\sum_{n=1}^{\infty}\frac{a_n}{n^s}=L(s,\chi)\zeta(s),\quad \operatorname{Re}s>1. \tag{2.1}$$

Since $L(s,\chi)$ converges and is analytic for $\operatorname{Re}s>0$ by Exercise 2.3.4 and $\zeta(s)$ has analytic continuation to $\operatorname{Re}s>0$ by Exercise 2.1.6 we can set $s=1/2+\epsilon$ in (2.1). The product on the right of (2.1) converges to $L(1/2,\chi)\zeta(1/2)$ as $\epsilon\to 0$, since $\zeta(s)$ has only a pole at $s=1$ by Exercise 2.1.6. On the other side of (2.1)

$$\sum_{n=1}^{\infty}\frac{a_n}{n^{1/2+\epsilon}}\geq\sum_{m=1}^{\infty}\frac{a_{m^2}}{m^{1+2\epsilon}}\geq\sum_{m=1}^{\infty}\frac{1}{m^{1+2\epsilon}}=\zeta(1+2\epsilon)$$

by 2.4.3. However, as $\epsilon\to 0$, $\zeta(1+2\epsilon)\to\infty$, since 1 is a pole of $\zeta(s)$. This gives a contradiction. □

2.4 Supplementary Problems

2.5.1 *Let $d_k(n)$ be the number of ways of writing n as a product of k numbers. Show that*

$$\sum_{n\leq x} d_k(n) = \frac{x(\log x)^{k-1}}{(k-1)!} + O(x(\log x)^{k-2})$$

for every natural number $k \geq 2$.

For $k = 2$, this is Exercise 2.1.4. We will prove the result by induction on k. Recall that

$$d_k(n) = \sum_{\delta|n} d_{k-1}(\delta),$$

so that

$$\begin{aligned}\sum_{n\leq x} d_k(n) &= \sum_{n\leq x}\sum_{\delta|n} d_{k-1}(\delta) \\ &= \sum_{\delta\leq x} d_{k-1}(\delta)\left[\frac{x}{\delta}\right] \\ &= x\sum_{\delta\leq x}\frac{d_{k-1}(\delta)}{\delta} + O(x(\log x)^{k-2})\end{aligned}$$

by the induction hypothesis. Also by the same, and by Theorem 2.1.1,

$$(k-2)!\sum_{\delta\leq x}\frac{d_{k-1}(\delta)}{\delta} = \int_1^x \frac{(\log t)^{k-2} + O((\log t)^{k-3})}{t}dt + O((\log x)^{k-2}),$$

which easily gives

$$\sum_{\delta\leq x}\frac{d_{k-1}(\delta)}{\delta} = \frac{(\log x)^{k-1}}{(k-1)!} + O((\log x)^{k-2}).$$

Inserting this in the above calculation gives the desired result. □

2.5.2 *Show that*

$$\sum_{n\leq x}\log\frac{x}{n} = x + O(\log x).$$

By Exercise 2.1.2,

$$\sum_{n\le x}\log n = x\log x - x + O(\log x).$$

Thus,

$$\sum_{n\le x}\log x = [x]\log x = x\log x + O(\log x).$$

Subtracting gives the result. □

2.5.3 *Let $A(x)=\sum_{n\le x}a_n$. Show that for x a positive integer,*

$$\sum_{n\le x}a_n\log\frac{x}{n} = \int_1^x\frac{A(t)dt}{t}.$$

We write the left-hand side as

$$\begin{aligned}\sum_{n\le x}\{A(n)-A(n-1)\}\log\frac{x}{n} &= \sum_{n\le x}A(n)\log\frac{x}{n} \\ &\qquad - \sum_{n\le x-1}A(n)\log\frac{x}{n+1} \\ &= \sum_{n\le x-1}A(n)\log\frac{n+1}{n} \\ &= \sum_{n\le x-1}A(n)\int_n^{n+1}\frac{dt}{t} \\ &= \int_1^x\frac{A(t)dt}{t},\end{aligned}$$

since $A(t)$ is a step function. □

2.5.4 *Let $\{x\}$ denote the fractional part of x. Show that*

$$\sum_{n\le x}\left\{\frac{x}{n}\right\} = (1-\gamma)x + O(x^{1/2}),$$

where γ is Euler's constant.

We have

$$\sum_{n\le x}\left\{\frac{x}{n}\right\} = \sum_{n\le x}\left(\frac{x}{n} - \left[\frac{x}{n}\right]\right)$$

$$= x\sum_{n\le x}\frac{1}{n} - \sum_{n\le x}\sigma_0(n).$$

By Example 2.1.10 and Exercise 2.4.2, we find that this is

$$x\left(\log x + \gamma + O\Big(\frac{1}{x}\Big)\right) - \left(x\log x + (2\gamma - 1)x + O(\sqrt{x})\right),$$

which simplifies to

$$(1-\gamma)x + O\left(\sqrt{x}\right),$$

as required. □

2.5.5 *Prove that*

$$\sum_{n\le x}\log^k\frac{x}{n} = O(x)$$

for any $k > 0$.

Since $\log t$ is an increasing function of t, we have for $n \ge 2$,

$$\log^k\frac{x}{n} \le \int_{n-1}^{n}\left(\log^k\frac{x}{t}\right)dt.$$

Hence,

$$\sum_{n\le x}\log^k\frac{x}{n} \le \int_1^x\left(\log^k\frac{x}{t}\right)dt.$$

Set $u = x/t$ in the integral to deduce

$$\sum_{n\le x}\log^k\frac{x}{n} \le x\int_1^x\frac{\log^k u}{u^2}du = O(x),$$

since the latter integral converges for any $k > 0$. (This also gives another proof of Exercise 2.5.2 in the case $k = 1$.) □

2.5.6 *Show that for $x \ge 3$,*

$$\sum_{3\le n\le x}\frac{1}{n\log n} = \log\log x + B = O\Big(\frac{1}{x\log x}\Big).$$

We apply Theorem 2.1.9 with $f(t) = 1/(t \log t)$, $a = 3$, $b = x$, and $k = 0$. Then,

$$\sum_{2 \le n \le x} \frac{1}{n \log n} = \int_3^x \frac{dt}{t \log t} + \Big(\frac{1}{2x \log x} - \frac{1}{6 \log 3} \Big)$$

$$+ \int_3^x \frac{(\{x\} - \frac{1}{2})(1 + \log t)}{(t \log t)^2} dt.$$

the first integral is

$$\log \log x - \log \log 3.$$

For the second integral, observe that the integrand is

$$O\Big(\frac{1}{t^2 \log t} \Big),$$

so that

$$\int_3^\infty \frac{(\{t\} - \frac{1}{2})(1 + \log t) dt}{(t \log t)^2} = c < \infty.$$

Thus, the second integral can be written as

$$c - O\Big(\int_x^\infty \frac{dt}{t^2 \log t} \Big) = c + O\Big(\frac{1}{x \log x} \Big).$$

This completes the proof. □

2.5.7 *Let* χ *be a nonprincipal character* (mod q). *Show that*

$$\sum_{n \ge x} \frac{\chi(n)}{\sqrt{n}} = O\Big(\frac{1}{\sqrt{x}} \Big).$$

By Exercise 2.3.4, we know that

$$\sum_{n \le x} \chi(n) = O(1).$$

Thus, by partial summation,

$$\sum_{n \ge x} \frac{\chi(n)}{\sqrt{n}} = O\Big(\int_x^\infty \frac{dt}{t^{3/2}} \Big) = O\Big(\frac{1}{\sqrt{x}} \Big),$$

as required. □

2.5.8 *For any integer $k \geq 0$, show that*

$$\sum_{n \leq x} \frac{\log^k n}{n} = \frac{\log^{k+1} x}{k+1} + O(1).$$

We apply Theorem 2.1.1 with $a_n = 1/n$ and $f(n) = \log^k n$. Using Example 2.1.10, we have

$$\begin{aligned}\sum_{n \leq x} \frac{\log^k n}{n} &= (\log^k x)\left(\log x + \gamma + O\left(\frac{1}{x}\right)\right) \\ &\quad - \int_1^x \left(\log t + \gamma + O\left(\frac{1}{t}\right)\right) k(\log^{k-1} t)\frac{dt}{t}.\end{aligned}$$

The main term is now evident. The terms involving γ as a coefficient cancel. The remaining error terms are easily seen to be $O(1)$. In fact, this argument can easily be modified to show that

$$\sum_{n \leq x} \frac{\log^k n}{n} = \frac{\log^{k+1} x}{k+1} + c + O\Big(\frac{\log^k x}{x}\Big).$$

2.5.9 *Let $d(n)$ be the number of divisors of n. Show that for some constant c,*

$$\sum_{n \leq x} \frac{d(n)}{n} = \frac{1}{2}\log^2 x + 2\gamma \log x + c + O\Big(\frac{1}{\sqrt{x}}\Big)$$

for positive integers $x \geq 1$.

We apply Theorem 2.1.1 with $a_n = d(n)$ and $f(n) = 1/n$. Using Exercise 2.4.2, we get

$$\begin{aligned}\sum_{n \leq x} \frac{d(n)}{n} &= \int_1^x \frac{(t\log t + (2\gamma - 1)t + O(\sqrt{t}))dt}{t^2} \\ &\quad + \frac{(x\log x + (2\gamma - 1)x + O(\sqrt{x}))}{x}.\end{aligned}$$

The integral is

$$\frac{1}{2}\log^2 x + (2\gamma - 1)\log x + O\Big(\int_1^x \frac{dt}{t^{3/2}}\Big).$$

Since the integral in the error term converges, we can write it as

$$c_1 - O\Big(\int_x^\infty \frac{dt}{t^{3/2}}\Big) = c_1 + O\Big(\frac{1}{\sqrt{x}}\Big)$$

for some constant c_1. Combining these estimates gives the final result. □

2.5.10 *Let $\alpha \geq 0$ and suppose $a_n = O(n^\alpha)$ and*

$$A(x) := \sum_{n \leq x} a_n = O(x^\delta)$$

for some fixed $\delta < 1$. Define

$$b_n = \sum_{d|n} a_d.$$

Prove that

$$\sum_{n \leq x} b_n = cx + O\left(x^{\frac{(1-\delta)(1+\alpha)}{2-\delta}}\right),$$

for some constant c. By Dirichlet's hyperbola method,

$$\begin{aligned}
\sum_{n \leq x} b_n &= \sum_{d \leq y} a_d\Big[\frac{x}{d}\Big] + \sum_{d \leq y} A\Big(\frac{x}{d}\Big) - A(y)\Big[\frac{x}{y}\Big] \\
&= \sum_{d \leq y} a_d\Big[\frac{x}{d}\Big] + \sum_{d \leq y} O\Big(\frac{x^\delta}{d^\delta}\Big) + O(xy^{\delta-1}).
\end{aligned}$$

The sum $\sum_{d \leq y} d^{-\delta}$ is $O(y^{1-\delta})$, so that

$$\sum_{n \leq x} b_n = \sum_{d \leq y} a_d\Big[\frac{x}{d}\Big] + O(x^\delta y^{1-\delta} + xy^{\delta-1}).$$

We choose $y = x^{\frac{1-\delta}{2-\delta}}$ to minimize the error terms (which is the case when the two terms are equal). Thus

$$\sum_{n \leq x} b_n = \sum_{d \leq y} a_d\Big[\frac{x}{d}\Big] + O\left(x^{\frac{1-\delta-\delta^2}{2-\delta}}\right).$$

Also,

$$\sum_{d\le y} a_d\left[\frac{x}{d}\right] = x\sum_{d\le y}\frac{a_d}{d} + O(y^{1+\alpha})$$

and

$$\sum_{d\le y}\frac{a_d}{d} = \sum_{d=1}^{\infty}\frac{a_d}{d} - \sum_{d>y}\frac{a_d}{d}.$$

We have

$$c = \sum_{d=1}^{\infty}\frac{a_d}{d} < \infty$$

(by Exercise 2.1.5). By partial summation,

$$\sum_{d>y}\frac{a_d}{d} \ll \int_y^{\infty}\frac{A(t)}{t^2}dt \ll y^{\delta-1}.$$

Thus,

$$\sum_{d\le y} a_d\left[\frac{x}{d}\right] = cx + O(xy^{\delta-1} + y^{1+\alpha}).$$

With the choice of y given above, we get

$$\sum_{n\le x} b_n = cx + O\left(x^{\frac{(1-\delta)(1+\alpha)}{2-\delta}} + x^{\frac{1-\delta-\delta^2}{2-\delta}}\right).$$

Since $(1-\delta)(1+\alpha) \ge 1-\delta-\delta^2$, we get

$$\sum_{n\le x} b_n = cx + O\left(x^{\frac{(1-\delta)(1+\alpha)}{2-\delta}}\right)$$

as required. □

2.5.11 *Let* χ *be a nontrivial character* $\pmod q$ *and set*

$$f(n) = \sum_{d|n}\chi(d).$$

Show that

$$\sum_{n\le x} f(n) = xL(1,\chi) + O(q\sqrt{x}),$$

where the constant implied is independent of q.

We apply Dirichlet's hyperbola method with $y = \sqrt{x}$. Let $S(x) = \sum_{n \le x} \chi(n)$. Then

$$\sum_{n \le x} f(n) = \sum_{d \le \sqrt{x}} \chi(d) \left[\frac{x}{d}\right] + \sum_{d \le \sqrt{x}} S\left(\frac{x}{d}\right) - S\left(\sqrt{x}\right)\left[\sqrt{x}\right].$$

Since $|S(x)| \le q$, we have

$$\sum_{n \le x} f(n) = \sum_{d \le \sqrt{x}} \chi(d) \left[\frac{x}{d}\right] + O(q\sqrt{x}).$$

Now,

$$\sum_{d \le \sqrt{x}} \chi(d) \left[\frac{x}{d}\right] = x \sum_{d \le \sqrt{x}} \frac{\chi(d)}{d} + O(\sqrt{x})$$

and

$$\sum_{d \le \sqrt{x}} \frac{\chi(d)}{d} = L(1, \chi) + O\left(\frac{q}{\sqrt{x}}\right),$$

by partial summation. Putting this all together gives the desired result. If we use Exercise 5.5.6, we can replace q by $q^{1/2} \log q$. □

2.5.12 *Suppose that $a_n \ge 0$ and that for some $\delta > 0$, we have*

$$\sum_{n \le x} a_n \ll \frac{x}{(\log x)^\delta}.$$

Let b_n be defined by the formal Dirichlet series

$$\sum_{n=1}^{\infty} \frac{b_n}{n^s} = \left(\sum_{n=1}^{\infty} \frac{a_n}{n^s}\right)^2.$$

Show that

$$\sum_{n \le x} b_n \ll x(\log x)^{1-2\delta}.$$

We have

$$b_n = \sum_{d|n} a_d a_{n/d},$$

and so we can apply Dirichlet's hyperbola method with

$$A(x) = \sum_{n \le x} a_n,$$

to get

$$\sum_{n\leq x} b_n = 2\sum_{d\leq\sqrt{x}} a_d A\Big(\frac{x}{d}\Big) - A\left(\sqrt{x}\right)^2.$$

The last term is $O(x/(\log x)^{2\delta})$. The summation on the right hand side is bounded by

$$\ll \frac{x}{(\log x)^{\delta}} \sum_{d\leq\sqrt{x}} \frac{a_d}{d}.$$

By partial summation,

$$\sum_{2\leq d\leq\sqrt{x}} \frac{a_d}{d} \ll \frac{A(\sqrt{x})}{\sqrt{x}} + \int_1^{\sqrt{x}} \frac{A(t)}{t^2}dt,$$

which is easily seen to be $O\left(\log^{1-\delta} x\right)$, and this gives the stated result. □

2.5.13 *Let $\{a_n\}$ be a sequence of nonnegative numbers. Show that there exists $\sigma_0 \in \mathbb{R}$ (possibly infinite) such that*

$$f(s) = \sum_{n=1}^{\infty} \frac{a_n}{n^s}$$

converges for $\text{Re}(s) > \sigma_0$ *and diverges for* $\text{Re}(s) < \sigma_0$. *Moreover, show that the series converges uniformly in* $\text{Re}(s) \geq \sigma_0 + \delta$ *for any* $\delta > 0$ *and that*

$$f^{(k)}(s) = (-1)^k \sum_{n=1}^{\infty} \frac{a_n(\log n)^k}{n^s}$$

for $\text{Re}(s) > \sigma_0$. *(σ_0 is called* **the abscissa of convergence** *of the Dirichlet series $\sum_{n=1}^{\infty} a_n/n^s$.)*

If there is no real value of s for which the series converges, we take $\sigma_0 = \infty$, and there is nothing to prove in this case. Now suppose there is some real s_0 for which the series converges. By the comparison test, the series converges for all $\text{Re}(s) > s_0$, since the coefficients are real and nonnegative. Now let σ_0 be the infimum of all real s_0 for which the series converges. This establishes the existence of σ_0. The uniform convergence in $\text{Re}(s) > \sigma_0 + \delta$ for any

$\delta > 0$ is immediate. Thus, in this region, we can differentiate the series term by term to derive the formula

$$f^k(s) = (-1)^k \sum_{n=1}^{\infty} \frac{a_n (\log n)^k}{n^s}.$$

□

2.5.14 *Let $a_n \geq 0$ be a sequence of nonnegative numbers. Let σ_0 be the abscissa of convergence of*

$$f(s) = \sum_{n=1}^{\infty} \frac{a_n}{n^s}.$$

Show that $s = \sigma_0$ is a singular point of $f(s)$. (That is, $f(s)$ cannot be extended to define an analytic function at $s = s_0$.)

By the previous exercise, $f(s)$ is holomorphic in $\mathrm{Re}(s) > \sigma_0$. If f is not singular at $s = \sigma_0$, then there is a disk

$$D = \{s : |s - \sigma_1| < \delta\}$$

where $\sigma_1 > \sigma_0$ such that $|\sigma_0 - \sigma_1| < \delta$ and a holomorphic function g in D such that $g(s) = f(s)$ for $\mathrm{Re}(s) > s_0$, $s \in D$. By Taylor's formula,

$$\begin{aligned} g(s) &= \sum_{k=0}^{\infty} \frac{g^{(k)}(\sigma_1)}{k!}(s - \sigma_1)^k \\ &= \sum_{k=0}^{\infty} \frac{f^{(k)}(\sigma_1)}{k!}(s - \sigma_1)^k, \end{aligned}$$

since $g(s) = f(s)$ for s in a neighborhood of σ_1. Thus, the series

$$\sum_{k=0}^{\infty} \frac{(-1)^k f^{(k)}(\sigma_1)}{k!}(\sigma_1 - s)^k$$

converges absolutely for any $s \in D$. By the previous exercise, we can write this as a double series

$$\sum_{k=0}^{\infty} \frac{(\sigma_1 - s)^k}{k!} \sum_{n=1}^{\infty} \frac{a_n (\log n)^k}{n^{\sigma_1}}.$$

If $\sigma_1 - \delta < s < \sigma_1$, this convergent double series consists of nonnegative terms and we may interchange the summation to find

$$\sum_{n=1}^{\infty} \frac{a_n}{n^{\sigma_1}} \sum_{k=0}^{\infty} \frac{(\sigma_1 - s)^k (\log n)^k}{k!} = \sum_{n=1}^{\infty} \frac{a_n}{n^s} < \infty.$$

Since $\sigma_1 - \delta < \sigma_0 < \sigma_1$, this is a contradiction for $s = \sigma_0$. Thus, the abscissa of convergence is a singular point of $f(s)$. □

2.5.15 *Let χ be a nontrivial character* (mod q) *and define*

$$\sigma_{a,\chi} = \sum_{d|n} \chi(d) d^a.$$

If χ_1, χ_2 are two characters (mod q), *prove that for $a, b \in \mathbb{C}$,*

$$\sum_{n=1}^{\infty} \sigma_{a,\chi_1}(n) \sigma_{b,\chi_2}(n) n^{-s} =$$

$$\frac{\zeta(s) L(s-a, \chi_1) L(s-b, \chi_2) L(s-a-b, \chi_1\chi_2)}{L(2s-a-b, \chi_1\chi_2)}$$

as formal Dirichlet series.

We apply Ramanujan's identity (see Exercise 1.2.8)

$$\sum_{n=1}^{\infty} \Big(\frac{\alpha^{n+1} - \beta^{n+1}}{\alpha - \beta} \Big) \Big(\frac{\gamma^{n+1} - \delta^{n+1}}{\gamma - \delta} \Big) T^n$$
$$= \frac{1 - \alpha\beta\gamma\delta T^2}{(1-\alpha\gamma T)(1-\alpha\delta T)(1-\beta\gamma T)(1-\beta\delta T)}$$

to deduce that

$$\sum_{n=0}^{\infty} \sigma_{a,\chi_1}(p^n) \sigma_{b,\chi_2}(p^n) T^n$$
$$= \sum_{n=0}^{\infty} \Big(\frac{\chi_1(p)^{n+1} p^{a(n+1)} - 1}{\chi_1(p) p - 1} \Big) \Big(\frac{\chi_2(p)^{n+1} p^{b(n+1)} - 1}{\chi_2(p) p - 1} \Big) T^n$$
$$= \frac{1 - \chi_1(p)\chi_2(p) p^{a+b} T^2}{(1 - \chi_1(p) p^a \chi_2(p) p^b T)(1 - \chi_1(p) p^a T)(1 - \chi_2(p) p^b T)(1 - T)}.$$

Putting $T = p^{-s}$ and multiplying over the primes p gives

$$\frac{\zeta(s) L(s-a, \chi_1)(L(s-b, \chi_2) L(s-a-b, \chi_1\chi_2)}{L(2s-a-b, \chi_1\chi_2)}.$$ □

2.5.16 *Let χ be a nontrivial character* (mod q). *Set $a = \overline{b}$, $\chi_1 = \chi$, and $\chi_2 = \overline{\chi}$ in the previous exercise to deduce that*

$$\sum_{n=1}^{\infty} |\sigma_{a,\chi}(n)|^2 n^{-s} = \frac{\zeta(s)L(s-a,\chi)L(s-\overline{a},\overline{\chi})L(s-a-\overline{a},\chi_0)}{L(2s-a-\overline{a},\chi_0)}.$$

Observe that

$$\sigma_{a,\chi}(n)\sigma_{\overline{a},\overline{\chi}}(n) = |\sigma_{a,\chi}(n)|^2$$

and $\chi\overline{\chi} = \chi_0$, so that the result is now immediate. □

2.5.17 *Using Landau's theorem and the previous exercise, show that $L(1,\chi) \neq 0$ for any nontrivial real character* (mod q).

Set $a = 0$ in Exercise 2.5.16. Then

$$\sum_{n=1}^{\infty} \frac{|\sigma_{0,\chi}(n)|^2}{n^s} = \frac{\zeta(s)L(s,\chi)L(s,\overline{\chi})L(s,\chi_0)}{L(2s,\chi_0)}.$$

The right hand side is regular for $\mathrm{Re}(s) > 1/2$, except possibly $s = 1$. However, if $L(1,\chi) = 0$, then the right-hand side is regular at $s = 1$. Therefore, the Dirichlet series

$$\sum_{n=1}^{\infty} \frac{|\sigma_{0,\chi}(n)|^2}{n^s}$$

represents an analytic function for $\mathrm{Re}(s) > \sigma_0$, where σ_0 is the abscissa of convergence. We must have $\sigma_0 < 1$. However, for χ real and $n = m^2$,

$$\sigma_{0,\chi}(m^2) \geq 1,$$

so that the Dirichlet series diverges for $s = 1/2$. Hence $1/2 \leq \sigma_0 < 1$. Since $L(2s,\chi_0)^{-1}$ is regular for $s \geq 1/2$, we have a contradiction because

$$\frac{\zeta(s)L(s,\chi)L(s,\overline{\chi})L(s,\chi_0)}{L(2s,\chi_0)}$$

is regular for any real $s \geq 1/2$. □

2.5.18 *Show that $\zeta(s) \neq 0$ for* $\mathrm{Re}(s) > 1$. We have for $\sigma = \mathrm{Re}(s)$,

$$\zeta(s) = \prod_p \left(1 + \frac{1}{p^s} + \frac{1}{p^{2s}} + \cdots\right),$$

so that

$$|\zeta(s)| \geq \prod_p \left(1 - \frac{1}{p^\sigma} - \frac{1}{p^{2\sigma}} - \cdots\right),$$

so that

$$|\zeta(s)| \geq \prod_p \left(1 - \frac{1}{p^\sigma - 1}\right) \neq 0,$$

and the infinite product converges because $\sigma > 1$. □

2.5.19 (Landau's theorem for integrals)*Let $A(x)$ be right continuous for $x \geq 1$ and of bounded finite variation on each finite interval. Suppose that*

$$f(s) = \int_1^\infty \frac{A(x)}{x^{s+1}} dx,$$

with $A(x) \geq 0$. Let σ_0 be the infimum of all real s for which the integral converges. Show that $f(s)$ has a singularity at $s = \sigma_0$. This is similar to Exercise 2.5.14, and so we merely indicate the modifications needed in the solution of that problem to obtain the required result. As before, we can differentiate under the integral sign to get

$$f^{(k)}(s) = (-1)^k \int_1^\infty \frac{A(x)(\log x)^k}{x^{s+1}} dx.$$

If σ_0 is not a singularity, we deduce that

$$\sum_{k=0}^\infty \frac{(\sigma_1 - k)^k}{k!} \int_1^\infty \frac{A(x)(\log x)^k}{x^{\sigma_1+1}} dx,$$

using the notation in the solution to Exercise 2.5.14. Interchanging the summation and integration gives

$$\int_1^\infty \frac{A(x)}{x^{s+1}} dx < \infty$$

for s satisfying $\sigma_1 - \delta < s < \sigma_1$. For $s = \sigma_0$, this is a contradiction. □

2.5.20 *Let λ denote Liouville's function and set*

$$S(x) = \sum_{n \leq x} \lambda(n).$$

Show that if $S(x)$ is of constant sign for all x sufficiently large, then $\zeta(s) \neq 0$ for $\mathrm{Re}(s) > 1/2$. (The hypothesis is an old conjecture of Pólya. It was shown by Haselgrove in 1958 that $S(x)$ changes sign infinitely often.)

We have by Exercise 1.2.5 and partial summation that

$$\frac{\zeta(2s)}{\zeta(s)} = s\int_1^\infty \frac{S(x)dx}{x^{s+1}}.$$

If $S(x) \geq 0$ for all x, then the integral represents an analytic function for $\text{Re}(s) > \sigma_0$, where σ_0 is the abscissa of convergence. However, by Exercise 2.1.6,

$$(s-1)\zeta(s) = s - s(s-1)\int_1^\infty \frac{\{x\}}{x^{s+1}}dx,$$

so if $\zeta(s) = 0$ for some s satisfying $1/2 < s < 1$, we get

$$1 = (s-1)\int_1^\infty \frac{\{x\}}{x^{s+1}}dx,$$

a contradiction because the right-hand side is negative. Thus $\zeta(s) \neq 0$ for $1/2 < s < 1$. We find that $\zeta(2s)/\zeta(s)$ has its first real singularity at $s = 1/2$. Therefore, $\sigma_0 = 1/2$. Therefore, $\zeta(2s)/\zeta(s)$ is regular for $\text{Re}(s) > 1/2$, which means that $\zeta(s) \neq 0$ for $\text{Re}(s) > 1/2$. (This is the celebrated Riemann hypothesis, which still remains unproved, as of the year 2000.) □

2.5.21 *Prove that*

$$b_n(x) = \sum_{k=0}^{n} \binom{n}{k} B_{n-k}x^k,$$

where $b_n(x)$ is the nth Bernoulli polynomial and B_n denotes the nth Bernoulli number.

We have from Exercise 2.1.7 that

$$F(x,t) = e^{xt}F(0,t).$$

As power series, this equation is

$$\sum_{r=0}^{\infty} \frac{b_r(x)t^r}{r!} = \Big(\sum_{r=0}^{\infty} \frac{x^r t^r}{r!}\Big)\Big(\sum_{r=0}^{\infty} \frac{B_r t^r}{r!}\Big),$$

so that comparing the coefficients of t^n on both sides gives the result. □

2.5.22 *Prove that*

$$b_n(1-x) = (-1)^n b_n(x),$$

where $b_n(x)$ denotes the nth Bernoulli polynomial.

We have from Exercise 2.1.7 that

$$\begin{aligned} F(1-x,t) &= \sum_{r=0}^{\infty} b_r(1-x)\frac{t^r}{r!} = \frac{te^{(1-x)t}}{e^t-1} \\ &= \frac{(-t)e^{x(-t)}}{e^{-t}-1} = \sum_{r=0}^{\infty} \frac{b_r(x)(-1)^r t^r}{r!}, \end{aligned}$$

from which the result follows. □

2.5.23 *Let*

$$s_k(n) = 1^k + 2^k + 3^k + \cdots + (n-1)^k.$$

Prove that for $k \geq 1$,

$$(k+1)s_k(n) = \sum_{i=0}^{k} \binom{k+1}{i} B_i n^{k+i-i}.$$

We consider the power series

$$\begin{aligned} \sum_{k=0}^{\infty} \frac{s_k(n)t^k}{k!} &= \sum_{k=0}^{\infty} \frac{t^k}{k!}\Big(\sum_{j=0}^{n-1} j^k\Big) \\ &= \sum_{j=0}^{n-1} e^{tj} = \frac{e^{nt}-1}{e^t-1}. \end{aligned}$$

Writing

$$\begin{aligned} \frac{e^{nt}-1}{e^t-1} &= \frac{e^{nt}-1}{t} \cdot \frac{t}{e^t-1} \\ &= \Big(\sum_{k=1}^{\infty} \frac{n^k t^{k-1}}{k!}\Big)\Big(\sum_{j=0}^{\infty} \frac{B_j t^j}{j!}\Big) \end{aligned}$$

and comparing coefficients of both sides gives the result. □

3

The Prime Number Theorem

3.1 Chebyshev's Theorem

3.1.1 *Let*

$$\theta(n) = \sum_{p \leq n} \log p,$$

where the summation is over primes. Prove that

$$\theta(n) \leq 4n \log 2.$$

Since every prime between n and $2n$ divides

$$\binom{2n}{n} \leq 2^{2n},$$

because it is one of the binomial coefficients occurring in the binomial expansion of $(1+1)^{2n}$, we see that

$$\theta(2n) - \theta(n) \leq 2n \log 2.$$

If $n = 2^r$, we obtain $\theta(2^{r+1}) - \theta(2^r) \leq 2^{r+1} \log 2$, valid for $r = 0, 1, 2, \ldots, m$ (say). Adding up these inequalities, we obtain

$$\begin{aligned} \theta(2^{m+1}) &\leq (2^{m+1} + 2^m + \cdots + 2 + 1) \log 2 \\ &\leq (2^{m+2} - 1) \log 2 \leq 2^{m+2} \log 2. \end{aligned}$$

If n satisfies $2^m \leq n < 2^{m+1}$, then

$$\begin{aligned}\theta(n) &= \theta(2^m) + (\theta(n) - \theta(2^m)) \\ &\leq 2^{m+1}\log 2 + (\theta(2^{m+1}) - \theta(2^m)) \\ &\leq 2^{m+1}\log 2 + 2^{m+1}\log 2 \leq 4n\log 2.\end{aligned}$$

□

3.1.2 *Prove that*

$$\theta(n) \leq 2n\log 2.$$

We induct on n. If n is not prime, then

$$\theta(n) = \theta(n-1) \leq 2(n-1)\log 2$$

by the induction hypothesis. If n is odd, write $n = 2m+1$, then notice that

$$\binom{2m+1}{m}$$

is divisible by all the primes between $m+1$ and $2m+1$. Notice that

$$\binom{2m+1}{m} + \binom{2m+1}{m+1} \leq 2^{2m+1},$$

so that

$$2\binom{2m+1}{m} \leq 2^{2m+1}.$$

Hence

$$\theta(2m+1) - \theta(m) \leq 2m\log 2$$

and induction gives $\theta(m) \leq 2m\log 2$, so that

$$\theta(2m+1) \leq 4m\log 2 \leq 2(2m+1)\log 2$$

as desired. □

3.1.3 *Let*

$$\psi(x) = \sum_{p^\alpha \leq x} \log p = \sum_{n \leq x} \Lambda(n),$$

where Λ is the von Mangoldt function. Show that

$$\text{lcm}[1, 2, \ldots, n] = e^{\psi(n)}.$$

Clearly, we can write

$$\text{lcm}[1, 2, \ldots, n] = \prod_{p \leq n} p^{e_p},$$

where e_p is the largest power of $p \leq n$. Thus

$$e_p = \left[\frac{\log n}{\log p}\right] = \sum_{p^\alpha \leq n} 1,$$

from which the result follows.

3.1.4 *Show that*

$$e^{\psi(2n+1)} \int_0^1 x^n (1-x)^n dx$$

is a positive integer. Deduce that $\psi(2n+1) \geq 2n \log 2$. (The method of deriving this is due to M. Nair.)

The integral

$$\begin{aligned} I = \int_0^1 x^n (1-x)^n dx &= \sum_{k=0}^{n} \binom{n}{k} (-1)^k \int_0^1 x^{n+k} dx \\ &= \sum_{k=0}^{n} \binom{n}{k} \frac{(-1)^k}{n+k+1} \end{aligned}$$

is a rational number. It is clear that $\text{lcm}[1, 2, \ldots, 2n+1]I$ is a positive integer. Since

$$x(1-x) \leq \frac{1}{4}$$

for $0 \leq x \leq 1$, we obtain

$$I \leq 2^{-2n}.$$

Hence, by Exercise 3.1.3, we obtain

$$e^{\psi(2n+1)} I \geq 1,$$

so that $e^{\psi(2n+1)} \geq 2^{2n}$, as required. □

3.1.5 *Prove that there are positive constants A and B such that*

$$\frac{Ax}{\log x} \leq \pi(x) \leq \frac{Bx}{\log x}$$

for all x sufficiently large. (This result was first proved by Chebycheff.)

By the previous exercises, we have

$$\begin{aligned} \theta(x) &\leq 2x \log 2, \\ \psi(2n+1) &\geq 2n \log 2. \end{aligned}$$

Hence,

$$\sum_{\sqrt{x}<p\leq x} \log p \leq 2x \log 2,$$

which implies

$$\left(\pi(x) - \pi\left(\sqrt{x}\right)\right) \frac{1}{2} \log x \leq 2x \log 2.$$

This yields

$$\pi(x) \leq \frac{4x \log 2}{\log x} + \pi\left(\sqrt{x}\right).$$

Since $\pi\left(\sqrt{x}\right) \leq \sqrt{x}$, we get

$$\pi(x) \leq \frac{4x \log 2}{\log x} + O\left(\sqrt{x}\right) = O\left(\frac{x}{\log x}\right).$$

For the lower bound, notice that

$$\psi(x) \gg x,$$

and that

$$\psi(x) - \theta(x) = \sum_{\substack{p^\alpha \leq x \\ \alpha \geq 2}} \log p = O\left(\sqrt{x} \log^2 x\right).$$

Hence

$$\theta(x) \gg x,$$

and as before,

$$\sum_{\sqrt{x}\leq p\leq x} \log p + O(\sqrt{x} \log x) \gg x,$$

so that $\pi(x)\log x \gg x$ for x sufficiently large. Thus, $\pi(x) \gg \frac{x}{\log x}$. □

3.1.6 *Prove that*

$$T(x) := \sum_{n\le x} \log n = x\log x - x + c + O(1/x)$$

for some constant c. (This improves Exercise 2.1.2.)

We apply Theorem 2.1.9 with $f(n) = \log n$, $a = 1$, $b = x$, and $k = 1$ to get

$$\sum_{n\le x} \log n = \int_1^x \log t dt + \frac{1}{2}\log x + \frac{1}{2}\int_1^x \frac{B_2(t)}{t^2} dt + O(1).$$

We have

$$\int_1^x \frac{B_2(t)dt}{t^2} = \int_1^\infty \frac{B_2(t)dt}{t^2} + O\Big(\frac{1}{x}\Big),$$

and the integral on the right-hand side converges because $B_2(t)$ is bounded. Since

$$\int_1^x \log t\, dt = x\log x - x + 1,$$

this completes the proof. □

3.1.7 *Using the fact*

$$\log n = \sum_{d|n} \Lambda(d),$$

prove that

$$\sum_{n\le x} \frac{\Lambda(n)}{n} = \log x + O(1).$$

We have

$$T(x) = \sum_{d\le x} \Lambda(d)\left[\frac{x}{d}\right] = x\sum_{d\le x}\frac{\Lambda(d)}{d} + O(\psi(x)).$$

Since $T(x) = x\log x + O(x)$ and $\psi(x) = O(x)$, we obtain the required result upon dividing by x. □

3.1.8 *Prove that*

$$\sum_{p\le x} \frac{1}{p} = \log\log x + O(1).$$

From Exercise 3.1.7, we deduce

$$\sum_{p \le x} \frac{\log p}{p} = \log x + O(1),$$

since the contribution from higher prime powers is bounded by a convergent sum. Thus, by partial summation,

$$\sum_{p \le x} \frac{\log p}{p} \cdot \frac{1}{\log p} = O(1) + \int_2^x \frac{\{\log t + O(1)\}}{(\log t)^2} \frac{dt}{t}.$$

Now,

$$\int_2^x \frac{dt}{t \log t} = \log \log x + O(1)$$

and

$$\int_2^x \frac{dt}{t(\log t)^2} = O(1).$$

The result is now immediate. □

3.1.10 *Suppose that $\{a_n\}_{n=1}^{\infty}$ is a sequence of complex numbers and set*

$$S(x) = \sum_{n \le x} a_n.$$

If

$$\lim_{x \to \infty} \frac{S(x)}{x} = \alpha,$$

show that

$$\sum_{n \le x} \frac{a_n}{n} = \alpha \log x + o(\log x)$$

as $x \to \infty$.

By partial summation

$$\sum_{n \le x} \frac{a_n}{n} = \frac{S(x)}{x} + \int_1^x \frac{S(t)}{t^2} dt = \alpha \log x + o(\log x).$$

The integral is divided into two parts:

$$\int_1^x \frac{S(t)dt}{t^2} = \int_1^y \frac{S(t)dt}{t^2} + \int_y^x \frac{S(t)dt}{t^2}.$$

We may use $S(t) = \alpha t + o(t)$ in the second integral if we choose $y = y(x) \to \infty$. The first integral is $O(\log y)$. Thus, choosing y such that $\log y = o(\log x)$ justifies the last step. □

3.1.11 *Show that*

$$\lim_{x\to\infty} \frac{\psi(x)}{x} = 1$$

if and only if

$$\lim_{x\to\infty} \frac{\pi(x)}{x/\log x} = 1.$$

We have

$$\begin{aligned} \sum_{2\le n\le x} \frac{\Lambda(n)}{\log n} &= \frac{\psi(x)}{\log x} + \int_2^x \frac{\psi(t)}{(\log t)^2}\frac{dt}{t} \\ &= \frac{x}{\log x} + o\Big(\frac{x}{\log x}\Big) + O\Big(\int_2^x \frac{dt}{\log^2 t}\Big). \end{aligned}$$

Now,

$$\sum_{n\le x} \frac{\Lambda(n)}{\log n} = \pi(x) + O\left(x^{1/2}\log x\right).$$

We have

$$\int_2^x \frac{dt}{\log^2 t} = O\left(\sqrt{x}\right) + \int_{\sqrt{x}}^x \frac{dt}{\log^2 t} = O\Big(\frac{x}{\log^2 x}\Big).$$

Thus, $\psi(x) = x + o(x)$ implies $\pi(x) = x/\log x + o(x/\log x)$. The converse is similarly deduced. Let $f(n) = 1$ if n is prime, and zero otherwise. Then

$$\begin{aligned} \theta(x) &= \sum_{n\le x} f(n)\log n = \pi(x)\log x - \int_2^x \frac{\pi(t)}{t}dt \\ &= x + o(x) + O\Big(\frac{x}{\log x}\Big). \end{aligned}$$

Therefore, $\theta(x) = x + o(x)$. Since $\psi(x) = \theta(x) + O\left(x^{1/2}\log^2 x\right)$, we deduce $\psi(x) = x + o(x)$ as required.

□

3.2 Nonvanishing of Dirichlet Series on $\mathrm{Re}(s) = 1$

3.1.12 *If*

$$\lim_{x\to\infty} \frac{\pi(x)}{x/\log x} = \alpha,$$

then show that

$$\sum_{p\le x} \frac{1}{p} = \alpha \log\log x + o(\log\log x).$$

Deduce that if the limit exists, it must be 1.

By partial summation,

$$\begin{aligned}\sum_{p\le x} \frac{1}{p} &= \frac{\pi(x)}{x} + \int_2^x \frac{\pi(t)}{t^2}dt \\ &= \alpha \log\log x + o(\log\log x).\end{aligned}$$

By Exercise 3.1.8, we know that α must be 1. □

3.2.1 *Show that*

$$\zeta(s) = \frac{s}{s-1} - s\int_1^\infty \frac{\{x\}}{x^{s+1}}dx$$

for $\mathrm{Re}(s) > 1$. *Since the right-hand side of the equation is analytic for* $\mathrm{Re}(s) > 0$, $s \neq 1$, *we obtain an analytic continuation of* $(s-1)\zeta(s)$. This

was already derived in Exercise 2.1.6. It remains only to observe that the integral on the right-hand side converges for $\mathrm{Re}(s) > 0$. Observe that $\zeta(s)$ has a simple pole at $s = 1$ with residue 1. □

3.2.2 *Show that* $\zeta(s) \neq 0$ *for* $\mathrm{Re}(s) > 1$.

We have

$$\zeta(s) = \prod_p \left(1 - \frac{1}{p^s}\right)^{-1}$$

for $\sigma = \mathrm{Re}(s) > 1$. Since

$$\left|\left(1 - \frac{1}{p^s}\right)^{-1}\right| = \left|1 + \frac{1}{p^s} + \frac{1}{p^{2s}} + \cdots\right| \geq \left|1 - \frac{1}{p^\sigma} - \frac{1}{p^{2\sigma}} - \cdots\right|$$

and

$$1 - \frac{1}{p^\sigma} - \frac{1}{p^{2\sigma}} - \cdots = 1 - \frac{1}{p^\sigma - 1} \geq 1 - \frac{1}{2^\sigma - 1} > 0$$

for $\sigma > 1$, we are done. □

3.2.3 *Prove that for $\sigma > 1$, $t \in \mathbb{R}$,*

$$\text{Re}\log\zeta(\sigma + it) = \sum_{n=2}^{\infty} \frac{\Lambda(n)}{n^\sigma \log n}\cos(t\log n).$$

We have

$$\begin{aligned}
\log\zeta(s) &= -\sum_p \log\left(1 - \frac{1}{p^s}\right) \\
&= \sum_p \sum_{k=1}^{\infty} \frac{1}{kp^{ks}} \\
&= \sum_{n=1}^{\infty} \frac{\Lambda(n)}{n^\sigma \log n}\left\{\cos(t\log n) - i\sin(t\log n)\right\},
\end{aligned}$$

from which the result follows. □

3.2.4 *Prove that*

$$\text{Re}(3\log\zeta(\sigma) + 4\log\zeta(\sigma + it) + \log\zeta(\sigma + 2it)) \geq 0,$$

for $\sigma > 1$, $t \in \mathbb{R}$.

By Exercise 3.2.3, we see that the left-hand side of the inequality is

$$\sum_{n=1}^{\infty} \frac{\Lambda(n)}{n^\sigma \log n}\left\{3 + 4\cos(t\log n) + \cos(2t\log n)\right\}.$$

Since $3 + 4\cos\theta + \cos 2\theta = 2(1 + \cos\theta)^2 \geq 0$, the result is now immediate. □

3.2.5 *Prove that for $\sigma > 1$, $t \in \mathbb{R}$,*

$$|\zeta(\sigma)^3\zeta(\sigma + it)^4\zeta(\sigma + 2it)| \geq 1.$$

Deduce that $\zeta(1 + it) \neq 0$ for any $t \in \mathbb{R}$, $t \neq 0$. Deduce in a similar way, by considering

$$\zeta(\sigma)^3 L(\sigma, \chi)^4 L(\sigma, \chi^2),$$

that $L(1, \chi) \neq 0$ for χ not real.

By Exercises 3.2.3 and 3.2.4 we obtain

$$\left|\zeta(\sigma)^3\zeta(\sigma+it)^4\zeta(\sigma+2it)\right| \geq 1.$$

Now, we know that

$$\lim_{\sigma\to 1^+}(\sigma-1)\zeta(\sigma)=1.$$

Suppose $\zeta(s)$ has zero of order m at $s=1+it, t\neq 0$. Then

$$\lim_{\sigma\to 1^+}\frac{\zeta(\sigma+it)}{(\sigma-1)^m}=c\neq 0.$$

Hence,

$$\left|(\sigma-1)^3\zeta(\sigma)^3(\sigma-1)^{-4m}\zeta(\sigma+it)^4\zeta(\sigma+2it)\right| \geq (\sigma-1)^{3-4m}.$$

Letting $\sigma \to 1^+$ gives us a finite limit on the left-hand side and infinity on the right-hand side if $m\geq 1$. Therefore, $\zeta(1+it)\neq 0$ for $t\in\mathbb{R}$, $t\neq 0$. If $\chi^2\neq\chi_0$, where χ_0 is the principal character $(\bmod\ q)$, then

$$\log L(\sigma,\chi)=\sum_p\sum_{\nu=1}^{\infty}\frac{\chi(p)^\nu}{p^{\sigma\nu}\nu}, \qquad \sigma>1,$$

and similarly for χ^2. Notice that if $\chi(p)=e^{2\pi i\theta_p}$, then $\chi^2(p)=e^{4\pi i\theta_p}$. Using the inequality $3+4\cos\theta+\cos(2\theta)\geq 0$ and Exercise 3.2.3 with $t=0$, we get by taking real parts that

$$3\log\zeta(\sigma)+4\operatorname{Re}\log L(\sigma,\chi)+\operatorname{Re}\log L(\sigma,\chi^2)\geq 0.$$

This gives

$$|\zeta(\sigma)^3L(\sigma,\chi)^4L(\sigma,\chi^2)|\geq 1,$$

similarly to the above. If $L(1,\chi)=0$ we get a fourth-order zero for $L(\sigma,\chi)^4$, while $\zeta(\sigma)^3$ gives a third-order pole. However, $L(\sigma,\chi^2)$ does not have a pole at $s=1$, since χ^2 is not the principal character. □

3.2.6 *Show that* $-\frac{\zeta'}{\zeta}(s)$ *has an analytic continuation to* $\operatorname{Re}(s)=1$, *with only a simple pole at* $s=1$, *with residue* 1.

Since $\zeta(s) \neq 0$ for $\text{Re}(s) \geq 1$, $s \neq 1$, $-\frac{\zeta'}{\zeta}(s)$ is analytic for $\text{Re}(s) \geq 1$, $s \neq 1$. Now,

$$(s-1)\zeta(s) = s - s(s-1)\int_1^\infty \frac{\{x\}}{x^{s+1}}dx$$

by Exercise 3.2.1. Thus, we can write

$$(s-1)\zeta(s) = sf(s),$$

where $f(s)$ is analytic for $\text{Re}(s) > 0$. Therefore, differentiating the equation, we get $\zeta(s) + (s-1)\zeta'(s) = sf'(s) + f(s)$, so that

$$1 + (s-1)\frac{\zeta'(s)}{\zeta(s)} = \frac{f(s)}{\zeta(s)} + s\frac{f'(s)}{\zeta(s)}.$$

Since $\lim_{s\to 1^+}\zeta(s) = +\infty$, we get $\lim_{s\to 1^+}(s-1)\frac{\zeta'}{\zeta}(s) = -1$. □

3.2.7 *Prove that*

$$\frac{1}{2} + \cos\theta + \cos 2\theta + \cdots + \cos n\theta = \frac{\sin(n+\frac{1}{2})\theta}{2\sin\frac{\theta}{2}}.$$

The left-hand side is the real part of

$$-\frac{1}{2} + \left(1 + e^{i\theta} + e^{2i\theta} + \cdots + e^{ni\theta}\right).$$

The term in the parentheses is the sum of a geometric progression and equals

$$\frac{e^{i(n+1)\theta} - 1}{e^{i\theta} - 1} = \frac{\left(e^{i(n+1)\theta} - 1\right)e^{-i\theta/2}}{2i\sin(\theta/2)}.$$

The real part is

$$\frac{\sin(n+\frac{1}{2})\theta}{2\sin\frac{\theta}{2}} + \frac{1}{2},$$

and the result is now immediate. □

3.2.8 *Prove that*

$$\cos\theta + \cos 3\theta + \cdots + \cos(2n-1)\theta = \frac{\sin 2n\theta}{2\sin\theta}.$$

By Exercise 3.2.7,

$$\frac{1}{2} + \cos\theta + \cos 2\theta + \cdots + \cos 2n\theta = \frac{\sin(2n+\frac{1}{2})\theta}{2\sin\frac{\theta}{2}}$$

and

$$\frac{1}{2} + \cos 2\theta + \cos 4\theta + \cdots + \cos 2n\theta = \frac{\sin(2n+1)\theta}{2\sin\theta},$$

putting first $2n$ instead of n and 2θ instead of θ, respectively. Subtracting gives

$$\cos\theta + \cos 3\theta + \cdots + \cos(2n-1)\theta =$$

$$= \frac{\sin(2n+\frac{1}{2})\theta}{2\sin\frac{\theta}{2}} - \frac{\sin(2n+1)\theta}{2\sin\theta}.$$

Now, $\sin\theta = 2\sin\frac{\theta}{2}\cos\frac{\theta}{2}$, so that the above is equal to

$$\frac{2\cos\frac{\theta}{2}\cdot\sin(2n+\frac{1}{2})\theta - \sin(2n+1)\theta}{4\sin\frac{\theta}{2}\cos\frac{\theta}{2}}.$$

Since

$$\sin(2n+1)\theta = \sin\left(2n+\frac{1}{2}\right)\theta\cos\frac{\theta}{2} + \sin\frac{\theta}{2}\cos\left(2n+\frac{1}{2}\right)\theta,$$

we deduce that the expression in question is

$$\frac{\cos\frac{\theta}{2}\sin(2n+\frac{1}{2})\theta - \sin\frac{\theta}{2}\cos(2n+\frac{1}{2})\theta}{2\sin\theta} = \frac{\sin 2n\theta}{2\sin\theta},$$

as desired. □

3.2.9 *Prove that*

$$1 + \frac{\sin 3\theta}{\sin\theta} + \frac{\sin 5\theta}{\sin\theta} + \cdots + \frac{\sin(2n-1)\theta}{\sin\theta} = \left(\frac{\sin n\theta}{\sin\theta}\right)^2.$$

We prove this by induction on n. For $n = 1$, it is clear. Assuming that it is true for $n \le m$, we must show it for $n = m+1$. After a simple calculation, we are led to prove that

$$\sin^2(n+1)\theta = \sin^2 n\theta + (\sin(2n+1)\theta)\sin\theta,$$

or equivalently,

$$(\sin(n+1)\theta - \sin n\theta)(\sin(n+1)\theta + \sin n\theta) = (\sin(2n+1)\theta)\sin\theta.$$

Using

$$\sin A + \sin B = 2\sin\frac{A+B}{2}\cos\frac{A-B}{2}$$

and

$$\sin A - \sin B = 2\cos\frac{A+B}{2}\sin\frac{A-B}{2}$$

we find that we must prove that

$$4\cos\left(n+\frac{1}{2}\right)\theta\sin\frac{\theta}{2}\sin\left(n+\frac{1}{2}\right)\theta\cos\frac{\theta}{2} = (\sin(2n+1)\theta)\sin\theta.$$

But the left-hand side is

$$\sin 2\left(n+\frac{1}{2}\right)\theta\cdot\sin\theta,$$

as desired. □

3.2.10 *Prove that*

$$(2m+1) + 2\sum_{j=0}^{2m-1}(j+1)\cos(2m-j)\theta = \left(\frac{\sin(m+\frac{1}{2})\theta}{\sin\frac{\theta}{2}}\right)^2,$$

for all integers $m \geq 0$.

We must prove

$$2m+1+2\sum_{j=1}^{2m}(2m-j+1)\cos j\theta = \left(\frac{\sin(m+\frac{1}{2})\theta}{\sin\frac{\theta}{2}}\right)^2.$$

Changing θ to 2φ, we must prove that

$$2m+1+2\sum_{j=1}^{2m}(2m-j+1)\cos 2j\varphi = \left(\frac{\sin(2m+1)\varphi}{\sin\varphi}\right)^2.$$

By Exercise 3.2.7, we know that

$$\frac{1}{2} + \cos 2\theta + \cos 4\theta + \cdots + \cos 2n\theta = \frac{\sin(2n+1)\theta}{2\sin\theta}.$$

That is,

$$1 + 2\sum_{j=1}^{n} \cos 2j\varphi = \frac{\sin(2n+1)\varphi}{\sin\varphi}.$$

Summing both sides over $0 \le n \le 2m$, we obtain

$$(2m+1) + 2\sum_{n=0}^{2m}\sum_{j=1}^{n} \cos 2j\varphi = \sum_{n=0}^{2m} \frac{\sin(2n+1)\varphi}{\sin\varphi}.$$

The left-hand side is

$$(2m+1)+2\sum_{j=1}^{2m} \cos 2j\varphi \sum_{j\le n\le 2m} 1 = (2m+1)+2\sum_{j=1}^{2m}(2m-j+1)\cos 2j\varphi,$$

and the right hand side is

$$\left(\frac{\sin(2m+1)\varphi}{\sin\varphi}\right)^2$$

by Exercise 3.2.9, as desired. □

3.2.11 *Let $f(s)$ be a complex-valued function satisfying*

1. *f is holomorphic in $\mathrm{Re}(s) > 1$ and non-zero there;*
2. *$\log f(s)$ can be written as a Dirichlet series*

$$\sum_{n=1}^{\infty} \frac{b_n}{n^s}$$

with $b_n \ge 0$ for $\mathrm{Re}(s) > 1$;

3. *on the line $\mathrm{Re}(s) = 1$, f is holomorphic except for a pole of order $e \ge 0$ at $s = 1$.*

If f has a zero on the line $\mathrm{Re}(s) = 1$, then the order of the zero is bounded by $e/2$. (This result is due to Kumar Murty.)

Suppose f has a zero at $1 + it_0$ of order $k > \frac{e}{2}$. Then $e \le 2k - 1$. Consider the function

$$\begin{aligned} g(s) &= f(s)^{2k+1} \prod_{j=1}^{2k} f(s + ijt_0)^{2(2k+1-j)} \\ &= f(s)^{2k+1} f(s+it_0)^{4k} f(s+2it_0)^{4k-2} \cdots f(s+2kit_0)^2. \end{aligned}$$

Then g is holomorphic for $\mathrm{Re}(s)>1$ and vanishing to at least first order at $s=1$, since

$$4k^2-(2k+1)e \geq 4k^2-(2k+1)(2k-1)=1.$$

However, for $\mathrm{Re}(s)>1$,

$$\log g(s)=\sum_{n=1}^{\infty}\frac{b_n}{n^s}\Big(2k+1+2\sum_{j=1}^{2k}2(2k+1-j)n^{-ijt_0}\Big).$$

Let $\theta=t_0\log n$. Then for $s=\sigma>1$,

$$\mathrm{Re}\log g(\sigma)=\log|g(\sigma)|=\sum_{n=1}^{\infty}\frac{b_n}{n^\sigma}\Big(2k+1+2\sum_{j=1}^{2k}2(2k+1-j)\cos j\theta\Big).$$

By Exercise 3.2.10, the quantity in the parentheses is greater than or equal to 0. Thus,

$$|g(\sigma)|\geq 1.$$

Letting $\sigma\to 1^+$ we get a contradiction, since $g(1)=0$. □

3.2.12 *Let $f(s)=\prod_\chi L(s,\chi)$, where the product is over Dirichlet characters* $\pmod q$. *Show that $f(s)$ is a Dirichlet series with nonnegative coefficients. Deduce that $L(s,\chi)\neq 0$ for* $\mathrm{Re}(s)=1$.

By the Euler product for each $L(s,\chi)$, we know that it does not vanish for $\mathrm{Re}(s)>1$. Also, for $\mathrm{Re}(s)>1$,

$$\log f(s)=\sum_\chi \log L(s,\chi)=\sum_{n,p}\frac{1}{np^{ns}}\sum_\chi \chi(p^n)$$

which by the orthogonality relations (see Exercise 2.3.8) is equal to

$$\varphi(q)\sum_{\substack{n,p\\ p^n\equiv 1 \pmod q}}\frac{1}{np^{ns}}.$$

This is patently a Dirichlet series with nonnegative coefficients. $L(s,\chi)$ is regular for $\mathrm{Re}(s)>0$ (by Exercise 2.3.4) for $\chi\neq\chi_0$. $L(s,\chi_0)$ has a simple pole at $s=1$. Applying Exercise 3.2.11 gives the desired result. □

3.3 The Ikehara - Wiener Theorem

3.3.3 *Suppose*

$$f(s) = \sum_{n=1}^{\infty} a_n/n^s$$

is a Dirichlet series with real coefficients absolutely convergent for $\mathrm{Re}(s) > 1$. *If* $f(s)$ *extends to a meromorphic function in the region* $\mathrm{Re}(s) \geq 1$, *with only a simple pole at* $s = 1$ *with residue* r, *and* $|a_n| \leq b_n$, *where* $F(s) = \sum_{n=1}^{\infty} b_n/n^s$ *satisfies the hypotheses of Theorem* 3.3.1, *show that*

$$\sum_{n \leq x} a_n = rx + o(x)$$

as $x \to \infty$.

The series $G(s) = F(s) - f(s)$ is a Dirichlet series satisfying the hypotheses of Theorem 3.3.1, and therefore

$$\sum_{n \leq x} (b_n - a_n) = (R - r)x + o(x)$$

as $x \to \infty$. On the other hand,

$$\sum_{n \leq x} b_n = Rx + o(x),$$

so that

$$\sum_{n \leq x} a_n = rx + o(x),$$

as required. □

3.3.4 *Show that the conclusion of the previous exercise is still valid if* $a_n \in \mathbb{C}$.

Define

$$f^*(s) = \sum_{n=1}^{\infty} \overline{a}_n/n^s$$

and observe that

$$f = \frac{1}{2}(f + f^*) + i\left(\frac{f - f^*}{2i}\right).$$

Furthermore, $(f+f^*)/2$ and $(f-f^*)/2i$ are represented by Dirichlet series with real coefficients, absolutely convergent in $\text{Re}(s) > 1$. Since

$$f^*(\overline{s}) = \overline{f(s)},$$

we have $f^*(s) = \overline{f(\overline{s})}$.

We leave it to the reader to show that $f^*(s)$ satisfies the Cauchy - Riemann equations and thus both $(f+f^*)/2$ and $(f-f^*)/2i$ satisfy the condition of the previous exercise. The result is now immediate. □

3.3.5 *Let q be a natural number. Suppose $(a,q)=1$. Show that*

$$\psi(x;q,a) := \sum_{\substack{n\leq x \\ n\equiv a \ (\text{mod } q)}} \Lambda(n)$$

satisfies

$$\lim_{x\to\infty} \frac{\psi(x)}{x/\varphi(q)} = 1.$$

We apply the previous exercise to the function

$$f(s) = \frac{1}{\varphi(q)} \sum_{\chi \ (\text{mod } q)} \overline{\chi(a)}\Big(-\frac{L'}{L}(s,\chi)\Big)$$

which is

$$\sum_{n\equiv a \ (\text{mod } q)} \frac{\Lambda(n)}{n^s}.$$

Since $L(s,\chi)\neq 0$ on $\text{Re}(s)=1$, and the only character contributing a pole to the sum is the principal character, we see that

$$\text{Res}_{s=1} f(s) = \frac{1}{\varphi(q)},$$

from which the result is immediate. □

3.3.6 *Suppose $F(s) = \sum_{n=1}^{\infty} b_n/n^s$ is a Dirichlet series with non-negative coefficients and is convergent for $\text{Re}(s) > c > 0$. If $F(s)$ extends to a* *meromorphic function in the region* $\text{Re}(s) \geq c$ *with only a simple pole at* $s = c$ *with residue* R*, show that*

$$\sum_{n \leq x} b_n = \frac{Rx^c}{c} + o(x^c)$$

as $x \to \infty$.

The Dirichlet series $G(s) = F(s + c - 1)$ satisfies the conditions of Theorem 3.3.1. Therefore,

$$A(x) := \sum_{n \leq x} \frac{b_n}{n^{c-1}} = Rx + o(x)$$

as $x \to \infty$. Now, by partial summation,

$$\begin{aligned}
\sum_{n \leq x} b_n &= A(x)x^{c-1} - \int_1^x A(t)(c-1)t^{c-2}dt \\
&= Rx^c - (c-1)\frac{Rx^c}{c} + o(x^c) \\
&= \frac{Rx^c}{c} + o(x^c),
\end{aligned}$$

as required. □

3.3.7 *Suppose* $f(s) = \sum_{n=1}^{\infty} a_n/n^s$ *is a Dirichlet series with complex coefficients absolutely convergent for* $\text{Re}(s) > c$. *If* $f(s)$ *extends to a meromorphic function in the region* $\text{Re}(s) \geq c$ *with only a simple pole at* $s = c$ *and residue* r, *and* $|a_n| \leq b_n$, *where* $f(s) = \sum_{n=1}^{\infty} b_n/n^s$ *satisfies the hypothesis of Exercise 3.3.6, show that*

$$\sum_{n \leq x} a_n = \frac{rx^c}{c} + o(x^c)$$

as $x \to \infty$.

If we write $g(s) = f(s + c - 1)$, then $g(s)$ satisfies the conditions of Exercises 3.3.3 and 3.3.4. Thus,

$$\sum_{n \leq x} \frac{a_n}{n^{c-1}} = rx + \underline{o}(x)$$

as $x \to \infty$. By partial summation (as in the previous exercise), the result is now immediate. $\square$

3.3.8 *Let $a(n)$ be a multiplicative function defined by $a(1) = 1$ and*

$$a(p^\alpha) = \begin{cases} p + c_p & \text{if} \quad \alpha = 1, \\ 0 & \text{otherwise}, \end{cases}$$

where $|c_p| \leq p^\theta$ with $\theta < 1$. Show that as $x \to \infty$,

$$\sum_{n \leq x} a(n) = \frac{rx^2}{2} + o(x^2)$$

for some nonzero constant r.

The Dirichlet series $f(s) = \sum_{n=1}^\infty a(n)/n^s$ is

$$\prod_p \left(1 + \frac{p + c_p}{p^s}\right) = \prod_p \left(1 + \frac{1}{p^{s-1}} + \frac{c_p}{p^s}\right).$$

We can factor

$$\begin{aligned} \left(1 + \frac{1}{p^{s-1}} + \frac{c_p}{p^s}\right) &= \left(1 + \frac{1}{p^{s-1}}\right)\left(1 + \frac{c_p}{p^s}\left(1 + \frac{1}{p^{s-1}}\right)^{-1}\right) \\ &= \left(1 + \frac{1}{p^{s-1}}\right)\left(1 + \frac{c_p}{p^s} - \frac{c_p}{p^{2s-1}} + \cdots\right). \end{aligned}$$

It is easy to see that

$$h(s) := \prod_p \left(1 + \frac{c_p}{p^s}\left(1 + \frac{1}{p^{s-1}}\right)^{-1}\right)$$

converges absolutely for $\text{Re}(s) > 1 + \theta$. Moreover, $h(s)$ does not vanish in this half-plane. Also,

$$g(s) := \prod_p \left(1 + \frac{1}{p^{s-1}}\right) = \frac{\zeta(s-1)}{\zeta(2s-2)}$$

by Exercise 1.2.7. Thus,

$$f(s) = \frac{\zeta(s-1)}{\zeta(2s-2)} h(s)$$

can be continued analytically to $\text{Re}(s) \geq 2$ with only a simple pole at $s = 2$ and residue $r = h(2)/\zeta(2) \neq 0$. We can now apply the previous exercise with $c = 2$ to deduce the result. □

3.3.9 *Suppose $c_n \geq 0$ and that*

$$\sum_{n \leq x} c_n = Ax + o(x).$$

Show that

$$\sum_{n \leq x} \frac{c_n}{n} = A \log x + o(\log x)$$

as $x \to \infty$.

Let $s(x) = \sum_{n \leq x} c_n$. By partial summation, we get

$$\begin{aligned} \sum_{n \leq x} \frac{c_n}{n} &= \frac{s(x)}{x} + \int_1^x \frac{s(t)}{t^2} dt \\ &= A \log x + o(\log x) \end{aligned}$$

as required. □

3.4 Supplementary Problems

3.4.1 *Show that*

$$\sum_{n \leq x} \Lambda(n) \log n = \psi(x) \log x + O(x).$$

By partial summation,

$$\sum_{n \leq x} \Lambda(n) \log n = \psi(x) \log x - \int_1^x \frac{\psi(t) dt}{t}.$$

Using Chebyshev's estimate that $\psi(x) = O(x)$ in the integral gives the result. □

3.4.2 *Show that*

$$\sum_{d|n} \Lambda(d) \Lambda\left(\frac{n}{d}\right) = \Lambda(n) \log n + \sum_{d|n} \mu(d) \log^2 d.$$

By Exercise 1.1.6, we have

$$\Lambda(n) = -\sum_{d|n} \mu(d) \log d,$$

so that

$$\begin{aligned}
\sum_{d|n} \Lambda(d)\Lambda\left(\frac{n}{d}\right) &= \sum_{de=n} \Lambda(d)\Lambda(e) \\
&= -\sum_{de=n} \Lambda(d) \sum_{t\delta=e} \mu(\delta) \log \delta \\
&= -\sum_{\delta td=n} \mu(\delta) \log \delta \Lambda(d).
\end{aligned}$$

Since

$$\sum_{td=n/\delta} \Lambda(d) = \log \frac{n}{\delta},$$

we have

$$\begin{aligned}
\sum_{d|n} \Lambda(d)\Lambda\left(\frac{n}{d}\right) &= -\sum_{d|n} \mu(d) \log d \log \frac{n}{d} \\
&= \Lambda(n) \log n + \sum_{d|n} \mu(d) \log^2 d
\end{aligned}$$

as required. □

3.4.3 *Show that*

$$\sum_{d|n} \mu(d) \log^2 \frac{x}{d} = \begin{cases} \log^2 x & \textit{if} \quad n = 1, \\ 2\Lambda(n) \log x - \Lambda(n) \log n + \sum_{hk=n} \Lambda(h)h(k) & \textit{if} \quad n > 1. \end{cases}$$

If $n = 1$, the result is clear. For $n > 1$, recall that $\sum_{d|n} \mu(d) = 0$ (Exercise 1.1.1) and that $-\sum_{d|n} \mu(d) \log d = \Lambda(n)$ (Exercise 1.1.6), so that

$$\begin{aligned}\sum_{d|n} \mu(d) \log^2 \frac{x}{d} &= \sum_{d|n} \mu(d)(\log^2 d - 2 \log x \log d) \\ &= 2\Lambda(n) \log x + \sum_{d|n} \mu(d) \log^2 d.\end{aligned}$$

By the previous exercise, we have

$$\sum_{d|n} \mu(d) \log^2 d = \sum_{hk=n} \Lambda(h)\Lambda(k) - \Lambda(n) \log n,$$

which completes the proof. □

3.4.4 *Let*

$$S(x) = \sum_{n \leq x} \left(\sum_{d|n} \mu(d) \log^2 \frac{x}{d} \right).$$

Show that

$$S(x) = \psi(x) \log x + \sum_{n \leq x} \Lambda(n) \psi\left(\frac{x}{n}\right) + O(x).$$

We sum the result of Exercise 3.4.3 to get

$$S(x) = \log^2 x + 2\psi(x) \log x - \sum_{n \leq x} \Lambda(n) \log n + \sum_{mn \leq x} \Lambda(m)\Lambda(n).$$

The first sum, by Exercise 3.4.1, is $\psi(x) \log x + O(x)$. The second sum is

$$\sum_{n \leq x} \Lambda(n) \psi\left(\frac{x}{n}\right).$$

Putting all this together gives the desired result. □

3.4.5 *Show that*

$$S(x) - \gamma^2 = \sum_{d \leq x} \mu(d) \left[\frac{x}{d}\right] \left\{ \log^2 \frac{x}{d} - \gamma^2 \right\},$$

where γ *is Euler's constant.*

We have

$$S(x) - \gamma^2 = \sum_{n \le x} \sum_{d|n} \mu(d) \left\{ \log^2 \frac{x}{d} - \gamma^2 \right\}$$

since $\sum_{d|n} \mu(d) = 1$ if $n = 1$, and 0 otherwise. Interchanging the sums now gives the required result. □

3.4.6 *Show that*

$$S(x) = x \sum_{d \le x} \frac{\mu(d)}{d} \left\{ \log^2 \frac{x}{d} - \gamma^2 \right\} + O(x).$$

Recall that (Exercise 2.5.5)

$$\sum_{d \le x} \log^2 \frac{x}{d} = O(x),$$

so that when we remove the square brackets in $[x/d]$ in Exercise 3.4.5, the error term is $O(x)$. □

3.4.7 *Using the fact*

$$\sum_{n \le x} \frac{1}{n} = \log x + \gamma + O\left(\frac{1}{x}\right),$$

deduce that

$$\frac{S(x)}{x} = \sum_{de \le x} \frac{\mu(d)}{de} \left(\log \frac{x}{d} - \gamma \right) + O(1).$$

By the previous exercise, we can write

$$\begin{aligned} \frac{S(x)}{x} &= \sum_{d \le x} \frac{\mu(d)}{d} \left\{ \log^2 \frac{x}{d} - \gamma^2 \right\} + O(1) \\ &= \sum_{d \le x} \frac{\mu(d)}{d} \left\{ \log \frac{x}{d} - \gamma \right\} \left\{ \log \frac{x}{d} + \gamma \right\} + O(1). \end{aligned}$$

Writing

$$\log \frac{x}{d} + \gamma = \sum_{e \le x/d} \frac{1}{e} + O\left(\frac{d}{x}\right)$$

gives

$$\frac{S(x)}{x} = \sum_{d \le x} \frac{\mu(d)}{d} \Big(\log \frac{x}{d} - \gamma \Big) \sum_{e \le x/d} \frac{1}{e} + O\Big(\frac{1}{x} \sum_{d \le x} \log \frac{x}{d} \Big),$$

and the error term is $O(1)$ by Exercise 2.5.5, which proves the result. □

3.4.8 *Prove that*

$$\frac{S(x)}{x} = 2 \log x + O(1).$$

By the previous exercise,

$$\begin{aligned} \frac{S(x)}{x} &= \sum_{n \le x} \frac{1}{n} \sum_{d|n} \mu(d) \Big(\log \frac{x}{d} - \gamma \Big) \\ &= \sum_{n \le x} \frac{\Lambda(n)}{n} + \log x - \gamma \\ &= 2 \log x + O(1) \end{aligned}$$

by Exercise 3.1.7.

3.4.9 (Selberg's identity) *Prove that*

$$\psi(x) \log x + \sum_{n \le x} \Lambda(n) \psi\Big(\frac{x}{n}\Big) = 2x \log x + O(x).$$

By Exercise 3.4.4,

$$S(x) = \psi(x) \log x + \sum_{n \le x} \Lambda(n) \psi\Big(\frac{x}{n}\Big) + O(x).$$

By the previous exercise,

$$S(x) = 2x \log x + O(x).$$

Putting these facts together gives the result. □

3.4.10 *Show that*

$$\nu(n) = O\Big(\frac{\log n}{\log \log n} \Big),$$

where $\nu(n)$ denotes the number of distinct prime factors of n.

In the interval $[1, n]$, the number with the largest number of prime factors is

$$N = \prod_{p \le t} p,$$

where t is chosen as large as possible so that $N \le n$. Hence

$$\nu(n) \le \pi(t),$$

and by Chebyshev's theorem (Exercise 3.1.4) we have $\log N \gg t$. By Exercise 3.1.5,

$$\pi(t) \ll \frac{t}{\log t},$$

so that $\nu(n) \ll (\log N)/\log t$. Also, $n \le \prod_{p \le t+1} p$, by our choice of t. Again by Chebyshev's theorem,

$$\log n \ll t,$$

so that

$$\nu(n) \ll \log n / \log \log n$$

as required. □

3.4.11 *Let $\nu(n)$ be as in the previous exercise. Show that*

$$\sum_{n \le x} \nu(n) = x \log \log x + O(x).$$

We have

$$\nu(n) = \sum_{p|n} 1,$$

so that

$$\begin{aligned} \sum_{n \le x} \nu(n) &= \sum_{p \le x} \left[\frac{x}{p}\right] = x \sum_{p \le x} \frac{1}{p} + O(x) \\ &= x \log \log x + O(x) \end{aligned}$$

by Exercise 3.1.8.

3.4.12 *Let $\nu(n)$ be as in the previous exercise. Show that*

$$\sum_{n\leq x}\nu^2(n) = x(\log\log x)^2 + O(x\log\log x).$$

We have

$$\sum_{n\leq x}\nu^2(n) = \sum_{n\leq x}\sum_{p,q|n} 1 = \sum_{\substack{pq\leq x\\ p\neq q}}\left[\frac{x}{pq}\right] + \sum_{p\leq x}\left[\frac{x}{p}\right].$$

The second sum is $O(x\log\log x)$ by the previous exercise. The first sum is

$$\begin{aligned}\sum_{\substack{pq\leq x\\ p\neq q}}\frac{x}{pq} + O(x) &= \sum_{pq\leq x}\frac{x}{pq} - \sum_{p^2\leq x}\frac{x}{p^2} + O(x)\\ &= \sum_{pq\leq x}\frac{x}{pq} + O(x).\end{aligned}$$

Now,

$$\sum_{pq\leq x}\frac{1}{pq} = \left(\sum_{p\leq x}\frac{1}{p}\right)^2 - \sum_{\substack{p,q\leq x\\ pq>x}}\frac{1}{pq},$$

and the first sum on the right-hand side is

$$(\log\log x + O(1))^2$$

by Exercise 3.1.8. The second sum is bounded by

$$\sum_{\sqrt{x}<p<x}\sum_{q\leq x}\frac{1}{pq} + \sum_{\sqrt{x}<q<x}\sum_{p\leq x}\frac{1}{pq},$$

since $p, q \leq x$ and $pq > x$ imply either $p > \sqrt{x}$ or $q > \sqrt{x}$. But

$$\sum_{\sqrt{x}<p<x}\frac{1}{p} \ll \int_{\sqrt{x}}^{x}\frac{\pi(t)dt}{t^2} \ll 1$$

by partial summation and Chebyshev's estimate for $\pi(t)$. Thus, the second sum in question is

$$O(\log\log x),$$

which completes the proof. □

3.4.13 *Prove that*

$$\int_{-\infty}^{\infty} \frac{\sin^2 \lambda x}{\lambda x^2} dx = \pi.$$

Let

$$f(t) = \begin{cases} 1 & \text{if} \quad |t| \leq \lambda, \\ 0 & \text{otherwise.} \end{cases}$$

Then

$$\hat{f}(x) = \frac{2 \sin \lambda x}{\sqrt{2\pi} x}.$$

By Parseval's theorem

$$\frac{2}{\pi} \int_{\infty}^{\infty} \frac{\sin^2 \lambda x}{x^2} dx = 2\lambda,$$

as desired. □

3.4.14 *Let*

$$T(x) := \sum_{n \leq x} \log n.$$

Show that for $x > 1$,

$$|T(x) - (x \log x - x)| \leq 4 + \log(x+1).$$

By the inequalities of the integral test, we have

$$T(x) \geq \int_1^x \log t \, dt = x \log x - x + 1.$$

Also,

$$T(x) \leq \int_1^{x+1} \log t \, dt = (x+1)\log(x+1) - (x+1) - 2\log 2 + 2.$$

Hence

$$T(x) - (x \log x - x) \leq x \log \left(\frac{x+1}{x} \right) + \log(x+1) + 3 - 2\log 2.$$

Since $\log(1+t) \leq t$, for $|t| < 1$ we deduce

$$|T(x) - (x\log x - x)| \leq 4 + \log(x+1)$$

as required. □

3.4.15 *Show that*

$$\psi(x) - \psi\left(\frac{x}{2}\right) \leq (\log 2)x + 12 + 3\log(x+1).$$

Deduce that

$$\psi(x) \leq 2(\log 2)x + \frac{12\log x}{\log 2} + \frac{3\log(x+1)\log x}{\log 2}.$$

We have (by the proof of Theorem 3.1.9)

$$\psi(x) - \psi\left(\frac{x}{2}\right) \leq T(x) - 2T\left(\frac{x}{2}\right).$$

By Exercise 3.4.14, we have

$$T(x) - 2T\left(\frac{x}{2}\right) \leq (\log 2)x + 12 + 3\log(x+1).$$

By iteration we obtain

$$\psi\left(\frac{x}{2}\right) - \psi\left(\frac{x}{4}\right) \leq (\log 2)\frac{x}{2} + 12 + 3\log(x+1)$$

and so on. Adding these up gives the stated inequality. □

3.4.16 *Show that*

$$\psi(x) - \psi\left(\frac{x}{2}\right) + \psi\left(\frac{x}{3}\right) \geq (\log 2)x - 2\log(x+1) - 7.$$

We have (as in the previous exercise or by the proof of Theorem 3.1.9)

$$\psi(x) - \psi\left(\frac{x}{2}\right) + \psi\left(\frac{x}{3}\right) \geq T(x) - 2T\left(\frac{x}{2}\right).$$

Using Exercise 3.4.14 now gives the result. □

3.4.17 *Prove that for $x \geq e^{12}$,*

$$\psi(x) - \psi\left(\frac{x}{2}\right) \geq \frac{1}{3}(\log 2)x - \frac{5(\log x)\log(x+1)}{\log 2} - 7.$$

By the previous two exercises

$$\psi(x)-\psi\left(\frac{x}{2}\right) \geq \frac{1}{3}(\log 2)x - 2(\log(x+1)) - 7$$

$$-\frac{12\log(x+1)}{\log 2} - \frac{3(\log x)\log(x+1)}{\log 2}.$$

If $x \geq e^{12}$, we can replace $2\log(x+1)$ and $2\log(x+1)/\log 2$ by $(\log x)(\log(x+1))/\log 2$. □

3.4.18 *Find an explicit constant c_0 such that for $x \geq c_0$,*

$$\psi(x)-\psi\left(\frac{x}{2}\right) > \frac{(\log 2)x}{6} - 7.$$

Since $\log x < \log(x+1)$, we may write by the previous exercise

$$\psi(x)-\psi\left(\frac{x}{2}\right) \geq \frac{1}{3}(\log 2)x - \frac{5(\log(x+1))^2}{\log 2} - 7.$$

Now let $c = (\log 2)^2/30$, so that we have $(\log(x+1))^2 < cx$, provided that

$$1+x < 1+\frac{(cx)^{3/2}}{6}$$

or, equivalently, $x \geq 36/c^3$. This yields

$$\psi(x)-\psi\left(\frac{x}{2}\right) \geq \frac{1}{6}(\log 2)x - 7,$$

provided that $x \geq c_0 = 36/c^3$. □

3.4.19 *With c_0 as in the previous exercise, show that for $x \geq c_0$,*

$$\theta(x)-\theta\left(\frac{x}{2}\right) > \frac{(\log 2)x}{6} - \frac{\sqrt{x}(\log x)^2}{\log 2} - 7.$$

Let

$$\theta^*(x) = \sum_{\substack{p^\alpha \leq x \\ \alpha \geq 2}} \log p.$$

Then, by the previous exercise,

$$\theta(x) - \theta\left(\frac{x}{2}\right) + \theta^*(x) > \frac{(\log 2)x}{6} - 7$$

for $x > c_0$. Also,

$$\theta^*(x) \leq \sqrt{x}(\log x)^2/\log 2,$$

from which the result follows. □

3.4.20 *Find an explicit constant c_1 such that for $x \geq c_1$,*

$$\theta(x) - \theta\left(\frac{x}{2}\right) > \frac{(\log 2)x}{12} - 7.$$

We have

$$\frac{\sqrt{x}(\log x)^2}{\log 2} < \frac{(\log 2)x}{12}$$

iff $x < \exp\left((\log 2)x^{1/4}/\sqrt{12}\right)$. This is certainly the case if

$$x < \frac{1}{8!}\left(\frac{\log 2}{\sqrt{12}}x^{1/4}\right)^8,$$

or in other words, if $x \geq c_2 = 12^4 \cdot 8!/(\log 2)^8$. Therefore,

$$\theta(x) - \theta\left(\frac{x}{2}\right) > \frac{(\log 2)x}{12} - 7$$

if $x \geq \max(c_0, c_2)$, with c_0 as in Exercise 3.4.18. We set $c_1 = \max(c_0, c_2)$ to deduce the stated inequality. □.

3.4.21 *Find an explicit constant c_3 such that for $x \geq c_3$, $\theta(x) - \theta(x/2) \geq 1$. Deduce that for $x \geq c_3$, there is always a prime between $x/2$ and x.*

By the previous exercise, we may set $c_3 = \max(c_1, 96/\log 2)$ to deduce that $\theta(x) - \theta(x/2) \geq 1$ for $x \geq c_3$. □

3.4.22 *Let*

$$F(x) = \sum_{n \leq x} f\left(\frac{x}{n}\right)$$

be a function of bounded variation in every finite interval $[1, x]$. Suppose that as $x \to \infty$,

$$F(x) = x\log x + Cx + O(x^\beta)$$

with C, β constant and $0 \leq \beta < 1$. Show that if $M(x) := \sum_{n\leq x} \mu(n) = o(x)$ as $x \to \infty$, then

$$f(x) = x + o(x).$$

By replacing $f(x)$ by

$$f_0(x) = f(x) - x - (C - \gamma),$$

we find that

$$F_0(x) := \sum_{n\leq x} f_0\left(\frac{x}{n}\right)$$

satisfies $F_0(x) = O(x^\beta)$. It suffices to show that $f_0(x) = o(x)$. By Möbius inversion,

$$f_0(x) = \sum_{n\leq x} \mu(n) F_0\left(\frac{x}{n}\right).$$

It is clear that F_0 is also of bounded variation. We write

$$\begin{aligned} f_0(x) &= \sum_{n\leq \epsilon x} \mu(n) F_0\left(\frac{x}{n}\right) + \sum_{\epsilon x < n \leq x} \mu(n) F_0\left(\frac{x}{n}\right) \\ &= \sum\nolimits_1 + \sum\nolimits_2 \quad \text{(say).} \end{aligned}$$

We estimate $\sum_1$ trivially:

$$\left|\sum\nolimits_1\right| \ll \sum_{n\leq \epsilon x} \left(\frac{x}{n}\right)^\beta \ll x^\beta \int_1^{\epsilon x} t^{-\beta} dt,$$

which is $O(\epsilon^{1-\beta} x)$.

For $\sum_2$, we may write $F_0(x) = P(x) - Q(x)$ with P and Q positive monotonic increasing functions, since F_0 is of bounded variation. Thus

$$\begin{aligned} \sum\nolimits_2 &= \sum_{\epsilon x \leq n \leq x} \mu(n) F_0\left(\frac{x}{n}\right) \\ &= \sum_{\epsilon x \leq n \leq x} \mu(n) P\left(\frac{x}{n}\right) - \sum_{\epsilon x \leq n \leq x} \mu(n) Q\left(\frac{x}{n}\right). \end{aligned}$$

We estimate $\sum_{\epsilon x \leq n \leq x} \mu(n) P\left(\frac{x}{n}\right)$ as follows. By partial summation, for x a positive integer,

$$\sum_{\epsilon x \leq n x} \mu(n) P\left(\frac{x}{n}\right) = \sum_{\epsilon x \leq n \leq x} \{M(n) - M(n-1)\} P\left(\frac{x}{n}\right)$$

$$= M(x)P(1) + \sum_{\epsilon x \leq n \leq x-1} M(n)\left(P\left(\frac{x}{n}\right) - P\left(\frac{x}{n+1}\right)\right).$$

Thus,

$$\left| \sum_{\epsilon x < n \leq x} \mu(n) P\left(\frac{x}{n}\right) \right| \leq 2P\left(\frac{1}{\epsilon}\right) \max_{\epsilon x < n \leq x} |M(n)|,$$

and a similar estimate holds for the Q-term. For any fixed $\epsilon > 0$,

$$\lim_{x \to \infty} \max_{\epsilon x \leq n \leq x} \frac{|M(n)|}{x} \leq \lim_{x \to \infty} \max_{\epsilon x \leq n \leq x} \frac{|M(n)|}{n} = 0,$$

so that for x sufficiently large,

$$f_0(x) = o(x)$$

as required. □

3.4.23 *Assuming $M(x) = o(x)$ as in the previous exercise, deduce that*

$$\lim_{x \to \infty} \frac{\psi(x)}{x} = 1.$$

We know that

$$T(x) := \sum_{n \leq x} \log n = x \log x - x + O(\log x)$$

and

$$T(x) = \sum_{n \leq x} \psi\left(\frac{x}{n}\right).$$

We may apply the previous exercise with $c = -1$ and any $0 < \beta < 1$. We deduce that $\psi(x) = x + o(x)$, which is the prime number theorem. □

4

The Method of Contour Integration

4.1 Some Basic Integrals

4.1.1 *If $x > 1$, show that*

$$\frac{1}{2\pi i}\int_{(c)} \frac{x^s}{s} ds = 1$$

for any $c > 0$.

Consider the integral

$$\frac{1}{2\pi i}\int_{c-iR}^{c+iR} \frac{x^s}{s} ds,$$

with $R > c$, and the contour ζ_R described by the line segment joining $c - iR$ to $c + iR$ and the semicircle S_R of radius R centered at c and enclosing the origin. By Cauchy's theorem,

$$\frac{1}{2\pi i}\int_{\zeta_R} \frac{x^s}{s} ds = \text{Res}_{s=0}\frac{x^s}{s} = 1.$$

Thus

$$\frac{1}{2\pi i}\int_{c-iR}^{c+iR} \frac{x^s}{s} ds + \frac{1}{2\pi i}\int_{S_R} \frac{x^s}{s} ds = 1.$$

The second integral satisfies

$$\left|\frac{1}{2\pi i}\int_{S_R}\frac{x^s}{s}ds\right| \ll \frac{x^c}{2\pi}\int_{\pi/2}^{3\pi/2} x^{R\cos\varphi}d\varphi.$$

Since $x>1$ and $\cos\phi$ is negative in $[\pi/2/3\pi/2]$, we see that $|x^{r\cos\phi}| \leq 1$. We split the integral by writing

$$\int_{\pi/2}^{3\pi/2} x^{R\cos\phi}d\phi \leq \int_{\pi/2}^{\pi/2+\delta} x^{R\cos\phi}d\phi + \int_{\pi/2+\delta}^{3\pi/2-\delta} x^{R\cos\phi}d\phi$$

$$+\int_{3\pi/2-\delta}^{3\pi/2} x^{R\cos\phi}d\phi,$$

for some arbitrarily small $\delta > 0$. The first and last integrals are bounded by δ. The middle integral is bounded by

$$\pi x^{-R\sin\delta}.$$

As R tends to infinity, the middle integral tends to zero. The other two integrals are bounded by δ and since δ can chosen to be arbitrarily small, the integral in question tends to zero as R tends to infinity. □

4.1.2 *If* $0 < x < 1$, *show that*

$$\frac{1}{2\pi i}\int_{(c)}\frac{x^s}{s}ds = 0, \qquad c > 0.$$

We proceed as in Exercise 4.1.1 and consider

$$\frac{1}{2\pi i}\int_{c-iR}^{c+iR}\frac{x^s}{s}ds.$$

However, the contour we choose will be $\mathcal{D}_R$ described as the line segment joining $c - iR$ to $c + iR$ and the semicircle s_R to the right of the line segment, of radius R, centered at c and **not** enclosing the origin.

By Cauchy's theorem,

$$\frac{1}{2\pi i}\int_{\mathcal{D}_R}\frac{x^s}{s}ds = 0.$$

Thus

$$\frac{1}{2\pi i}\int_{c-iR}^{c+iR}\frac{x^s}{s}ds + \frac{1}{2\pi i}\int_{s_R}\frac{x^s ds}{s} = 0.$$

The second integral can be estimated as before by

$$\left|\frac{1}{2\pi i}\int_{S_R}\frac{x^s}{s}ds\right| \ll \frac{x^c}{2\pi}\int_{\pi/2}^{-\pi/2} x^{R\cos\varphi}d\varphi.$$

The integral is easily estimated as in the previous exercise and so the integralgoes to zero as $R \to \infty$. □

4.1.3 *Show that*

$$\frac{1}{2\pi i}\int_{(c)}\frac{ds}{s} = \frac{1}{2}, \qquad c > 0.$$

We have

$$\begin{aligned}\frac{1}{2\pi i}\int_{c-iR}^{c+iR}\frac{ds}{s} &= \frac{1}{2\pi i}\int_{-R}^{R}\frac{i\,dt}{c+it}\\ &= \frac{1}{2\pi}\int_{-R}^{R}\frac{c-it}{c^2+t^2}dt.\end{aligned}$$

The imaginary part of the integral vanishes, since the range of integration varies from $-R$ to R. Thus, the integral is

$$\frac{c}{\pi}\int_0^R\frac{dt}{c^2+t^2} = \frac{1}{\pi}\int_0^{R/c}\frac{du}{1+u^2}.$$

The latter integral tends to $\arctan\infty = \pi/2$, so that the final result is $1/2$. □

4.1.5 *Let*

$$f(s) = \sum_{n=1}^{\infty}\frac{a_n}{n^s}$$

be a Dirichlet series absolutely convergent in $\mathrm{Re}(s) > c - \epsilon$. *Show that if* x *is not an integer,*

$$\sum_{n<x} a_n = \frac{1}{2\pi i}\int_{(c)} f(s)\frac{x^s}{s}ds.$$

(The integral is taken in the sense of Cauchy's principal value.)

We can integrate term by term in the expression

$$\frac{1}{2\pi i}\int_{(c)} f(s)\frac{x^s}{s}ds = \frac{1}{2\pi i}\int_{(c)}\sum_{n=1}^{\infty} a_n\left(\frac{x}{n}\right)^s\frac{ds}{s},$$

since the function $f(s)$ is uniformly convergent in this half-plane. By Theorem 4.1.4, we get, letting $T \to \infty$,

$$\sum_{n=1}^{\infty} a_n\delta\left(\frac{x}{n}\right) = \sum_{n<x} a_n.$$

□

4.1.6 *Prove that for* $c > 0$,

$$\frac{1}{2\pi i}\int_{(c)}\frac{x^s}{s^{k+1}}ds = \begin{cases} \frac{1}{k!}(\log x)^k & \text{if} \quad x \geq 1, \\ 0 & \text{if} \quad x \leq 1, \end{cases}$$

for every integer $k \geq 1$.

When $x \geq 1$, we choose our contour ζ_R as in Exercise 4.1.1. By Cauchy's theorem,

$$\frac{1}{2\pi i}\int_{\zeta_R}\frac{x^s}{s^{k+1}}ds = \text{Res}_{s=0}\frac{x^s}{s^{k+1}} = \frac{1}{k!}(\log x)^k.$$

Thus

$$\frac{1}{2\pi i}\int_{c-iR}^{c+iR}\frac{x^s}{s^{k+1}}ds + \frac{1}{2\pi i}\int_{S_R}\frac{x^s}{s^{k+1}}ds = \frac{1}{k!}(\log x)^k.$$

The second integral is bounded by

$$\frac{x^c}{2R^{k+1}},$$

which goes to zero as $R \to \infty$.

If $x < 1$, we choose our contour $\mathcal{D}_R$ as in Exercise 4.1.2. By Cauchy's theorem,

$$\frac{1}{2\pi i}\int_{\mathcal{D}_R}\frac{x^s}{s^{k+1}}ds = 0.$$

Thus

$$\frac{1}{2\pi i}\int_{c-iR}^{c+iR}\frac{x^s}{s^{k+1}}ds + \frac{1}{2\pi i}\int_{S_R}\frac{x^s}{s^{k+1}}ds = 0.$$

The second integral is easily estimated by

$$\frac{x^c}{2R^{k+1}},$$

which goes to zero as $R \to \infty$. □

4.1.7 *Let*

$$f(s) = \sum_{n=1}^{\infty} \frac{a_n}{n^s}$$

be a Dirichlet series absolutely convergent in $\text{Re}(s) > c - \epsilon$. *For* $k \geq 1$, *show that*

$$\frac{1}{k!} \sum_{n \leq x} a_n \left(\log \frac{x}{n} \right)^k = \frac{1}{2\pi i} \int_{(c)} f(s) \frac{x^s}{s^{k+1}} ds.$$

This is straightforward from the previous exercise. The proof is analogous to that of Exercise 4.1.5. □

4.1.8 *If* k *is any positive integer,* $c > 0$, *show that*

$$\frac{1}{2\pi i} \int_{(c)} \frac{x^s ds}{s(s+1)\cdots(s+k)} = \begin{cases} \frac{1}{k!}\left(1 - \frac{1}{x}\right)^k & \text{if } \ x \geq 1, \\ 0 & \text{if } \ 0 \leq x \leq 1 \end{cases}$$

The method is identical to that of the previous exercises. If $x \geq 1$, we choose our contour as in Exercise 4.1.1. We choose $R > 2k$ such that by Cauchy's theorem,

$$\frac{1}{2\pi i} \int_{c-iR}^{c+iR} \frac{x^s ds}{s(s+1)\cdots(s+k)} + \frac{1}{2\pi i} \int_{S_R} \frac{x^s ds}{s(s+1)\cdots(s+k)}$$

$$= \sum_{j=0}^{k} \text{Res}_{s=-j} \frac{x^s}{s(s+1)\cdots(s+k)}.$$

The residues are easily calculated:

$$\text{Res}_{s=-j} \frac{x^s}{s(s+1)\cdots(s+k)} = \frac{x^{-j}}{(-j)(-j+1)\cdots(-1)(1)(2)\cdots(k-j)},$$

which is

$$\frac{(-1)^j x^{-j}}{j!(k-j)!},$$

and the sum of these residues is

$$\frac{1}{k!}\sum_{j=0}^{k}\binom{k}{j}\left(-\frac{1}{x}\right)^j = \frac{1}{k!}\left(1-\frac{1}{x}\right)^k.$$

On S_R we obtain

$$|s+j| \geq R-k \geq \frac{R}{2},$$

and hence

$$\frac{1}{2\pi i}\int_{S_R}\frac{x^s ds}{s(s+1)\cdots(s+k)} = O\left(\frac{x^c}{R^k}\right),$$

which goes to zero as $R \to \infty$. In the case $0 \leq x \leq 1$, we use the contour as in Exercise 4.1.2, and since the integrand is analytic inside this contour, Cauchy's theorem gives

$$\frac{1}{2\pi i}\int_{c-iR}^{c+iR}\frac{x^s ds}{s(s+1)\cdots(s+k)} = -\frac{1}{2\pi i}\int_{S_R}\frac{x^s ds}{s(s+1)\cdots(s+k)}$$

with $R > 2k$, as before. The integral on the right is

$$O\left(\frac{x^c}{R^k}\right),$$

which tends to zero as $R \to \infty$. □

4.1.9 *Let*

$$f(s) = \sum_{n=1}^{\infty}\frac{a_n}{n^s}$$

be a Dirichlet series absolutely convergent in $\mathrm{Re}(s) > c - \epsilon$. *Show that*

$$\frac{1}{x^k}\sum_{n\leq x} a_n(x-n)^k = \frac{k!}{2\pi i}\int_{c-i\infty}^{c+i\infty}\frac{f(s)x^s ds}{s(s+1)\cdots(s+k)}$$

for any $k \geq 1$.

Substituting the Dirichlet series for $f(s)$ in the expression

$$\frac{1}{2\pi i}\int_{(c)}\frac{f(s)x^s ds}{s(s+1)\cdots(s+k)}$$

and integrating term by term using the previous exercises, we obtain the result. □

4.2 The Prime Number Theorem

4.2.1 *Using the Euler - Maclaurin summation formula (Theorem 2.1.9), prove that for $\sigma = \mathrm{Re}(s) > 0$,*

$$\zeta(s) = \sum_{m=1}^{n-1} \frac{1}{m^s} + \frac{n^{-s}}{2} + \frac{n^{1-s}}{s-1} - s\int_n^\infty \frac{x - [x] - \frac{1}{2}}{x^{s+1}} dx$$

where $[x]$ denotes the greatest integer function.

In Theorem 2.1.9, we take $f(t) = 1/t^s$ and $k = 1$ to get

$$\sum_{m=n}^{B} \frac{1}{m^s} = \int_n^B \frac{dt}{t^s} + \frac{1}{2}\left(\frac{1}{n^s} + \frac{1}{B^s}\right) - s\int_n^B \frac{x - [x] - 1/2}{x^{s+1}} dx.$$

Let $B \to \infty$. Then,

$$\sum_{m=n}^{\infty} \frac{1}{m^s} = \frac{1}{2n^s} - \frac{n^{1-s}}{1-s} - s\int_n^\infty (x - [x] - \frac{1}{2})\frac{dx}{x^{s+1}}.$$

Thus,

$$\zeta(s) = \sum_{m=1}^{n-1} \frac{1}{m^s} + \frac{1}{2n^s} - \frac{n^{1-s}}{1-s} - s\int_n^\infty \frac{x - [x] - \frac{1}{2}}{x^{s+1}} dx$$

as desired. □

4.2.2 *Using the previous exercise, show that*

$$\zeta(s) - \frac{1}{s-1} = O(\log T)$$

for $s \in R_T$.

We have

$$\zeta(s) - \frac{1}{s-1} = \sum_{m=1}^{n-1} \frac{1}{m^s} + \frac{1}{2n^s} + \frac{n^{1-s} - 1}{s-1} - s\int_n^\infty \frac{(x - [x] - 1/2)dx}{x^{s+1}},$$

and we observe that writing $s = \sigma + it$,

$$\left|\zeta(s) - \frac{1}{s-1}\right| \leq \sum_{m=1}^{n-1} \frac{1}{m^\sigma} + \frac{1}{2n^\sigma} + \int_1^n x^{-\sigma} dx + \frac{|s|}{2}\int_n^\infty \frac{dx}{x^{\sigma+1}},$$

since

$$\frac{n^{1-s}-1}{s-1} = -\int_1^n \frac{dx}{x^s}$$

and $|x - [x] - \frac{1}{2}| \leq \frac{1}{2}$. Also, by the integral test,

$$\sum_{m=1}^{n-1} \frac{1}{m^\sigma} < 1 + \int_1^n \frac{dx}{x^\sigma},$$

which gives an estimate of

$$\begin{aligned} 1 + 2\int_1^n \frac{dx}{x^\sigma} + \frac{|s|}{2}\int_n^\infty \frac{dx}{x^{\sigma+1}} &\leq 1 + \frac{2(n^{1-\sigma}-1)}{1-\sigma} + \frac{|s|n^{-\sigma}}{2\sigma} \\ &\leq \frac{2n^{1-\sigma}}{1-\sigma} + \frac{|s|n^{-\sigma}}{2\sigma}. \end{aligned}$$

We are free to choose n optimally to minimize this quantity. Let $n = [T]$. In R_T, $|s| < 2 + T$ and for $\sigma > 1/2$,

$$\frac{|s|}{2\sigma} < \frac{2+T}{2\sigma} < 2 + T,$$

which leads to a final estimate of

$$T^{1-\sigma}\left(\frac{2}{1-\sigma} + \frac{2+T}{T}\right).$$

Since

$$\sigma \geq \sigma_0 = 1 - \frac{1}{\log T},$$

we have $1 - \sigma \leq 1/\log T$, from which we get from above

$$\left|\zeta(s) - \frac{1}{s-1}\right| \leq 1 + 2\int_1^T \frac{dt}{t^\sigma} + \frac{|s|}{2}\int_T^\infty \frac{dt}{t^{\sigma+1}}.$$

By monotonicity, we get

$$\begin{aligned} \left|\zeta(s) - \frac{1}{s-1}\right| &\leq 1 + 2\int_1^T \frac{dt}{t^{\sigma_0}} + |s|\int_T^\infty \frac{dt}{t^{\sigma_0+1}} \\ &\ll \log T, \end{aligned}$$

for $s \in R_T$, as required. □

4.2.3 *Show that*

$$\zeta(s) = O(\log T)$$

for s on the boundary of R_T.

Since

$$\left|\frac{1}{s-1}\right| = \left|\frac{1}{\sigma - 1 + iT}\right| \leq \min\left(\frac{1}{\sigma-1}, \frac{1}{T}\right)$$

and $\sigma \geq 1 - 1/\log T$ for s on the boundary of R_T, we get the desired result. □

4.2.4 *Show that for $\sigma \geq 1/2$, $\zeta(s) = O(T^{1/2})$, where $T = |\operatorname{Im}(s)| \to \infty$.*

By Exercise 4.2.1, we get with $n = [T]$,

$$|\zeta(s)| \leq O\left(T^{1/2}\right)$$

by an easy estimation of the quantities in that formula. □

4.2.5 *For $s \in R_T$, show that*

$$\zeta'(s) + \frac{1}{(s-1)^2} = O(\log^2 T).$$

We use Exercise 4.2.1 again and differentiate the formula there with respect to s. Thus

$$\begin{aligned}\zeta'(s) + \frac{1}{(s-1)^2} &= -\sum_{m=1}^{n-1} \frac{\log m}{m^s} - \frac{\log n}{2n^s} + \int_1^n \frac{(\log x)dx}{x^s} \\ &\quad - \int_n^\infty \frac{x - [x] - 1/2}{x^{s+1}} dx \\ &\quad + s\int_n^\infty \frac{x - [x] - 1/2}{x^{s+1}} (\log x) dx.\end{aligned}$$

Estimating all of the terms on the right-hand side as in Exercise 4.2.2, we get with $n = [T]$ the desired estimate. □

4.2.6 *Show that*

$$\zeta'(s) = O(\log^2 T),$$

where $T = |\operatorname{Im}(s)|$ and s is on the boundary of R_T.

We proceed as before:

$$\sigma \geq 1 - \frac{1}{\log T},$$

which implies $(1-\sigma)^{-2} \leq \log^2 T$, and this gives us the stated result. □

4.2.8 *Let $s = \sigma + it$, with $1 \leq |t| \leq T$. There is a constant $c > 0$ such that*

$$\frac{\zeta'(s)}{\zeta(s)} = O(\log^9 T)$$

for

$$1 - \frac{c}{(\log T)^9} \leq \sigma \leq 2.$$

Combining Theorem 4.2.7 and Exercise 4.2.5 gives the result. □

4.3 Further Examples

4.3.2 *Suppose that for any $\epsilon \geq 0$, we have $a_n = O(n^\epsilon)$. Prove that for any $c > 1$, and x not an integer,*

$$\sum_{n \leq x} a_n = \frac{1}{2\pi i} \int_{c-iR}^{c+iR} \frac{f(s)x^s}{s} ds + O\Big(\frac{x^{c+\epsilon}}{R}\Big) + O\Big(\frac{x^\epsilon \log x}{R}\Big),$$

where

$$f(s) = \sum_{n=1}^{\infty} \frac{a_n}{n^s}.$$

By Theorem 4.1.4, we have

$$\frac{1}{2\pi i} \int_{c-iR}^{c+iR} \frac{f(s)x^s}{s} ds = \sum_{n \leq x} a_n + O\Big(\sum_{\substack{n=1 \\ n \neq x}}^{\infty} |a_n| \Big(\frac{x}{n}\Big)^c \min\Big(1, \frac{1}{R|\log \frac{x}{n}|}\Big)\Big).$$

The analysis of the error term is handled as in the proof of Theorem 4.2.9. We split the sum into three parts: $n < x/2$, $x/2 < n < 2x$, and $n > 2x$. For the first and last parts $|\log x/n| \geq \log 2$, so that

$$\sum_{n \leq x/2} |a_n| \Big(\frac{x}{n}\Big)^c \ll x^{c+\epsilon}.$$

Also,

$$\sum_{n \geq 2x} |a_n| \left(\frac{x}{n}\right)^c \ll x^c \sum_{n \geq 2x} \frac{1}{n^{c-\epsilon}} \ll x^{1+\epsilon}.$$

Finally, for the middle part (x/n) is bounded so that

$$\sum_{x/2 \leq n \leq 2x} |a_n| \min\left(1, \frac{1}{R|\log \frac{x}{n}|}\right) \ll \frac{x^\epsilon}{R} \log x.$$

Putting all this together gives the desired result. □

4.3.3 *Assuming the Lindelöf hypothesis, prove that for any* $\epsilon > 0$,

$$\sum_{n \leq x} d_k(n) = x P_{k-1}(\log x) + O(x^{1/2+\epsilon}),$$

where $d_k(n)$ *denotes the number of ways of writing* n *as a product of* k *natural numbers.*

By Exercise 1.5.5, we know that

$$\zeta^k(s) = \sum_{n=1}^{\infty} \frac{d_k(n)}{n^s}.$$

By Exercise 1.3.2 and the fact that $d_k(n) \leq d(n)^k$, we see that $d_k(n) = O(n^\epsilon)$ for any $\epsilon > 0$. Applying the previous exercise, we obtain

$$\sum_{n \leq x} d_k(n) = \frac{1}{2\pi i} \int_{c-iR}^{c+iR} \frac{\zeta^k(s) x^s}{s} ds + O\Big(\frac{x^{c+\epsilon}}{R}\Big)$$

for any $c > 1$. If C is the rectangular contour joining $c - iR$, $c + iR$, $\frac{1}{2} + iR$, $\frac{1}{2} - iR$, we have by Cauchy's theorem

$$\frac{1}{2\pi i} \int_C \frac{\zeta^k(s)}{s} x^s ds = \text{Res}_{s=1} \frac{\zeta^k(s) x^s}{s} = x P_{k-1}(\log x)$$

for some polynomial P_{k-1} of degree $k - 1$. Also

$$\frac{1}{2\pi i} \int_C \frac{\zeta^k(s) x^s}{s} ds =$$

$$\frac{1}{2\pi i} \int_{c-iR}^{c+iR} \frac{\zeta^k(s) x^s}{s} ds + O\Big(\frac{x^c}{R^{1-\epsilon} \log x}\Big) + O\Big(x^{1/2} R^\epsilon\Big)$$

where the horizontal and vertical integrals in the contour have been estimated using $\zeta(s) = O(t^\epsilon)$. Choosing $R = x$ and $c = 1 + 1/\log x$ gives the desired result. $\square$

4.3.4 *Show that*

$$M(x) := \sum_{n \leq x} \mu(n) = O(x \exp(-c(\log x)^{1/10}))$$

for some positive constant c.

By Exercise 4.3.2 with $\epsilon = 0$,

$$\sum_{n \leq x} \mu(n) = \frac{1}{2\pi i} \int_{c-iR}^{c+iR} \frac{x^s ds}{s\zeta(s)} + O\Big(\frac{x^{c+\epsilon}}{R}\Big).$$

By Theorem 4.2.7, $|1/\zeta(s)| = O(\log^7 R)$ for $1 \leq |\mathrm{Im}(s)| \leq R$ and

$$\delta = 1 - \frac{c_1}{\log^9 R} \leq \sigma \leq 2.$$

We choose C to be the rectangular contour joining $c - iR$, $c + iR$, $\delta + iR$ and $\delta - iR$. Then, by Cauchy's theorem,

$$\frac{1}{2\pi i} \int_C \frac{x^s ds}{s\zeta(s)} = 0.$$

Therefore,

$$\frac{1}{2\pi i} \int_{c-iR}^{c+iR} \frac{x^s ds}{s\zeta(s)} = -\frac{1}{2\pi i} \left(\int_{c+iR}^{\delta+iR} + \int_{\delta+iR}^{\delta-iR} + \int_{\delta-iR}^{c-iR} \right) \left(\frac{x^s ds}{s\zeta(s)} \right).$$

We use the estimate provided by Theorem 4.2.7 to estimate these integrals:

$$\frac{1}{2\pi i} \left(\int_{c+iR}^{\delta+iR} + \int_{\delta-iR}^{c-iR} \right) \left(\frac{x^s ds}{s\zeta(s)} \right) \ll \frac{x^c \log^7 R}{R \log x},$$

if $R \geq 1$. For the vertical integral, we can use the same technique to bound the integrand, observing that $1/\zeta(s)$ is regular at $s = 1$ and thus is bounded in $0 \leq |\mathrm{Im}(s)| \leq 1$, $\delta \leq \mathrm{Re}(s) \leq 2$. Therefore, the vertical integral is

$$\ll x^\delta \log^8 R.$$

Putting this all together gives

$$\sum_{n\le x}\mu(n) \ll \frac{x^{c+\epsilon}}{R} + \frac{x^c \log^7 R}{R\log x} + x^\delta \log^8 R.$$

Put $c = 1 + 1/\log x$. The optimal choice of R is obtained hy equating error terms. We choose $R = \exp(c_1(\log x)^{1/10})$ to get for some constant $c > 0$,

$$M(x) \ll x\exp\left(-c(\log x)^{1/10}\right)$$

as required. □

4.3.5 *Let $E(x)$ be the number of square-free $n \le x$ with an even number of prime factors. Prove that*

$$E(x) = \frac{3}{\pi^2}x + O\left(x\exp\left(-c(\log x)^{1/10}\right)\right)$$

for some constant $c > 0$.

The function $a_n = \mu^2(n)(1+\mu(n))/2$ is 1 if n is squarefree and has an even number of prime factors, and 0 otherwise. Thus,

$$E(x) = \sum_{n\le x} a_n = Q(x)/2 + O(M(x)),$$

where $Q(x)$ is the number of square-free numbers less than or equal to x. Now apply Exercise 1.4.4 to deduce the behavior of the main term. By the previous exercise, $M(x) = O\left(x\exp\left(-c(\log x)^{1/10}\right)\right)$, so that the result is now immediate. □

4.4 Supplementary Problems

4.4.1 *Let $\lambda(n)$ be the Liouville function defined by $\lambda(n) = (-1)^{\Omega(n)}$, where $\Omega(n)$ is the total number of prime factors of n, counted with multiplicity. Show that*

$$\sum_{n\le x}\lambda(n) = O\left(x\exp\left(-c(\log x)^{1/10}\right)\right)$$

for some constant $c > 0$.

Recall that (see Exercise 1.2.5)

$$\sum_{n=1}^{\infty} \frac{\lambda(n)}{n^s} = \frac{\zeta(2s)}{\zeta(s)}.$$

One can apply the method of contour integration as in Exercise 4.3.4 and derive the result. An alternative approach is to make use of Exercise 4.3.4 in the following way. We have from the above Dirichlet series

$$\lambda(n) = \sum_{d^2 e = n} \mu(e),$$

so that

$$\sum_{n \le x} \lambda(n) = \sum_{d^2 e \le x} \mu(e) = \sum_{d^2 \le x} M(x/d^2)$$

in the notation of Exercise 4.3.4. By that exercise, we have

$$M(x) = O\left(x \exp\left(-c(\log x)^{1/10}\right)\right)$$

for some constant $c > 0$. Inserting this estimate above gives

$$\begin{aligned} \sum_{n \le x} \lambda(n) &= \sum_{d \le \sqrt{x}} O\left(\frac{x}{d^2} \exp\left(-c\left(\log \frac{x}{d^2}\right)^{1/10}\right)\right) \\ &= \sum_{d \le x^{1/4}} + \sum_{x^{1/4} < d \le x^{1/2}} \qquad \text{(say)}. \end{aligned}$$

The first sum is easily seen to be

$$O\left(x \exp\left(-c_1(\log x)^{1/10}\right)\right)$$

for some constant $c_1 > 0$. The second sum is bounded by

$$\sum_{d > x^{1/4}} \frac{x}{d^2} \ll x^{3/4},$$

and this completes the proof. □

4.4.2 *Show that*

$$\sum_{n=1}^{\infty} \frac{\mu(n)}{n^s}$$

converges for every s with $\text{Re}(s) = 1$.

Let $s = 1 + it$. By partial summation

$$\sum_{n \le N} \mu(n) n^{-1-it} = \frac{M(N)}{N^{1+it}} + (1+it) \int_1^N M(w) w^{-2-it} dw.$$

The first term on the right-hand side is, by Exercise 4.3.4,

$$O\left(\exp\left(-c(\log N)^{1/10}\right)\right).$$

The second term can be written as

$$(1+it)\int_1^\infty M(w) w^{-2-it} dw - (1+it)\int_N^\infty M(w) w^{-2-it} dw.$$

Since $M(w) = O(w/\log^2 w)$, the first integral above converges to a limit L (say). The second integral is bounded by

$$\ll \int_N^\infty \exp(-c(\log w)^{1/10}) dw/w$$

$$\ll \exp(-\frac{c}{2}(\log N)^{1/10}) \int_N^\infty \exp\left(-\frac{c}{2}(\log w)^{1/10}\right) \frac{dw}{w},$$

which is

$$O\left(\exp\left(-\frac{c}{2}(\log N)^{1/10}\right)\right),$$

since the integral converges. Letting $N \to \infty$ shows that the series converges to L. □

4.4.3 *Show that*

$$\sum_{n \le x} \frac{\Lambda(n)}{n} = \log x + B + O\left(\exp\left(-c(\log x)^{1/10}\right)\right)$$

for some constants B and c, with $c > 0$. The summation is over prime numbers. (This improves upon Exercise 3.1.7.)

We have

$$\begin{aligned}\sum_{n \le x} \frac{\Lambda(n)}{n} &= \frac{\psi(x)}{x} + \int_1^x \frac{\psi(t)dt}{t^2} \\ &= 1 + O(\exp(-c(\log x)^{1/10})) + \log x \\ &\quad + \int_1^x O(t^{-1}\exp(-c(\log t)^{1/10}))dt\end{aligned}$$

The integral is easily seen to converge. Accordingly, we split the integral into two parts as

$$\int_1^\infty - \int_x^\infty$$

and estimate the second integral as in the previous exercise. This shows that

$$\sum_{n \leq x} \frac{\Lambda(n)}{n} = \log x + B_1 + O\left(\exp\left(-c_1(\log x)^{1/10}\right)\right)$$

for some constants B_1, c_1 with $c_1 > 0$, as desired. □

4.4.4 *Let* $f(s) = \sum_{n=1}^\infty A_n/n^s$ *be a Dirichlet series absolutely convergent for* $\mathrm{Re}(s) > 1$. *Show that for any* $c > 1$,

$$\sum_{n \leq x} A_n = O(x^c).$$

We have

$$\left|\sum_{n \leq x} A_n\right| \leq \sum_{n \leq x} |A_n| \leq \sum_{n \leq x} |A_n| \left(\frac{x}{n}\right)^c \ll x^c,$$

as required. □

4.4.5 *Define* a_n *for* $n \geq 1$ *by*

$$\sum_{n=1}^\infty \frac{a_n}{n^s} = \frac{1}{\zeta^2(s)}.$$

Prove that

$$\sum_{n \leq x} a_n = O\left(x \exp\left(-c(\log x)^{1/10}\right)\right)$$

for some positive constant c.

We have

$$a_n = \sum_{de=n} \mu(d)\mu(e).$$

Applying Dirichlet's hyperbola method (Theorem 2.4.1), we have

$$\sum_{n \leq x} a_n = \sum_{d \leq y} \mu(d) M\left(\frac{x}{d}\right) + \sum_{d \leq x/y} \mu(d) M\left(\frac{x}{d}\right) - M(y) M\left(\frac{x}{y}\right).$$

We choose $y = \sqrt{x}$ and apply Exercise 4.3.4 to get

$$\sum_{n\leq x} a_n = O\left(\sum_{d\leq\sqrt{x}} \frac{x}{d} \exp\left(-c\left(\log\frac{x}{d}\right)^{1/10}\right)\right)$$

$$= O\left(x \exp\left(-c_1(\log x)^{1/10}\right)\right)$$

for some positive constant c_1 as required. □

4.4.6 *Prove that*

$$\sum_{n\leq x} \mu(n)d(n) = O\left(x \exp\left(-c(\log x)^{1/10}\right)\right)$$

for some constant $c > 0$.

We have

$$f(s) := \sum_{n=1}^{\infty} \frac{\mu(n)d(n)}{n^s} = \prod_p \left(1 - \frac{2}{p^s}\right).$$

We may write

$$\left(1 - \frac{2}{p^s}\right) = \left(1 - \frac{1}{p^s}\right)^2 \left(1 - \frac{1}{p^{2s}}\left(1 - \frac{1}{p^s}\right)^{-2}\right),$$

so that

$$f(s) = \frac{g(s)}{\zeta^2(s)},$$

where $g(s)$ is a Dirichlet series absolutely convergent for $\text{Re}(s) > 1/2$. Writing

$$g(s) = \sum_{n=1}^{\infty} \frac{b_n}{n^s},$$

let us note that

$$\mu(n)d(n) = \sum_{de=n} a_d b_e,$$

where a_n is as in the previous exercise. Applying Dirichlet's hyperbola method with

$$A(x) := \sum_{n\leq x} a_n,$$

$$B(x) := \sum_{n \le x} b_n,$$

we get

$$\sum_{n \le x} \mu(n) d(n) = \sum_{d \le y} b_d A\Big(\frac{x}{d}\Big) + \sum_{d \le x/y} a_d B\Big(\frac{x}{d}\Big) - A(y) B\Big(\frac{x}{y}\Big).$$

We choose $y = \sqrt{x}$ and note that for some positive constants c,

$$A(x) = O\left(x \exp\left(-c(\log x)^{1/10}\right)\right)$$

(by Exercise 4.4.5). Also, by Exercise 4.4.4, $B(x) = O\left(x^{1/2+\epsilon}\right)$. Thus

$$\sum_{n \le x} \mu(n) d(n)$$

$$\ll x \sum_{d \le \sqrt{x}} \frac{|b_d|}{d} \exp(-c(\log x)^{1/10}) + \sum_{d \le \sqrt{x}} \frac{|a_d|}{d^{1/2+\epsilon}} x^{1/2+\epsilon} + O(x^{3/4+\epsilon}).$$

The series

$$\sum_d \frac{|b_d|}{d}$$

is finite, and $a_n = O(n^\epsilon)$, and the second sum is

$$O\left(x^{3/4+\epsilon}\right).$$

The final contribution is

$$O\left(x \exp\left(-c(\log x)^{1/10}\right)\right)$$

as required. □

4.4.7 *If* $f(s) = \sum_{n=1}^\infty a_n/n^s$ *is a Dirichlet series converging absolutely for* $\sigma = \mathrm{Re}(s) = \sigma_a$, *show that*

$$\lim_{T \to \infty} \frac{1}{2T} \int_{-T}^{T} f(\sigma + it) m^{\sigma+it} dt = a_m.$$

We have

$$\frac{1}{2T} \int_{-T}^{T} \Big(\sum_{n=1}^\infty \frac{a_n}{n^{\sigma+it}}\Big) m^{\sigma+it} dt = \frac{m^\sigma}{2T} \sum_{n=1}^\infty \frac{a_n}{n^\sigma} \int_{-T}^{T} \Big(\frac{m}{n}\Big)^{it} dt.$$

Now,

$$\int_{-T}^{T}\left(\frac{m}{n}\right)^{it}dt = \begin{cases} 2T & \text{if} \quad m = n, \\ \frac{2\sin(T\log m/n)}{\log m/n} & \text{otherwise.} \end{cases}$$

The sum

$$\sum_{\substack{n=1 \\ n\neq m}} \frac{|a_n|}{n^{\sigma}|\log m/n|}$$

converges. Indeed, if $n < m/2$ or $n > 2m$, this is clear, since $|\log m/n|$ is then bounded. If $m/2 < n < 2m$, then the finite sum is clearly bounded. The result is now immediate. □

4.4.8 *Suppose*

$$f(s) := \sum_{n=1}^{\infty} a_n/n^s,$$

$$g(s) := \sum_{n=1}^{\infty} b_n/n^s,$$

and $f(s) = g(s)$ *in a half-plane of absolute convergence. Then* $a_n = b_n$ *for all* n.

We apply the previous exercise:

$$\begin{aligned} a_m &= \lim_{T\to\infty}\frac{1}{2T}\int_{-T}^{T} f(\sigma+it)m^{\sigma+it}dt \\ &= \lim_{T\to\infty}\frac{1}{2T}\int_{-T}^{T} g(\sigma+it)m^{\sigma+it}dt = b_m. \end{aligned}$$

□

4.4.9 *If*

$$f(s) = \sum_{n=1}^{\infty} a_n/n^s$$

converges absolutely for $\sigma = \mathrm{Re}(s) > \sigma_a$, *show that*

$$\lim_{T\to\infty}\frac{1}{2T}\int_{-T}^{T}|f(\sigma+it)|^2dt = \sum_{n=1}^{\infty}\frac{|a_n|^2}{n^{2\sigma}}.$$

We have

$$|f(\sigma+it)|^2 = \sum_{m,n} \frac{a_m \overline{a_n}}{m^{\sigma+it} n^{\sigma-it}}$$

$$= \sum_{n=1}^{\infty} \frac{|a_n|^2}{n^{2\sigma}} + \sum_{m \neq n} \frac{a_m \overline{a_n}}{m^\sigma n^\sigma} \left(\frac{n}{m}\right)^{it},$$

so that

$$\frac{1}{2T} \int_{-T}^{T} |f(\sigma+it)|^2 dt = \sum_{n=1}^{\infty} \frac{|a_n|^2}{n^{2\sigma}} + \sum_{m \neq n} \frac{a_m \overline{a_n}}{m^\sigma n^\sigma} \frac{2\sin(T\log n/m)}{2T(\log n/m)}.$$

The double series is analyzed as before. For fixed m, the ranges $n < m/2$ and $n > 2m$ are easily handled, and the remaining range is finite. Thus, for fixed m, the summation over n is bounded. The summation over m is also bounded, since $\sigma > \sigma_a$. Thus the double sum is $O(1/T)$ and the result follows. □

4.4.10 *Let $Q(x)$ be the number of squarefree numbers less than or equal to x. Show that*

$$Q(x) = \frac{x}{\zeta(2)} + O\left(x^{1/2} \exp\left(-c(\log x)^{1/10}\right)\right)$$

for some positive constant c.

We have

$$Q(x) = \sum_{d^2 e \leq x} \mu(d) = \sum_{d^2 \leq x} \mu(d) \left[\frac{x}{d^2}\right]$$

as in Exercise 1.1.9. Writing $[x/d^2] = x/d^2 + E(x,d)$, we observe that $|E(x,d)| \leq 1$. Now,

$$Q(x) = \sum_{d \leq \sqrt{x}} \mu(d) \frac{x}{d^2} + \sum_{d \leq \sqrt{x}} \mu(d) E(x,d).$$

Let us analyze the first term. We have

$$\sum_{d \leq \sqrt{x}} \frac{\mu(d)}{d^2} = \sum_{d=1}^{\infty} \frac{\mu(d)}{d^2} - \sum_{d > \sqrt{x}} \frac{\mu(d)}{d^2}$$

$$= \frac{1}{\zeta(2)} + O\left(\int_{\sqrt{x}}^{\infty} \frac{M(t)}{t^3} dt\right).$$

By Exercise 4.3.4,

$$M(x) = O\left(x \exp\left(-c(\log x)^{1/10}\right)\right),$$

so that

$$\int_{\sqrt{x}}^{\infty} \frac{M(t)dt}{t^3} \ll x^{-1/2} \exp\left(-c(\log x)^{1/10}\right).$$

For the second term, we write

$$\sum_{d \leq \sqrt{x}} \mu(d) E(x,d) = \sum_{d \leq \sqrt{x}} (M(d) - M(d-1)) E(x,d)$$

$$= M([\sqrt{x}]) E(x, [\sqrt{x}]) + \sum_{d \leq \sqrt{x}-1} M(d) \big\{E(x,d) - E(x,d+1)\big\}.$$

Using the estimate for $M(x)$ and the fact $|E(x,d)| \leq 1$ gives the result. $\square$

4.4.11 *Let $\gamma(n) = \prod_{p|n} p$. Show that*

$$\sum_{n \leq x} \frac{1}{n\gamma(n)} < \infty.$$

Clearly, $\gamma(n)$ is multiplicative. Also,

$$\sum_{n \leq x} \frac{1}{n\gamma(n)} \ll \prod_{p \leq x} \left(1 + \frac{1}{p^2} + \frac{1}{p^3} + \cdots\right),$$

from which the result follows. $\square$

4.4.12 *Show that*

$$\sum_{n \leq x} \frac{n}{\phi(n)} \ll x.$$

We have

$$\sum_{n \leq x} \frac{n}{\phi(n)} = \sum_{n \leq x} \prod_{p|n} \left(1 - \frac{1}{p}\right)^{-1} = \sum_{n \leq x} \sum_{\gamma(d)|n} \frac{1}{d}$$

$$\leq \sum_{\gamma(d) \leq x} \frac{1}{d} \cdot \frac{x}{\gamma(d)} \leq \sum_{d} \frac{x}{d\gamma(d)} \ll x,$$

by the previous exercise. □

4.4.13 *Deduce by partial summation from the previous exercise that*

$$\sum_{n \leq x} \frac{1}{\phi(n)} \ll \log x.$$

By partial summation,

$$\sum_{n \leq x} \frac{n}{\phi(n)} \cdot \frac{1}{n} \ll \int_1^x \frac{dt}{t} \ll \log x$$

as desired. □

4.4.14 *Prove that*

$$\sum_{n \leq x} \frac{1}{\phi(n)} \sim c \log x$$

for some positive constant c.

We consider the Dirichlet series

$$f(s) = \sum_{n \leq x} \frac{1}{\phi(n)} \cdot \frac{1}{n^s} = \prod_p \left(1 + \frac{1}{p^s(p-1)} + \frac{1}{p^{2s+1}(p-1)} + \cdots\right)$$
$$= \prod_p \left(1 + \frac{p}{(p-1)(p^{s+1}-1)}\right).$$

The quotient $f(s)/\zeta(s+1)$ is easily seen to be regular for $\mathrm{Re}\, s > -1$, simply by long division of the Euler factors. We may write

$$f(s) = \zeta(s+1)h(s),$$

so that

$$f(s-1) = \sum_{n=1}^{\infty} \frac{n}{\phi(n)} \cdot \frac{1}{n^s}$$

is $\zeta(s)h(s-1)$, with $h(s-1)$ regular for $\mathrm{Re} s > 0$. We therefore have by contour integration (or by an application of a Tauberian theorem) that

$$\sum_{n \leq x} \frac{n}{\phi(n)} \sim c_1 x.$$

By partial summaton, we can deduce the desired result. □

4.4.15 (Perron's formula) *Let* $f(s) = \sum_{n=1}^{\infty} a_n/n^s$ *be a Dirichlet series absolutely convergent for* $\mathrm{Re}(s) > 1$. *Show that for* x *not an integer and* $\sigma > 1$,

$$\sum_{n \leq x} a_n = \frac{1}{2\pi i}\int_{\sigma-iT}^{\sigma+iT} f(s)\frac{x^s}{s}ds + O\Big(\sum_{n=1}^{\infty}\left(\frac{x}{n}\right)^{\sigma}|a_n|\min\left(1, \frac{1}{T|\log\frac{x}{n}|}\right)\Big).$$

This is just a straightforward application of Theorem 4.1.4. □

4.4.16 *Suppose* $a_n = O(n^{\epsilon})$ *for any* $\epsilon > 0$ *in the previous exercise. Show that for* x *not an integer,*

$$\sum_{n \leq x} a_n = \frac{1}{2\pi i}\int_{\sigma-iT}^{\sigma+iT} f(s)\frac{x^s}{s}ds + O\Big(\frac{x^{\sigma+\epsilon}}{T}\Big).$$

We estimate the error term in the previous exercise as in Theorem 4.2.9 for $n < x/2$ or $n > 3x/2$. In these cases, the log term is bounded absolutely from below. The series

$$\sum_{n=1}^{\infty}\frac{|a_n|}{n^{\sigma}}$$

converges. For $x/2 < n < 3x/2$, we have $|a_n| = O(x^{\epsilon})$, and we use this in the estimate. The log term for this range of n is handled as in the proof of Theorem 4.2.9. □

4.4.17 *Let* $f(s) = \sum_{n=1}^{\infty} a_n/n^s$, *with* $a_n = O(n^{\epsilon})$. *Suppose that*

$$f(s) = \zeta(s)^k g(s),$$

where k *is a natural number and* $g(s)$ *is a Dirichlet series absolutely convergent in* $\mathrm{Re}(s) > 1 - \delta$ *for some* $0 < \delta < 1$.

Show that

$$\sum_{n \leq x} a_n \sim g(1)x(\log x)^{k-1}/(k-1)!$$

as $x \to \infty$.

By the previous exercise,

$$\sum_{n \leq x} a_n = \frac{1}{2\pi i}\int_{\sigma-iT}^{\sigma+iT} \zeta^k(s)g(s)\frac{x^s}{s}ds + O\Big(\frac{x^{\sigma+\epsilon}}{T}\Big).$$

We move the line of integration to $\text{Re}(s) = a > 1 - \delta$. The pole at $s = 1$ of $\zeta(s)$ contributes

$$\frac{xg(1)(\log x)^{k-1}}{(k-1)!}.$$

By Exercise 4.2.4, $\zeta(s) = O(T^{1/2})$ for $T \to \infty$. Thus, the horizontal integral contributes

$$O\left(\frac{x^\delta T^{\frac{k}{2}-1}}{\log x}\right),$$

and the vertical integral contributes

$$O\left(x^a T^{k/2}\right)$$

on the line $\text{Re}(s) = a$. We choose $T = x^{2\sigma/k}$, and this gives an error term of

$$O\left(x^{a+\sigma} + x^{\sigma(1-\frac{2}{k})+\epsilon}\right).$$

We can choose σ such that $a + \sigma < 1$ and $\sigma(1 - 2/k) + \epsilon < 1$. This completes the proof. □

4.4.18 *Let $\nu(n)$ denote the number of distinct prime factors of n. Show that*

$$\sum_{n \leq x} 2^{\nu(n)} \sim \frac{x \log x}{\zeta(2)}$$

as $x \to \infty$.

We have

$$f(s) = \sum_{n=1}^{\infty} \frac{2^{\nu(n)}}{n^s} = \frac{\zeta^2(s)}{\zeta(2s)}$$

by Exercise 1.2.6. Also, $f(s)$ satisfies the hypotheses of the previous exercise. Hence

$$\sum_{n \leq x} 2^{\nu(n)} \sim \frac{x \log x}{\zeta(2)}$$

as $x \to \infty$. □

5
Functional Equations

5.1 Poisson's Summation Formula

5.1.1 *For* $\mathrm{Re}(c) > 0$, *let* $F(x) = e^{-c|x|}$. *Show that*

$$\hat{F}(u) = \frac{2c}{c^2 + 4\pi^2 u^2}.$$

We have

$$\begin{aligned}\hat{F}(u) &= \int_{-\infty}^{\infty} e^{-c|x|} e^{-2\pi i u x}\, dx \\ &= \int_{0}^{\infty} e^{-(c+2\pi i u)x}\, dx + \int_{0}^{-\infty} e^{-(c-2\pi i u)x}\, dx.\end{aligned}$$

Since

$$\int_0^{\infty} e^{-vx}\, dx = \frac{1}{v}$$

for $\mathrm{Re}(v) > 0$, we get

$$\hat{F}(u) = \frac{1}{c+2\pi i u} + \frac{1}{c - 2\pi i u} = \frac{2c}{c^2 + 4\pi^2 u^2}.$$

□

5.1.2 *For $F(x) = e^{-\pi x^2}$, show that $\hat{F}(u) = e^{-\pi u^2}$.*

We must show that

$$\int_{-\infty}^{\infty} e^{-\pi x^2} e^{-2\pi i x u} dx = e^{-\pi u^2},$$

which is the same as

$$\int_{-\infty}^{\infty} e^{-\pi (x+iu)^2} dx = 1.$$

But this is essentially the famous probability integral

$$I = \int_{-\infty}^{\infty} e^{-\pi x^2} dx = 1.$$

To see this, observe that for $u = 0$, we have

$$\begin{aligned} I^2 &= \int_{-\infty}^{\infty} e^{-\pi x^2} dx \int_{-\infty}^{\infty} e^{-\pi y^2} dy \\ &= \int_{-\infty}^{\infty} \int_{-\infty}^{\infty} e^{-\pi (x^2+y^2)} dx\, dy \\ &= \int_0^{\infty} r dr \int_0^{2\pi} e^{-\pi r^2} d\theta, \end{aligned}$$

where we have made the polar substitution $x = r\cos\theta$, $y = r\sin\theta$. Thus

$$I^2 = \int_0^{\infty} e^{-\pi r^2} (2\pi r) dr = 1.$$

Since $I > 0$, we conclude that $I = 1$. For the general case, notice that

$$\begin{aligned}
\frac{\partial}{\partial u}\int_{-\infty}^{\infty} e^{-\pi(x+iu)^2}\,dx &= \int_{-\infty}^{\infty}\left(\frac{\partial}{\partial u}e^{-\pi(x+iu)^2}\right)dx \\
&= 2\pi i\int_{-\infty}^{\infty}(x+iu)e^{-\pi(x+iu)^2}\,dx \\
&= i\int_{-\infty}^{\infty}\left(\frac{\partial}{\partial x}e^{-\pi(x+iu)^2}\right)dx \\
&= \left[ie^{-\pi(x+iu)^2}\right]_{x=-\infty}^{x=+\infty} = 0.
\end{aligned}$$

Thus, the value of the integral is independent of u. But for $u = 0$, the value is 1. Hence

$$\int_{-\infty}^{\infty} e^{-\pi(x+iu)^2}\,dx = 1.$$

□

5.1.5 *With F as in Theorem 5.1.3, show that*

$$\sum_{n\in\mathbb{Z}} F\left(\frac{v+n}{t}\right) = \sum_{n\in\mathbb{Z}} |t|\hat{F}(nt)e^{2\pi intv}.$$

Observe that the Fourier transform of $F(x/t)$ is $|t|\hat{F}(tu)$, so that the result is now immediate from Theorem 5.1.3. □

5.1.6 *Show that*

$$\frac{e^c+1}{e^c-1} = \sum_{-\infty}^{\infty}\frac{2c}{c^2+4\pi^2n^2}.$$

By Exercise 5.1.1 and Corollary 5.1.4, this result is immediate. □

5.1.7 *Show that*

$$\sum_{n\in\mathbb{Z}} e^{-(n+\alpha)^2\pi/x} = x^{1/2}\sum_{n\in\mathbb{Z}} e^{-n^2\pi x+2\pi in\alpha}$$

for any $\alpha \in \mathbb{R}$, and $x > 0$.

We have the pair of Fourier transforms

$$F(t) = e^{-\pi t^2} \quad \text{and} \quad \hat{F}(t) = e^{-\pi t^2}.$$

Thus, the function $e^{-\pi(t+a)^2}$ has transform $e^{2\pi iat}e^{-\pi t^2}$. Also, $e^{-\pi(a+t/\sqrt{x})^2}$ has transform

$$x^{1/2}e^{2\pi iat\sqrt{x}}e^{-\pi t^2 x}.$$

Applying the Poisson summation formula gives

$$\sum_{n\in\mathbb{Z}} e^{-\pi(a+n/\sqrt{x})^2} = x^{1/2}\sum_{n\in\mathbb{Z}} e^{-\pi n^2 x+2\pi ian\sqrt{x}}.$$

Setting $\alpha = a\sqrt{x}$ gives

$$\sum_{n\in\mathbb{Z}} e^{-\pi(n+\alpha)^2/x} = x^{1/2}\sum_{n\in\mathbb{Z}} e^{-\pi n^2 x+2\pi in\alpha}$$

as desired. □

5.2 The Riemann Zeta Function

5.2.1 *Show that*

$$\Gamma(s+1) = s\Gamma(s)$$

for Re$(s) > 0$ *and that this functional equation can be used to extend* $\Gamma(s)$ *as a meromorphic function for all* $s \in \mathbb{C}$ *with only simple poles at* $s = 0, -1, -2, \ldots$.

The equation

$$\Gamma(s+1) = s\Gamma(s)$$

is easily deduced by an integration by parts. Thus, for Re$(s) > -1$, we can define

$$\Gamma(s) = \frac{\Gamma(s+1)}{s}$$

from which we see that $\Gamma(s)$ has a simple pole at $s = 0$. Continuing in this way, we see that

$$\Gamma(s) = \frac{\Gamma(s+2)}{s(s+1)},$$

which gives a meromorphic continuation to $\text{Re}(s) > -2$ with again a simple pole at $s = -1$. The result is now clear. $\square$

5.2.3 *Show that $\zeta(s)$ has simple zeros at $s = -2n$, for n a positive integer.*

The integral on the right-hand side in Theorem 5.2.2 converges for all $s \in \mathbb{C}$. Thus $\pi^{-s/2}\Gamma(s/2)\zeta(s)$ is analytic for any $s = -2n$, with n a positive integer. Note that

$$1 + 2n(2n+1)\int_1^\infty W(x)\left(x^{-n} + x^{n+1/2}\right)\frac{dx}{x} > 0.$$

Since the Γ-function has a simple pole there, $\zeta(s)$ must have a simple zero at that point. $\square$

5.2.4 *Prove that $\zeta(0) = -1/2$.*

Since $\Gamma(s/2) \sim (s/2)^{-1}$ as $s \to 0$, multiplying the equation in Theorem 5.2.2 by $s/2$ and taking limits as $s \to 0$ gives the result. $\square$

5.2.5 *Show that $\zeta(s) \neq 0$ for any real s satisfying $0 < s < 1$.*

Since

$$\zeta(s) = \frac{s}{s-1} - s\int_1^\infty \frac{\{x\}}{x^{s+1}}dx,$$

we see that

$$\left|\zeta(s) - \frac{s}{s-1}\right| < s\int_1^\infty \frac{dx}{x^{s+1}} = 1.$$

Hence

$$\frac{1}{s-1} = -1 + \frac{s}{s-1} < \zeta(s) < 1 + \frac{s}{s-1} = \frac{2s-1}{s-1}.$$

Thus, for $1/2 < s < 1$, we have $(2s-1)/(s-1) < 0$, which shows that $\zeta(s) \neq 0$ for $1/2 < s < 1$. By the functional equation, we have it for the whole range $0 < s < 1$.

5.3 Gauss Sums

5.3.2 *If χ is a primitive nonprincipal character* $(\text{mod } q)$, *show that*

$$\chi(n)\tau(\overline{\chi}) = \sum_{m=1}^{q} \overline{\chi}(m)e\left(\frac{mn}{q}\right)$$

if $(n, q) > 1$.

Let us put

$$\frac{n}{q} = \frac{n_1}{q_1},$$

where $(n_1, q_1) = 1$ and $q_1 | q$, $q_1 < q$. If n is a multiple of q, the left-hand side is zero, and so is the right-hand side, since

$$\sum_{m=1}^{q} \overline{\chi}(m) = 0.$$

So, we may suppose $1 < q_1 < q$. We have to prove that

$$\sum_{m=1}^{q} \overline{\chi}(m) e\Big(\frac{mn_1}{q_1}\Big) = 0.$$

Write $q = q_1 q_2$ and put $m = aq_1 + b$, where $0 \leq a < q_2$, $1 \leq b \leq q_1$. Then, the above sum can be rewritten

$$\sum_{1 \leq b \leq q_1} e\left(\frac{bn_1}{q_1}\right) \sum_{0 \leq a < q_2} \overline{\chi}(aq_1 + b),$$

and it suffices to prove that the inner sum is zero. Let us write

$$S(b) = \sum_{0 \leq a < q_2} \overline{\chi}(aq_1 + b).$$

Observe that $S(b + q_1) = S(b)$. If c is any integer satisfying

$$(c, q) = 1, \qquad c \equiv 1 \pmod{q_1},$$

then

$$\begin{aligned} \chi(c)S(b) &= \sum_{0 \leq a < q_2} \overline{\chi}(caq_1 + cb) \\ &= \sum_{0 \leq a < q_2} \overline{\chi}(aq_1 + b) = S(b), \end{aligned}$$

since $S(b + q_1) = S(b)$. Since χ is a primitive character $(\mathrm{mod}\ q)$, it is not periodic to any modulus q_1 that is a proper factor of q. Thus, there are integers c_1, c_2 such that

$$(c_1, q) = (c_2, q) = 1, \quad c_1 \equiv c_2 \pmod{q_1},$$

and $\chi(c_1) \neq \chi(c_2)$. Hence, there exists $c \equiv c_1 c_2^{-1} \pmod{q_1}$, $(c, q) = 1$, such that $\chi(c) \neq 1$. Thus $S(b) = 0$, as desired. □

5.4 Dirichlet L-functions

5.4.2 *Suppose $\chi(-1) = 1$. Show that $L(s,\chi)$ has simple zeros at $s = -2, -4, -6, \ldots$.*

Since $L(1-s,\overline{\chi})$ has no zeros for $\mathrm{Re}(1-s) > 1$ and $\Gamma((1-s)/2)$ has no zeros at all, the only zeros of $L(s,\chi)$ for $\mathrm{Re}(s) < 0$ are at $s = -2, -4, -6, \ldots$ corresponding to the poles of $\Gamma(s/2)$. This is so because by the above, their product is entire. □

5.4.3 *Prove that*

$$\pi^{-(s+1)/2} q^{(s+1)/2} \Gamma\left(\frac{s+1}{2}\right) n^{-s} = \int_0^\infty n e^{-\pi n^2 x/q} x^{\frac{s+1}{2}} \frac{dx}{x},$$

and hence deduce that

$$\pi^{-(\frac{s+1}{2})} q^{(\frac{s+1}{2})} \Gamma\Big(\frac{s+1}{2}\Big) L(s,\chi) = \frac{1}{2}\int_0^\infty \theta_1(x,\chi) x^{\frac{s+1}{2}} \frac{dx}{x}$$

where

$$\theta_1(x,\chi) = \sum_{n=-\infty}^{\infty} n\chi(n) e^{-n^2\pi x/q}.$$

Changing s to $s+1$ in the formula

$$\pi^{-s/2} q^{s/2} \Gamma(s/2) n^{-s} = \int_0^\infty e^{-n^2\pi x/q} x^{\frac{s}{2}} \frac{dx}{x}$$

gives the first result. Then summing over n gives the second equation upon noting that since $\chi(-1) = -1$,

$$\theta_1(x,\chi) = \sum_{n=-\infty}^{\infty} n\chi(n) e^{-n^2\pi x/q} = 2\sum_{n=1}^{\infty} n\chi(n) e^{-n^2\pi x/q}.$$

□

5.4.4 *Prove that*

$$\sum_{n=-\infty}^{\infty} n e^{-n^2\pi x/q + 2\pi i m n/q} = i(q/x)^{3/2} \sum_{n=-\infty}^{\infty} \left(n + \frac{m}{q}\right) e^{-\pi(n+m/q)^2 q/x}.$$

This is immediate from Poisson's summation formula. Indeed, by Exercise 5.1.7, we have

$$\sum_{n=-\infty}^{\infty} e^{-n^2\pi y+2\pi i n\alpha} = y^{-1/2}\sum_{n=-\infty}^{\infty} e^{-\pi(n+\alpha)^2/y}.$$

Differentiating with respect to α gives

$$2\pi i\sum_{n=-\infty}^{\infty} ne^{-n^2\pi y+2\pi i n\alpha} = -2\pi y^{-\frac{3}{2}}\sum_{n=-\infty}^{\infty}(n+\alpha)e^{-(n+\alpha)^2\pi/y},$$

and substituting x/q for y and m/q for α gives the stated equation. □

5.4.5 *Prove that for $\chi(-1) = -1$, if we set*

$$\xi(s,\chi) = \pi^{-s/2}q^{s/2}\Gamma((s+1)/2)L(s,\chi),$$

then $\xi(s,\chi)$ is entire and

$$\xi(s,\chi) = w_\chi\xi(1-s,\overline{\chi}),$$

where $w_\chi = \tau(\chi)/(iq^{1/2})$.

By Exercises 5.4.3 and 5.4.4, we obtain

$$\pi^{-(\frac{s+1}{2})}q^{\frac{s+1}{2}}\Gamma\Big(\frac{s+1}{2}\Big)L(s,\chi)$$

$$= \frac{1}{2}\int_1^\infty \theta_1(x,\chi)x^{\frac{s+1}{2}}\frac{dx}{x} + \frac{1}{2}\int_1^\infty \theta_1(x^{-1},\chi)x^{-\frac{s+1}{2}}\frac{dx}{x}$$

$$= \frac{1}{2}\int_1^\infty \theta_1(x,\chi)x^{\frac{s}{2}}\frac{dx}{\sqrt{x}} + \frac{iq^{1/2}}{2\tau(\overline{\chi})}\int_1^\infty \theta_1(x,\overline{\chi})x^{\frac{1-s}{2}}\frac{dx}{\sqrt{x}}.$$

This gives the analytic continuation for $L(s,\chi)$ and establishes the functional equation, since the change of the right-hand side when s is replaced by $1-s$ is as stated. □

5.5.1 *Let*

$$f(y) = \sum_{n=1}^{\infty} a_n e^{-2\pi n y}$$

converge for $y > 0$. Suppose that for some $w \in \mathbb{Z}$,

$$f(1/y) = (-1)^w y^r f(y),$$

and that $a_n = O(n^c)$ for some constant $c > 0$. Let

$$L_f(s) = \sum_{n=1}^{\infty} a_n n^{-s}.$$

Show that $(2\pi)^{-s}\Gamma(s)L_f(s)$ extends to an entire function and satisfies the functional equation

$$(2\pi)^{-s}\Gamma(s)L_f(s) = (-1)^w(2\pi)^{-(r-s)}\Gamma(r-s)L_f(r-s).$$

We have

$$\begin{aligned}\int_0^\infty f(y)y^{s-1} &= \int_0^\infty \sum_{n=1}^{\infty} a_n e^{-2\pi n y} y^{s-1} dy \\ &= \sum_{n=1}^{\infty} a_n \int_0^\infty e^{-2\pi n y} y^{s-1} dy,\end{aligned}$$

the interchange being justified by the estimate $a_n = O(n^c)$ which implies the absolute convergence of the integral. Changing variables in the integral gives

$$(2\pi)^{-s}\Gamma(s)L_f(s) = \int_0^\infty f(y)y^{s-1}dy,$$

which converges absolutely for $\mathrm{Re}(s) > 0$. Now write the integral as

$$\int_0^1 f(y)y^{s-1}dy + \int_1^\infty f(y)y^{s-1}dy.$$

We make a change of variable $y = 1/t$ in the first integral:

$$\int_0^1 f(y)y^{s-1}dy = \int_1^\infty f(1/t)t^{-s-1}dt.$$

Using the fact that $f(1/t) = (-1)^w t^r f(t)$, we obtain

$$\int_0^1 f(y)y^{s-1}dy = (-1)^w \int_1^\infty f(t)t^{r-s-1}dt.$$

Hence

$$\begin{aligned}(2\pi)^{-s}\Gamma(s)L_f(s) &= \int_1^\infty f(y)\{y^s + (-1)^w y^{r-s}\}\frac{dy}{y} \\ &= (-1)^w(2\pi)^{-(r-s)}\Gamma(r-s)L_f(r-s),\end{aligned}$$

which establishes the functional equation. Note that the integral converges for all $s \in \mathbb{C}$. This gives the result. □

5.5 Supplementary Problems

5.5.2 *Let*

$$g(y) = \sum_{n=0}^\infty a_n e^{-2\pi n y}$$

converge for $y > 0$. Suppose that for some $w \in \mathbb{Z}$,

$$g(1/y) = (-1)^w y^r g(y)$$

and that $a_n = O(n^c)$ for some constant $c > 0$. Let

$$L_g(s) = \sum_{n=1}^\infty a_n n^{-s}.$$

Show that $(2\pi)^{-s}\Gamma(s)L_g(s)$ extends to a meromorphic function with at most simple poles at $s = 0$ and $s = r$ and satisfies the functional equation

$$(2\pi)^{-s}\Gamma(s)L_g(s) = (-1)^w(2\pi)^{r-s}\Gamma(r-s)L_g(r-s).$$

Set

$$h(y) = \sum_{n=1}^\infty a_n e^{-2\pi n y} = g(y) - a_0.$$

Note that the Dirichlet series $\sum_{n=1}^{\infty} a_n n^{-s}$ converges absolutely for $\text{Re}(s) > 1 + c$. Thus, in this half-plane,

$$\begin{aligned} (2\pi)^{-s}\Gamma(s)L_f(s) &= \int_0^\infty h(y)y^{s-1}dy \\ &= \int_0^\infty (g(y) - a_0)y^{s-1}dy, \end{aligned}$$

which converges for $\text{Re}(s) > 0$. Now,

$$\begin{aligned} h(1/y) &= g(1/y) - a_0 \\ &= (-1)^w y^r g(y) - a_0 \\ &= (-1)^w y^r h(y) - a_0(-1)^w y^r - a_0. \end{aligned}$$

We write the integral

$$\int_0^1 h(y)y^{s-1}dy + \int_1^\infty h(y)y^{s-1}dy$$

and change variables in the first integral by setting $y = 1/t$ to obtain

$$\begin{aligned} \int_0^1 h(y)y^{s-1}dy &= \int_1^\infty h(1/t)t^{-s-1}dt \\ &= \int_0^1 \{(-1)^w y^r h(y) - a_0(-1)^w y^r - a_0\} \, y^{-s-1}dy \end{aligned}$$

by the functional equation for h. Thus

$$\begin{aligned} &(2\pi)^{-s}\Gamma(s)L_f(s) \\ &= \int_1^\infty h(y)(y^{s-1} + (-1)^w y^{r-s-1})dy \\ &\quad -a_0 \int_1^\infty (-1)^w y^{r-s-1}dy - a_0 \int_1^\infty y^{-s-1}dy, \end{aligned}$$

and the two integrals are easily evaluated:

$$\int_1^\infty y^{-s-1}dy = \left[-\frac{y^{-s}}{s}\right]_1^\infty = \frac{1}{s},$$

$$\int_1^\infty y^{r-s-1}dy = \left[-\frac{y^{r-s}}{r-s}\right]_1^\infty = \frac{1}{r-s}$$

so that

$$\begin{aligned}(2\pi)^{-s}\Gamma(s)L_f(s) &= -a_0\left(\frac{(-1)^w}{r-s}+\frac{1}{s}\right)\\ &\quad+\int_1^\infty h(y)(y^{s-1}+(-1)^w y^{r-s-1})dy,\end{aligned}$$

and the right-hand side gives the meromorphic continuation with only simple poles at $s = 0, r$. Also, the functional equation is immediate, since

$$\begin{aligned}(2\pi)^{-(r-s)}\Gamma(r-s)L_f(r-s) &= -a_0\left(\frac{(-1)^w}{s}+\frac{1}{r-s}\right)\\ &\quad+\int_1^\infty h(y)(y^{r-s-1}+(-1)^w y^{s-1})dy\\ &= (-1)^w(2\pi)^{-s}\Gamma(s)L_f(s),\end{aligned}$$

as required. □

5.5.3 *Let*

$$\Psi(x) = \begin{cases} x-[x]-\frac{1}{2} & \text{if } x \notin \mathbb{Z} \\ 0 & \text{if } x \in \mathbb{Z} \end{cases}$$

Show that

$$\left|\Psi(x) + \sum_{0<|m|\le M} \frac{e(mx)}{2\pi im}\right| \le \frac{1}{2\pi M||x||}$$

where $e(t) = e^{2\pi it}$ *and* $||x||$ *denotes the distance from* x *to the nearest integer.*

The function $\Psi(x)$ is periodic with period 1. If $x \in \mathbb{Z}$, the result is clear, since in the sum we can pair m and $-m$ to get 0. Suppose first $0 < x \leq 1/2$. Then

$$\int_{1/2}^{x} e(mt)dt = \frac{e(mx)}{2\pi i m} - \frac{(-1)^m}{2\pi i m},$$

so that summing both sides of this equation for $0 < |m| \leq M$ gives

$$\int_{1/2}^{x} \Big(\sum_{0<|m|\leq M} e(mt) \Big) dt = \sum_{0<|m|\leq M} \frac{e(mx)}{2\pi i m},$$

since

$$\sum_{0<|m|\leq M} \frac{(-1)^m}{2\pi i m} = 0.$$

Thus

$$\int_{1/2}^{x} \Big(\sum_{0\leq|m|\leq M} e(mt) \Big) dt = \sum_{0<|m|\leq M} \frac{e(mx)}{2\pi i m} + \left(x - \frac{1}{2} \right).$$

The integrand is a geometric progression, which is easily summed to

$$\begin{aligned} e(-Mt) \left(\frac{e((2M+1)t) - 1}{e(t) - 1} \right) &= \frac{e((M+\frac{1}{2})t) - e(-(M+\frac{1}{2})t)}{e(t/2) - e(-t/2)} \\ &= \frac{\sin((2M+1)\pi t)}{\sin \pi t}. \end{aligned}$$

Recall the following mean value theorem for integrals: Let $f(x)$ be bounded, monotonic decreasing, nonnegative, and differentiable in $[a, b]$ and let $g(x)$ be a bounded integrable function. Then

$$\int_{a}^{b} f(x)g(x)dx = f(a) \int_{a}^{\xi} g(x)dx$$

for some $a \leq \xi \leq b$. Indeed, letting

$$G(t) = \int_{a}^{x} g(x)dx$$

we have by integration by parts,

$$\int_a^b f(x)g(x)dx = G(b)f(b) - \int_a^b f'(x)G(x)dx,$$

and the last integral is, by the familiar mean value theorem for integrals,

$$= G(\eta) \int_a^b f'(x)dx = G(\eta)(f(b) - f(a)),$$

for some $a \leq \eta \leq b$. Suppose now, without loss of generality, that $G(\eta) \leq G(b)$. Then, since $f(a) \geq f(b)$, we deduce that

$$G(\eta)f(a) \leq G(b)f(b) + G(\eta)(f(a) - f(b)) \leq G(b)f(a).$$

Since G is continuous, we must have

$$G(b)f(b) + G(\eta)(f(a) - f(b)) = G(\xi)f(a)$$

for some ξ satisfying $a \leq \xi \leq b$. Note that we apply this with $f(x) = 1/\sin \pi x$, $g(x) = \sin(2m+1)\pi x$, and $[a,b] = [x, 1/2]$. Then $f(x)$ is monotone decreasing, and we have

$$\begin{aligned} \int_x^{1/2} \frac{\sin(2m+1)\pi t}{\sin \pi t} dt &= \frac{1}{\sin \pi x} \int_x^{\xi} \sin(2m+1)\pi t dt \\ &= \frac{1}{\sin \pi x} \left[- \frac{\cos(2m+1)\pi t}{(2m+1)\pi} \right]_x^{\xi}. \end{aligned}$$

Thus,

$$\left| \int_{1/2}^{x} \frac{\sin(2m+1)\pi t}{\sin \pi t} dt \right| \leq \frac{1}{(2m+1)\pi x}$$

by the elementary inequality $\sin \pi x \geq 2x$, valid for $0 \leq x \leq 1/2$. The result is proved for $0 < x \leq 1/2$. We still need to treat the range $1/2 < x < 1$. Observe that $\Psi(1-x) = \Psi(-x)$ (because Ψ has period 1) and $\Psi(-x) = -\Psi(x)$ because for $x > 0$, $[-x] = -[x] - 1$. Thus by the above,

$$\left| \Psi(1-x) + \sum_{0<|m|\leq M} \frac{e(m(1-x))}{2\pi i m} \right| \leq \frac{1}{(2M+1)\pi(1-x)}.$$

Now, $1 - x = ||x||$ for $1/2 < x < 1$. Hence

$$\left| -\Psi(x) + \sum_{0<|m|\leq M} \frac{e(-mx)}{2\pi i m} \right| \leq \frac{1}{(2M+1)\pi||x||}.$$

which gives

$$\left| \Psi(x) + \sum_{0<|m|\leq M} \frac{e(mx)}{2\pi i m} \right| \leq \frac{1}{(2M+1)\pi||x||}$$

for $\frac{1}{2} \leq x < 1$. This completes the proof. □.

5.5.4 *Let $f(x)$ be a differentiable function on $[0,1]$ satisfying $|f'(x)| \leq K$. Show that*

$$\left| \sum_{|m|\leq M} \int_0^1 f(x)e(mx)dx - \frac{f(0)+f(1)}{2} \right| \ll \frac{K \log M}{M}.$$

Deduce that

$$\sum_{-\infty}^{\infty} \int_0^1 f(x)e(mx)dx = \frac{f(0)+f(1)}{2}.$$

By integrating by parts, we have for $m \neq 0$,

$$\int_0^1 f(x)e(mx)dx = \left[\frac{f(x)e(mx)}{2\pi i m} \right]_0^1 - \int_0^1 \frac{f'(x)e(mx)dx}{2\pi i m}.$$

Summing both sides over $0 < |m| \leq M$ gives

$$\sum_{0<|m|\leq M} \int_0^1 f(x)e(mx)dx = -\int_0^1 f'(x) \sum_{0<|m|\leq M} \frac{e(mx)}{2\pi i m} dx,$$

since

$$\left[f(x) \sum_{0<|m|\leq M} \frac{e(mx)}{2\pi i m} \right]_0^1 = 0,$$

as is easily seen by pairing m and $-m$ in the summation.

By the previous exercise,

$$-\sum_{0<|m|\leq M} \frac{e(mx)}{2\pi i m} = \Psi(x) + O\Big(\frac{1}{M||x||}\Big).$$

Before inserting this fact into the integral, let us note that

$$\left| \int_0^{1/M} f'(x) \sum_{0<|m|\leq M} \frac{e(mx)}{2\pi i m} dx \right| \leq \frac{K \log M}{M}.$$

Similarly,

$$\left| \int_{1-\frac{1}{M}}^{1} f'(x) \sum_{0<|m|\leq M} \frac{e(mx)}{2\pi i m} dx \right| \leq \frac{K \log M}{M}.$$

Thus,

$$-\int_{1/M}^{1-1/M} f'(x) \sum_{0<|m|\leq M} \frac{e(mx)}{2\pi i m} dx$$

$$= \int_{1/M}^{1-1/M} f'(x)\Psi(x)dx + O\left(\int_{1/M}^{1-1/M} \frac{f'(x)dx}{M||x||} \right).$$

The error term is easily estimated by breaking the interval into two parts: $[1/M, 1/2]$ and $[1/2, 1-1/M]$. The error is $O(K \log M/M)$.

Therefore,

$$\sum_{0<|m|\leq M} \int_0^1 f(x)e(mx)dx = \int_0^1 f'(x)\Psi(x)dx + O\left(\frac{K \log M}{M}\right).$$

The integral on the right-hand side is

$$\begin{aligned} \int_0^1 f'(x)\left(x - \frac{1}{2}\right) dx &= \left[f(x) \left(x - \frac{1}{2} \right) \right]_0^1 - \int_0^1 f(x)dx \\ &= \frac{f(1)+f(0)}{2} - \int_0^1 f(x)dx, \end{aligned}$$

which completes the proof. □

5.5.5 *By using the previous exercise with* $f(x) = x^2$, *deduce that*

$$\sum_{m=1}^{\infty} \frac{1}{m^2} = \frac{\pi^2}{6}.$$

We have for $m \neq 0$,

$$\int_0^1 f(x)e(mx)dx = \left[\frac{x^2 e(mx)}{2\pi i m}\right]_0^1 - 2\int_0^1 \frac{xe(mx)dx}{2\pi i m}$$

$$= \frac{1}{2\pi i m} + \frac{1}{2\pi^2 m^2},$$

by an easy integration by parts.

For $m = 0$,

$$\int_0^1 f(x)dx = \frac{1}{3}.$$

By Exercise 5.5.4,

$$\frac{1}{3} + \sum_{0<|m|\leq M} \left(\frac{1}{2\pi i m} + \frac{1}{2\pi^2 m^2}\right) = \frac{1}{2} + O\left(\frac{\log M}{M}\right).$$

Since

$$\sum_{0<|m|\leq M} \frac{1}{2\pi i m} = 0,$$

the result is now immediate upon letting $M \to \infty$. □

5.5.6 (Pólya - Vinogradov inequality) *Let* χ *be a primitive character* $\pmod q$. *Show that for* $q > 1$,

$$\left|\sum_{n\leq x} \chi(n)\right| \ll q^{1/2} \log q.$$

We use Gauss sums. By Example 5.3.1 and Exercise 5.3.2, we can write

$$\tau(\overline{\chi})\chi(n) = \sum_{m=1}^{q} \overline{\chi}(m) e\left(\frac{mn}{q}\right).$$

Since the summation is over any complete set of residues $\pmod q$, we can replace the range of summation by $-q/2 < m < q/2$. Thus,

$$\tau(\overline{\chi}) \sum_{n\leq x} \chi(n) = \sum_{0<|m|\leq q/2} \overline{\chi(m)} \sum_{n\leq x} e\left(\frac{mn}{q}\right).$$

The inner sum is bounded by $2/|e(m/q)-1|$. Writing $e(m/q)-1 = e(m/2q)(e(m/2q)-e(-m/2q))$ we obtain

$$|\tau(\overline{\chi})|\left|\sum_{n\le x}\chi(n)\right| \le \sum_{0<|m|<q/2}\frac{1}{|\sin \pi m/q|}.$$

Using the inequality $|\sin \pi x| \ge 2x$ for $0 \le x \le 1/2$, we get

$$|\tau(\overline{\chi})|\sum_{n\le x}\chi(n)| \ll q\log q.$$

Finally, by Theorem 5.3.3, $|\tau(\overline{\chi})| = q^{1/2}$, so that the result is now immediate. □

5.5.7 *Show that if χ is a primitive character* (mod q), *then*

$$L(1,\chi) = \sum_{n\le x}\frac{\chi(n)}{n} + O\left(\frac{q^{1/2}\log q}{x}\right)$$

for any $x \ge 1$ and $q > 1$.

We have

$$L(1,\chi) = \sum_{n=1}^{\infty}\frac{\chi(n)}{n} = \sum_{n\le x}\frac{\chi(n)}{n} + \sum_{n>x}\frac{\chi(n)}{n}.$$

By partial summation and the Pólya - Vinogradov inequality (Exercise 5.5.6), the second sum is

$$\ll \frac{q^{1/2}\log q}{x}$$

as required. □

5.5.8 *Prove that*

$$\sum_{\chi\ne\chi_0} L(1,\chi) = \varphi(q) + O(q^{1/2}\log q),$$

where the summation is over all nontrivial characters (mod q).

By Exercise 5.5.7,

$$L(1,\chi) = \sum_{n\le x}\frac{\chi(n)}{n} + O\Big(\frac{q^{1/2}\log q}{x}\Big)$$

for any nontrivial character χ because the conductor of χ is bounded by q. Summing this over $\chi \neq \chi_0 \pmod q$, we get

$$\sum_{\chi \neq \chi_0} L(1,\chi) = \sum_{n \leq x} \frac{1}{n} \Big(\sum_{\chi \neq \chi_0} \chi(n) \Big) + O\left(\frac{q^{3/2} \log q}{x} \right).$$

We choose $x = q$. Also,

$$\sum_{\chi \neq \chi_0} \chi(n) = \begin{cases} \varphi(q) - 1 & \text{if } n \equiv 1 \pmod q, \\ \\ -1 & \text{otherwise.} \end{cases}$$

Thus,

$$\sum_{\chi \neq \chi_0} L(1,\chi) = \varphi(q) + O\left(q^{1/2} \log q \right),$$

as desired. □

5.5.9 *For any $s \in \mathbb{C}$ with* $\mathrm{Re}(s) > 0$, *show that for any $x \geq 1$,*

$$L(s,\chi) = \sum_{n \leq x} \frac{\chi(n)}{n^s} + O\left(\frac{|s| q^{1/2} \log q}{\sigma x^\sigma} \right),$$

where χ is a nontrivial character $\pmod q$ *and* $\sigma = \mathrm{Re}(s)$.

By partial summation and the Pólya - Vinogradov inequality, we have

$$\sum_{n > x} \frac{\chi(n)}{n^s} \ll |s| \int_x^\infty \frac{O(q^{1/2} \log q)}{t^{\sigma+1}} dt,$$

from which the result is now immediate. □

5.5.10 *Prove that for any $\sigma > 1/2$,*

$$\sum_{\chi \neq \chi_0} L(\sigma,\chi) = \varphi(q) + O\left(q^{3/2-\sigma} \right)$$

where the sum is over all nontrivial characters $\pmod q$.

By the previous exercise,

$$L(\sigma,\chi) = \sum_{n \leq x} \frac{\chi(n)}{n^\sigma} + O\left(\frac{q^{1/2} \log q}{x^\sigma} \right).$$

Summing both sides over $\chi \neq \chi_0$, we get

$$\sum_{\chi\neq\chi_0} L(\sigma,\chi) = \sum_{n\leq x} \frac{1}{n^\sigma}\left(\sum_{\chi\neq\chi_0}\chi(n)\right) + O\left(\frac{q^{3/2}\log q}{x^\sigma}\right).$$

We treat the inner sum as in Exercise 5.5.8 and choose $x = q$, to obtain

$$\begin{aligned}\sum_{\chi\neq\chi_0} L(\sigma,\chi) &= \varphi(q) + O\left(q^{1-\sigma} + q^{3/2-\sigma}\right) \\ &= \varphi(q) + O\left(q^{3/2-\sigma}\right),\end{aligned}$$

as required. □

5.5.11 *Let $B_n(x)$ denote the nth Bernoulli polynomial introduced in Chapter 2. For $n \geq 2$, show that*

$$\frac{B_n(x)}{n!} = \sum_{m\neq 0} \frac{e(mx)}{(2\pi i m)^n}.$$

For $n \geq 2$, the function defined by the series is uniformly continuous. Let us denote it by $\tilde{B}_n(x)/n!$. Then $\tilde{B}_n'/n! = \tilde{B}_{n-1}(x)/(n-1)!$, so that $\tilde{B}_n'(x) = n\tilde{B}_{n-1}(x)$. Also,

$$\int_0^1 \tilde{B}_n(x)dx = 0 \quad \text{for} \quad n \geq 2.$$

Exercise 5.5.3 shows that the formula stated in the exercise holds for $n = 1$. These must therefore coincide with the Bernoulli polynomials. This completes the proof. □

5.5.12 *Let $f(x)$ be differentiable on $[A, B]$ satisfying $|f'(x)| \leq K$ for all $x \in [A, B]$. Show that*

$$\sum_{n=A}^{B}{}' f(n) = \sum_{m=-\infty}^{\infty} \int_A^B f(x)e(mx)dx,$$

where the dash on the summation means that the end-terms are replaced by $f(A)/2$ and $f(B)/2$.

By Exercise 5.5.4, we have

$$\left| \frac{f(n)+f(n+1)}{2} - \sum_{|m|\le M} \int_n^{n+1} f(x)e(mx)dx \right| \le \frac{K\log M}{M}.$$

Adding this result over $n \in [A, B]$ gives

$$\left| \sum_{n=A}^{B}{}' f(n) - \sum_{|m|\le M} \int_A^B f(x)e(mx)dx \right| \le \frac{(B-A)K\log M}{M}.$$

Now let $M \to \infty$ to deduce the result. □

5.5.13 *Apply the previous exercise to each of the functions* $f(x) = \cos(2\pi x^2/N)$ *and* $f(x) = \sin(2\pi x^2/N)$ *to deduce that*

$$S = \sum_{n=0}^{N-1} e\left(\frac{n^2}{N}\right) = \begin{cases} (1+i)N^{1/2} & \text{if } N \equiv 0 \pmod 4, \\ N^{1/2} & \text{if } N \equiv 1 \pmod 4, \\ 0 & \text{if } N \equiv 2 \pmod 4, \\ iN^{1/2} & \text{if } N \equiv 3 \pmod 4. \end{cases}$$

By Exercise 5.5.12, we have to evaluate

$$\sum_{n=0}^{N}{}' e(n^2/N) = \sum_{m=-\infty}^{\infty} \int_0^N e\left(\frac{x^2}{N} + mx\right) dx.$$

We change variables in the integrand: put $x = Nt$ so that the integral is

$$N\int_0^1 e(Nt^2 + mNt)dt = Ne\left(-\frac{Nm^2}{4}\right)\int_0^1 e\left(N(t+m/2)^2\right) dt.$$

We must therefore evaluate

$$\int_0^1 e(N(t+m/2)^2)dt = \int_{m/2}^{m/2+1} e(Ny^2)dy.$$

Thus, we have

$$\sum_{n=0}^{N}{}' e(n^2/N) = N\sum_{m=-\infty}^{\infty} e\left(-\frac{Nm^2}{4}\right)\int_{m/2}^{m/2+1} e(Ny^2)dy.$$

Now $e(-Nm^2/4)$ is 1 if m is even, and i^{-N} if m is odd. This suggests we divide the infinite sum into two parts, m even and m odd:

$$\sum_{n=0}^{N}{}' e(n^2/N) = N \sum_{j=-\infty}^{\infty} \int_j^{j+1} e(Ny^2)dy + Ni^{-N} \sum_{j=-\infty}^{\infty} \int_{j-1/2}^{j+1/2} e(Ny^2)dy$$
$$= N(1+i^{-N}) \int_{-\infty}^{\infty} e(Ny^2)dy.$$

If we put $y = N^{-1/2}u$, then the integral becomes

$$N^{-1/2} \int_{-\infty}^{\infty} e(u^2)du = N^{-1/2}c$$

for some constant c. This constant is easily evaluated upon setting $N = 1$. Then

$$\sum_{n=0}^{N}{}' e(n^2/N) = 1,$$

so that $c = (1+i^{-1})^{-1} = (1-i)^{-1}$. Therefore,

$$\sum_{n=0}^{N}{}' e(n^2/N) = \left(\frac{1+i^{-N}}{1-i}\right) N^{1/2}.$$

Notice that the left-hand side is equal to S and the right-hand side takes the four values stated according as N belongs to the various classes (mod 4).

5.5.14 *Let χ be a nontrivial quadratic character* (mod p) *with p prime. Show that*

$$\tau(\chi) = \sum_{m=1}^{p-1} \chi(m) e\left(\frac{m}{p}\right) = \begin{cases} \sqrt{p} & \textit{if} \quad p \equiv 1 \pmod 4, \\ i\sqrt{p} & \textit{if} \quad p \equiv 3 \pmod 4. \end{cases}$$

Clearly,

$$\tau(\chi) - 1 = \sum_{m=1}^{p-1} (1+\chi(m)) e\left(\frac{m}{p}\right),$$

since $\sum_{m=1}^{p-1} e(m/p) = -1$.

Now, $1+\chi(m) = 2$ or 0 according as m is a square $(\bmod p)$ or not for $(m,p)=1$. Thus,

$$\begin{aligned}\tau(\chi) &= \sum_{m=1}^{p-1} e\left(\frac{m^2}{p}\right) + 1 \\ &= \sum_{m=0}^{p-1} e\left(\frac{m^2}{p}\right).\end{aligned}$$

By the previous exercise, the sum is $p^{1/2}$ if $p \equiv 1 \pmod 4$ and $ip^{1/2}$ if $p \equiv 3 \pmod 4$, and this completes the proof.

5.5.15 *Let $\phi(s) = (2\pi)^{-s}\Gamma(s)\zeta(s)\zeta(s+1)$. Show that $\phi(-s) = \phi(s)$.*

By Legendre's duplication formula (see Exercise 6.3.6) we have

$$\Gamma(s)\Gamma\left(\frac{1}{2}\right) = 2^{s-1}\Gamma\left(\frac{s}{2}\right)\Gamma\left(\frac{s+1}{2}\right).$$

Also, by Exercise 6.3.5, $\Gamma(1/2) = \sqrt{\pi}$. Therefore,

$$\begin{aligned}\phi(s) &= (2\pi)^{-s}\pi^{-1/2}2^{s-1}\Gamma\left(\frac{s}{2}\right)\Gamma\left(\frac{s+1}{2}\right)\zeta(s)\zeta(s+1) \\ &= 2^{-1}(\pi^{-s/2}\Gamma(s/2)\zeta(s))\pi^{-(s+1)/2}\Gamma\left(\frac{s+1}{2}\right)\zeta(s+1).\end{aligned}$$

By the functional equation of the ζ-function, we see that

$$\begin{aligned}\phi(s) &= 2^{-1}(\pi^{-(1-s)/2}\Gamma\left(\frac{1-s}{2}\right)\zeta(1-s))\pi^{s/2}\Gamma\left(\frac{-s}{2}\right)\zeta(-s) \\ &= \phi(-s)\end{aligned}$$

by another application of the duplication formula.

5.5.16 *Show that $\phi(s)$ in Exercise 5.5.15 has a double pole at $s=0$ and simple poles at $s = \pm 1$. Show further that* $\mathrm{Res}_{s=1}\phi(s) = \pi/12$ *and* $\mathrm{Res}_{s=-1}\phi(s) = -\pi/12$.

Since $\Gamma(s)$ has a simple pole at $s=0$ and $\zeta(s+1)$ has a simple pole at $s=0$, it is clear that $\phi(s)$ has a double pole at $s=0$. It is also clear that $\phi(s)$ has simple poles at $s=1$ and $s=-1$, the latter pole arising from the Γ-function. We have

$$\mathrm{Res}_{s=1}\phi(s) = \lim_{s\to 1}(s-1)\phi(s) = (2\pi)^{-1}\zeta(2).$$

By Exercise 5.5.5, this is equal to $\pi/12$. Also,

$$\begin{aligned}\mathrm{Res}_{s=-1}\phi(s) &= \lim_{s\to -1}(s+1)\phi(s) \\ &= \lim_{s\to -1}(2\pi)^{-s}\frac{\Gamma(s+2)}{s}\zeta(s)\zeta(s+1) \\ &= -(2\pi)\zeta(-1)\zeta(0).\end{aligned}$$

By Exercise 5.2.4, $\zeta(0) = -1/2$. Also, by the functional equation for the Riemann zeta function, we have

$$\pi^{1/2}\Gamma\Big(-\frac{1}{2}\Big)\zeta(-1) = \pi^{-1}\zeta(2) = \pi/6.$$

Now $(-1/2)\Gamma(-1/2) = \Gamma(1/2)$, since $s\Gamma(s) = \Gamma(s+1)$. By Exercise 6.3.5, $\Gamma(1/2) = \sqrt{\pi}$. Therefore

$$\zeta(-1) = -1/12.$$

Therefore,

$$\mathrm{Res}_{s=-1}\phi(s) = -(2\pi)(-1/12)(-1/2) = \pi/12.$$

□

5.5.17 *Show that if* $\sigma(n) = \sum_{d|n} d$, *then*

$$\sum_{n=1}^{\infty}\frac{\sigma(n)}{n^{s+1}} = \zeta(s)\zeta(s+1),$$

and that

$$\sum_{n=1}^{\infty}\frac{\sigma(n)}{n}e^{-nx} = \frac{1}{2\pi i}\int_{2-i\infty}^{2+i\infty} x^{-s}\Gamma(s)\zeta(s)\zeta(s+1)ds.$$

The first part is clear. The second part follows from Exercise 6.6.3. □

5.5.18 *Show that*

$$\sum_{n=1}^{\infty}\frac{\sigma(n)}{n}e^{-2\pi nx} = \frac{\pi}{12x} - \frac{\pi x}{12} + \frac{1}{2}\log x + \sum_{n=1}^{\infty}\frac{\sigma(n)}{n}e^{-2\pi n/x}.$$

By Exercise 5.5.17, we can move the line of integration to $\text{Re}(s) = -2$ to deduce

$$\sum_{n=1}^{\infty} \frac{\sigma(n)}{n} e^{-2\pi n x} = \frac{\pi}{12x} - \frac{\pi x}{12} + \frac{1}{2}\log x + \frac{1}{2\pi i}\int_{-2-i\infty}^{-2+i\infty} (2\pi x)^{-s}\Gamma(s)\zeta(s)\zeta(s+1)ds$$

by an application of Exercise 5.5.16. By Exercise 5.5.15, the integrand becomes

$$x^s\phi(-s) = x^s\phi(s)$$

upon changing s to $-s$. Moreover,

$$\frac{1}{2\pi i}\int_{2-i\infty}^{2+i\infty} x^s(2\pi)^{-s}\Gamma(s)\zeta(s)\zeta(s+1)ds = \sum_{n=1}^{\infty}\frac{\sigma(n)}{n}e^{-2\pi n/x},$$

as desired. □

5.5.19 *For a and b coprime integers and $b > 0$, define*

$$C\left(\frac{a}{b}\right) = \sum_{j=0}^{b-1} e^{2\pi i j^2 a/b}.$$

Let q be prime and $(p, q) = 1$. Show that

$$\lim_{t\to 0}\sqrt{t}\theta\left(t + \frac{2pi}{q}\right) = \frac{1}{q}C\left(-\frac{p}{q}\right).$$

Observe that

$$\theta\left(t + \frac{2pi}{q}\right) = \sum_{n=-\infty}^{\infty} e^{-\pi n^2 t}e^{-2\pi i n^2 p/q} = \sum_{b=0}^{q-1} e^{-2\pi i b^2 p/q}\Big(\sum_{n\equiv b\,(\text{mod } q)} e^{-\pi n^2 t}\Big).$$

We now write $n = qm + b$ in the inner sum:

$$\sum_{n\equiv b(\text{mod } q)} e^{-\pi n^2 t} = \sum_{m=-\infty}^{\infty} e^{-\pi t(qm+b)^2}.$$

Let $\theta(t,a) = \sum_{n=-\infty}^{\infty} e^{2\pi i n a - n^2\pi t}$. Then, by Exercise 5.1.7, we have

$$\sum_{m=-\infty}^{\infty} e^{-\pi t q^2 (m+b/q)^2} = \left(tq^2\right)^{-1/2} \theta\left(\frac{1}{tq^2}, \frac{b}{q}\right).$$

Hence,

$$\lim_{t\to 0} \sqrt{t}\theta\left(t + \frac{2pi}{q}\right) = \lim_{t\to 0} \sqrt{t}\sum_{b=0}^{q-1} e^{-2\pi i b^2 p/q}(tq^2)^{-1/2}\theta\left(\frac{1}{tq^2}, \frac{b}{q}\right).$$

As $t \to 0$, $1/tq^2 \to \infty$, and the θ-term goes to 1. The result now follows. □

5.5.20 *Let $r = p/q$. Show that*

$$\lim_{t\to 0} \sqrt{\frac{t}{t+2ir}}\theta\left(\frac{1}{t+2ir}\right) = \frac{(1-i)}{4\sqrt{pq}} C\left(\frac{q}{4p}\right),$$

with notation as in the previous exercise.

Write

$$\frac{1}{t+2ir} = \tau - \frac{i}{2r},$$

where

$$\tau = \frac{it^2 + 2rt}{2r(t^2+4r^2)}.$$

Then the limit in question is

$$\lim_{t\to 0} \sqrt{\frac{t}{t+2ir}} \sum_{n=-\infty}^{\infty} e^{-\pi n^2(\tau - i/2r)}$$

$$= \lim_{t\to 0} \sqrt{\frac{t}{t+2ir}} \sum_{b=0}^{4p-1} e^{2\pi i b^2 q/4p} \sum_{n\equiv b \,(\mathrm{mod}\ 4p)} e^{-\pi n^2 \tau},$$

which is treated as in the previous exercise. The limit is easily evaluated to be

$$\sqrt{\frac{2p}{qi}} \frac{1}{4p} C\left(\frac{q}{4p}\right).$$

Since

$$\Theta(z) = \sum_{-\infty}^{\infty} e^{\pi i n^2 z}$$

is analytic for $\text{Im}(z) > 0$, the functional equation of the θ-function extends to Θ:

$$\Theta(-1/z) = \sqrt{\frac{z}{i}}\Theta(z).$$

Now, $\sqrt{z}$ is well-defined on the cut plane $\mathbb{C} \setminus (-\infty, 0]$. This means that $i = e^{i\pi/2}$ and

$$\frac{1}{\sqrt{i}} = \frac{\sqrt{2}}{2}(1-i)$$

in the above limit that was evaluated. This completes the proof. □

5.5.21 *Deduce from the previous exercise the law of quadratic reciprocity:*

$$\left(\frac{p}{q}\right)\left(\frac{q}{p}\right) = (-1)^{\frac{p-1}{2}\cdot\frac{q-1}{2}}$$

for odd primes p and q, and (a/b) denotes the Legendre symbol.

The limits in the two previous exercises are equal by the functional equation of the θ-function. Therefore,

$$\frac{1}{q}C\left(-\frac{p}{q}\right) = \frac{1-i}{4\sqrt{pq}}C\left(\frac{q}{4p}\right).$$

We have

$$C\left(-\frac{p}{q}\right) = (-1)^{(q-1)/2}\left(\frac{p}{q}\right)C\left(\frac{1}{q}\right)$$

and it is easily checked that

$$C\left(\frac{q}{4p}\right) = C\left(\frac{pq}{4}\right)C\left(\frac{4q}{p}\right).$$

Also

$$C\left(\frac{pq}{4}\right) = 2(1 + i^{pq}).$$

We use Exercise 5.5.13 (or put $p = 1$ in the above identity relating $C(-p/q)$ with $C(q/4p)$) to deduce

$$C(1/q) = \begin{cases} \sqrt{q} & \text{if } q \equiv 1 \pmod 4, \\ i\sqrt{q} & \text{if } q \equiv 3 \pmod 4. \end{cases}$$

Moreover, $C(4q/p) = (q/p)C(1/p)$, so that

$$\frac{(-1)^{\frac{q-1}{2}}}{q}\left(\frac{p}{q}\right)C\left(\frac{1}{q}\right) = \frac{(1-i)(1+i^{pq})}{2\sqrt{pq}}\left(\frac{q}{p}\right)C\left(\frac{1}{p}\right)$$

from which the result easily follows. □

5.5.22 *Suppose that $f(s)$ is an entire function satisfying the functional equation*

$$A^s\Gamma(s)f(s) = A^{1-s}\Gamma(1-s)f(1-s).$$

Show that if $f(1/2) \neq 0$, then

$$f'\left(\frac{1}{2}\right) = -f(1/2)\left(\log A + \frac{\Gamma'(1/2)}{\Gamma(1/2)}\right).$$

We logarithmically differentiate the functional equation and set $s = 1/2$ to get the desired result. □

6
Hadamard Products

6.1 Jensen's theorem

6.1.4 *Show that*

$$\int_0^R \frac{n(r)dr}{r} \leq \max_{|z|=R} \log |f(z)| - \log |f(0)|,$$

with f as in Jensen's theorem.

Let us order the z_i so that

$$0 = |z_0| < |z_1| \leq |z_2| \leq \cdots \leq |z_n| < |z_{n+1}| = R.$$

Then

$$\begin{aligned}
\int_0^R \frac{n(r)}{r} dr &= \sum_{i=0}^{n} \int_{|z_i|}^{|z_{i+1}|} \frac{n(r)}{r} dr \\
&= \log \frac{|z_2|}{|z_1|} + 2 \log \frac{|z_3|}{|z_2|} + \cdots + n \log \frac{R}{|z_n|} \\
&= \log \left(\frac{R^n}{|z_1||z_2| \cdots |z_n|} \right).
\end{aligned}$$

The result is now clear from Jensen's theorem. □

6.1.5 *If $f(z)$ is of order β, show that $n_f(r) = O(r^{\beta+\epsilon})$, for any $\epsilon > 0$.*

Since

$$\max_{|z|=R} |f(z)| \ll \exp\left(R^{\beta+\epsilon}\right),$$

we get by Exercise 6.1.4 that

$$\int_0^{2R} \frac{n(r)}{r} dr \ll R^{\beta+\epsilon}.$$

But then

$$\int_R^{2R} \frac{n(r)}{r} dr \ll R^{\beta+\epsilon},$$

so that $n(R)\log 2 \ll R^{\beta+\epsilon}$, as desired. □

6.1.6 *Let $f(z)$ be an entire function of order β. Show that*

$$\sum_{n=1}^{\infty} |z_n|^{-\beta-\epsilon}$$

converges for any $\epsilon > 0$ (Here, we have indexed the zeros z_i so that $|z_1| \le |z_2| \le \cdots$).

By partial summation,

$$\sum_{n=1}^{\infty} |z_n|^{-\beta-\epsilon} \ll \int_1^{\infty} \frac{n(r)dr}{r^{\beta+1+\epsilon}}.$$

By Exercise 6.1.5, $n(r) \ll r^{\beta+\epsilon/2}$, and therefore the integral converges.

6.2 The Gamma Function

6.3.1 *Show that*

$$\int_0^{\infty} \frac{v^{x-1}dv}{1+v} = \frac{\pi}{\sin \pi x}$$

for $0 < x < 1$.

Consider the integral

$$\int_{C_\epsilon} \frac{z^{x-1}dz}{1+z},$$

where C_ϵ is the contour taken along the real axis from ϵ to R, then in the positive direction along the circle c_1 of radius R centered at the origin, and then back along the real axis to $z = \epsilon$ and finally around the circle c_2 of radius ϵ centered at the origin, taken in the negative direction.

The function

$$\frac{z^{x-1}}{1+z}$$

is regular except at $z = -1$, where it has a simple pole with residue

$$e^{\pi i(x-1)}.$$

We will take $\epsilon < 1 < R$ so that integrating the function along the contour indicated above shows by Cauchy's theorem

$$\int_\epsilon^R \frac{u^{x-1}du}{1+u} + \int_{c_1} \frac{z^{x-1}dz}{1+z} + \int_R^\epsilon \frac{(ue^{2\pi i})^{x-1}du}{1+u} + \int_{c_2} \frac{z^{x-1}dz}{1+z}$$
$$= (2\pi i)e^{\pi i(x-1)}.$$

The two integrals along the real axis together give

$$(1 - e^{2\pi i(x-1)}) \int_\epsilon^R \frac{u^{x-1}du}{1+u} = -2ie^{\pi ix}(\sin \pi x) \int_\epsilon^R \frac{u^{x-1}du}{1+u}.$$

The other two integrals tend to 0 as $R \to \infty$ because on c_1,

$$\left|\frac{z^{x-1}}{1+z}\right| \leq \frac{R^{x-1}}{R-1},$$

so that

$$\left|\int_{c_1} \frac{z^{x-1}dz}{1+z}\right| \leq \frac{R^{x-1}}{R-1} 2\pi R = \frac{2\pi R^x}{R-1},$$

which tends to 0, since $x < 1$.

Similarly,

$$\left|\int_{c_2} \frac{z^{x-1}dz}{1+z}\right| \leq \frac{2\pi \epsilon^x}{1-\epsilon},$$

which tends to 0, as $\epsilon \to 0$ since $x > 0$. Therefore,

$$-2ie^{\pi ix}(\sin \pi x) \int_0^\infty \frac{u^{x-1}du}{1+u} = -2\pi i e^{\pi ix},$$

which gives

$$\int_0^\infty \frac{u^{x-1}}{1+u} = \frac{\pi}{\sin \pi x}.$$

□

6.3.2 *Show that*

$$\Gamma(x)\Gamma(y) = 2\Gamma(x+y) \int_0^{\pi/2} (\cos\theta)^{2x-1}(\sin\theta)^{2y-1}d\theta$$

for $x, y > 0$.

For $x, y > 0$, we have

$$\Gamma(x)\Gamma(y) = \left(\int_0^\infty e^{-t}t^{x-1}dt\right)\left(\int_0^\infty e^{-u}u^{y-1}du\right).$$

Putting $u = tv$ and inverting the order of integration, we obtain

$$\begin{aligned}
\Gamma(x)\Gamma(y) &= \int_0^\infty e^{-t}t^{x-1}dt \int_0^\infty t^y v^y e^{-tv}\frac{dv}{v} \\
&= \int_0^\infty v^{y-1}dv \int_0^\infty e^{-t(1+v)}t^{x+y-1}dt \\
&= \Gamma(x+y)\int_0^\infty \frac{v^{y-1}dv}{(1+v)^{x+y}}.
\end{aligned}$$

The interchanging of integrals is easily justified by Fubini's theorem. This last integral is

$$2\int_0^{\pi/2} (\cos\theta)^{2x-1}(\sin\theta)^{2y-1}d\theta,$$

where we have put $v = \tan^2\theta$.

Again, making a substitution of $\lambda = \cos^2\theta$ transforms the integral to

$$\int_0^1 \lambda^{x-1}(1-\lambda)^{y-1}d\lambda,$$

which is the familiar beta function $B(x, y)$.

For $0 < x < 1$, we obtain

$$\Gamma(x)\Gamma(1-x) = \int_0^1 \lambda^{x-1}(1-\lambda)^{-x} d\lambda,$$

Putting

$$v = \frac{\lambda}{1-\lambda}$$

in the integral gives

$$\Gamma(x)\Gamma(1-x) = \int_0^\infty \frac{v^{x-1} dv}{1+v},$$

which by Exercise 6.3.1 is

$$\frac{\pi}{\sin \pi x},$$

which gives the desired result. □

6.3.3 *Show that*

$$\Gamma(x)\Gamma(y) = \Gamma(x+y) \int_0^1 \lambda^{x-1}(1-\lambda)^{y-1} dy.$$

(The integral is denoted by $B(x, y)$ and called the beta function.)

Making the substitution of $\lambda = \cos^2 \theta$ in the integral of Exercise 6.3.2 gives

$$\int_0^1 \lambda^{x-1}(1-\lambda)^{y-1} d\lambda,$$

which is the familiar beta function $B(x, y)$. □

6.3.4 *Prove that*

$$\Gamma(x)\Gamma(1-x) = \frac{\pi}{\sin \pi x}$$

for $0 < x < 1$.

This is clear from the solution to Exercise 6.3.2. □

6.3.5 *Prove that*

$$\Gamma\left(\frac{1}{2}\right) = \sqrt{\pi}.$$

In Exercise 6.3.4, put $x = y = \frac{1}{2}$ to obtain

$$\Gamma\left(\frac{1}{2}\right)^2 = 2\Gamma(1) \int_0^{\pi/2} d\theta = \pi.$$

Since $\Gamma(1/2)$ is positive, we obtain $\Gamma(1/2) = \sqrt{\pi}$. □

6.3.6 *Show that*

$$\Gamma(2x)\Gamma\left(\frac{1}{2}\right) = 2^{2x-1}\Gamma(x)\Gamma\left(x+\frac{1}{2}\right)$$

for $x > 0$.

In Exercise 6.3.3, put $x = y$ to obtain

$$\begin{aligned}\Gamma(x)^2 &= \Gamma(2x)\int_0^1 \lambda^{x-1}(1-\lambda)^{x-1}d\lambda \\ &= 2\Gamma(2x)\int_0^{1/2} \lambda^{x-1}(1-\lambda)^{x-1}d\lambda.\end{aligned}$$

Since $\lambda(1-\lambda) \leq \frac{1}{4}$, we may write $\lambda(1-\lambda) = \frac{1}{4} - \frac{t}{4}$, so that $\lambda = \frac{1-\sqrt{t}}{2}$. This substitution gives

$$\begin{aligned}\Gamma(x)^2 &= \frac{\Gamma(2x)}{2}\int_0^1 \left(\frac{1-t}{4}\right)^{x-1}\frac{dt}{\sqrt{t}} \\ &= 2^{1-2x}\Gamma(2x)\int_0^1 (1-t)^{x-1}t^{-1/2}dt.\end{aligned}$$

The latter integral is, by Exercise 6.3.3,

$$\frac{\Gamma(x)\Gamma\left(\frac{1}{2}\right)}{\Gamma\left(x+\frac{1}{2}\right)}.$$

Inserting this in the penultimate step gives the desired result. □

6.3.7 *Let c be a positive constant. Show that as $x \to \infty$,*

$$\Gamma(x+c) \sim x^c\Gamma(x).$$

Suppose first that $c > 1$. Then, by Exercise 6.3.3,

$$\begin{aligned}\frac{\Gamma(x)\Gamma(c)}{\Gamma(x+c)} &= \int_0^1 (1-\lambda)^{c-1}\lambda^{x-1}d\lambda \\ &= \int_0^\infty (1-e^{-t})^{c-1}e^{-xt}dt \\ &= \int_0^\infty t^{c-1}e^{-xt}dt - \int_0^\infty \left\{t^{c-1} - (1-e^{-t})^{c-1}\right\} e^{-xt}dt.\end{aligned}$$

The first integral is $\Gamma(c)x^{-c}$. The second integral is easily estimated as follows. Notice that $1 - e^{-t} < t$ for $t > 0$, and that

$$1 - e^{-t} > t - \frac{1}{2}t^2$$

for $0 < t < 1$. Thus, the second integral is positive and less than

$$\int_0^1 \left\{1 - (1-t/2)^{c-1}\right\} t^{c-1}e^{-xt}dt + \int_1^\infty t^{c-1}e^{-xt}dt.$$

For $0 < x < 1$, we have for $c > 1$,

$$1 - \left(1 - \frac{t}{2}\right)^{c-1} \leq t,$$

as is easily checked by elementary calculus. Thus, the second integral is less than

$$\int_0^1 t^c e^{-xt}dt + \int_1^\infty t^c e^{-xt}dt = \frac{\Gamma(c+1)}{x^{c+1}}.$$

This proves the result for $c > 1$. For $0 < c < 1$, we can use the formula

$$\Gamma(x+1) = x\Gamma(x)$$

to deduce the result. □

6.3.8 *Show that*

$$\Gamma(x) \sim e^{-x}x^{x-1/2}\sqrt{2\pi}$$

as $x \to \infty$.

By Exercise 2.1.12 we know that for a natural number n,

$$\log \Gamma(n) = \log(n-1)! = \left(n - \frac{1}{2}\right)\log n - n + c_1 + o(1)$$

as $n \to \infty$ (and with c_1 an absolute constant). If x is not an integer, let us write $x = n + c$ for some $0 < c < 1$. By Exercise 6.3.7, we have

$$\Gamma(n+c) \sim n^c \Gamma(n),$$

so that

$$\begin{aligned} \log \Gamma(x) &= \log \Gamma(n) + c \log n + o(1) \\ &= \left(x - \frac{1}{2}\right)\log n - n + c_1 + o(1). \end{aligned}$$

Also,

$$\log\left(\frac{n+c}{n}\right) = \log\left(1 + \frac{c}{n}\right) = \frac{c}{n} + O\left(\frac{1}{n^2}\right),$$

so that

$$\log x = \log n + \frac{c}{n} + O\left(\frac{1}{x^2}\right).$$

Inserting this observation above gives

$$\log \Gamma(x) = \left(x - \frac{1}{2}\right)\log x - x + c_1 + o(1).$$

We can use the duplication formula to evaluate c_1. Indeed, on the one hand we have from above

$$\log \Gamma(2x) = \left(2x - \frac{1}{2}\right)\log 2x - 2x + c_1 + o(1).$$

On the other hand, by the duplication formula (Exercise 6.3.6) we have

$$\log \Gamma(2x) = (2x-1)\log 2 + \log \Gamma(x) + \log \Gamma\left(x + \frac{1}{2}\right) - \frac{1}{2}\log \pi,$$

which is equal to

$$\left(2x - \frac{1}{2}\right)\log 2x - 2x - \frac{1}{2}\log 2 + 2c_1 - \frac{1}{2}\log \pi + o(1),$$

so that

$$c_1 = 2c_1 - \frac{1}{2}\log \pi - \frac{\log 2}{2}.$$

Thus, as required

$$c_1 = \log \sqrt{2\pi}. \qquad \square$$

6.3.9 *Show that* $1/\Gamma(z)$ *is an entire function with simple zeros at* $z = 0, -1, -2, \ldots$.

From the functional equation

$$\Gamma(z)\Gamma(1-z) = \frac{\pi}{\sin \pi z},$$

we see that $\Gamma(z)\Gamma(1-z)$ is regular except when z is an integer, in which case it has a simple pole.

We also see from this functional equation that since $\Gamma(z)$ is regular in $\mathrm{Re}(z) > 0$, $\Gamma(1-z)$ has simple poles at $z = 1, 2, 3, \ldots$. Therefore,

$$1/\Gamma(z) = \Gamma(1-z)(\sin \pi z)/\pi$$

is regular in $\mathrm{Re}(1-z) \geq 0$. If $\mathrm{Re}(z) \leq 0$, then $\mathrm{Re}(1-z) \geq 1$ and the right-hand side of the above equation is regular. This completes the proof. $\square$

6.3.10 *Show that for some constant* K,

$$\frac{\Gamma'(z)}{\Gamma(z)} = \int_0^1 \left\{1 - (1-t)^{z-1}\right\} \frac{dt}{t} - K.$$

By Exercise 6.3.3, we have

$$\begin{aligned}\frac{\Gamma(z-h)\Gamma(h)}{\Gamma(z)} &= \int_0^1 t^{h-1}(1-t)^{z-h-1} dt \\ &= \frac{1}{h} + \int_0^1 \left\{(1-t)^{z-h-1} - 1\right\} t^{h-1} dt.\end{aligned}$$

The Taylor expansion of the left-hand side with respect to h is

$$\begin{aligned}&\frac{1}{\Gamma(z)}\left\{\Gamma(z) - \Gamma'(z)h + \cdots\right\}\left\{\frac{1}{h} + K + \cdots\right\} \\ &= \frac{1}{h} - \frac{\Gamma'(z)}{\Gamma(z)} + K + O(h).\end{aligned}$$

The Taylor expansion of the right-hand side is

$$= \frac{1}{h} + \int_0^1 \left\{ (1-t)^{z-1} - 1 \right\} \frac{dt}{t} + O(h),$$

so that by equating the constant terms we get the desired result. □

6.3.11 *Show that for z not equal to a negative integer,*

$$\frac{\Gamma'(z)}{\Gamma(z)} = \sum_{n=0}^{\infty} \left(\frac{1}{n+1} - \frac{1}{n+z} \right) - K$$

for some constant K.

First, for $z > 1$, we use Exercise 6.3.10 and expand

$$\frac{1}{t} = \sum_{n=0}^{\infty} (1-t)^n$$

in the integrand and integrate term by term to obtain the result. The step is valid for $z > 1$ and by analytic continuation for all z unequal to a negative integer. □

6.3.12 *Derive the Hadamard factorization of $1/\Gamma(z)$:*

$$1/\Gamma(z) = e^{\gamma z} z \prod_{n=1}^{\infty} \left(1 + \frac{z}{n} \right) e^{-z/n},$$

where γ denotes Euler's constant.

We integrate the formula

$$\frac{\Gamma'(z)}{\Gamma(z)} = \sum_{n=0}^{\infty} \left(\frac{1}{n+1} - \frac{1}{n+z} \right) - K$$

from $z = 1$ to $z = w$ and take exponentials, to obtain

$$\frac{1}{\Gamma(z)} = e^{Bz} \prod_{n=1}^{\infty} \left(1 + \frac{z}{n} \right) e^{-z/n}$$

for some constant B. Putting $z = 1$ gives

$$\begin{aligned} 0 &= B + \sum_{n=1}^{\infty} \left\{ \log\left(1 + \frac{1}{n}\right) - \frac{1}{n} \right\} \\ &= B + \lim_{N\to\infty} \sum_{n=1}^{N} \left\{ \log\left(1 + \frac{1}{n}\right) - \frac{1}{n} \right\} \\ &= B - \gamma. \qquad \square \end{aligned}$$

6.3.13 *Show that*

$$\log \Gamma(z) = \left(z - \frac{1}{2}\right) \log z - z + \frac{1}{2} \log 2\pi + \int_0^{\infty} \frac{[u] - u + \frac{1}{2}}{u + z} du.$$

By Exercise 6.3.12,

$$\log \Gamma(z) = \sum_{n=1}^{\infty} \left\{ \frac{z}{n} - \log\left(1 + \frac{z}{n}\right) \right\} - \gamma z - \log z$$

with each logarithm having its principal value.

By Exercise 2.1.12, we see that

$$\begin{aligned} \sum_{n=1}^{N-1} \left\{ \frac{z}{n} - \log\left(1 + \frac{z}{n}\right) \right\} &= \log(N-1)! + z\left(1 + \frac{1}{2} + \cdots + \frac{1}{N-1}\right) \\ &\quad -\left(z + \frac{1}{2}\right) \log z \\ &\quad -\left(N - \frac{1}{2} + z\right) \log(N+z) - N \\ &\quad + \int_0^N \frac{B_1(u)du}{u+z}. \end{aligned}$$

Letting $N \to \infty$, and using

$$1 + \frac{1}{2} + \cdots + \frac{1}{N-1} = \log N + \gamma + o(1)$$

as well as

$$\log(N+z) = \log N + \frac{z}{N} + O\Big(\frac{1}{N^2}\Big),$$

we obtain the desired result by an application of Stirling's formula. This completes the proof. □

6.3.14 *For any $\delta > 0$, show that*

$$\log\Gamma(z) = \left(z - \frac{1}{2}\right)\log z - z + \frac{1}{2}\log 2\pi + O\left(\frac{1}{|z|}\right)$$

uniformly for $-\pi + \delta \leq \arg z \leq \pi - \delta$.

By the previous exercise, it suffices to estimate

$$\int_0^\infty \frac{B_1(u)du}{u+z}.$$

Let us write $f(v) = \int_0^v \left([u] - u + \frac{1}{2}\right)du$. Then f is bounded, since $f(v+1) = f(v)$ for any integer v. Thus,

$$\int_0^\infty \frac{f'(u)}{u+z}du = \int_0^\infty \frac{f(u)du}{(u+z)^2}.$$

Writing $z = re^{i\varphi}$, we see that

$$\begin{aligned} |u+z|^2 &= (u+re^{i\varphi})(u+re^{-i\varphi}) \\ &= (u+r\cos\varphi)^2 + r^2\sin^2\varphi \\ &= u^2 + 2ur\cos\varphi + r^2. \end{aligned}$$

We break the integral into three parts,

$$\int_0^{r/2} + \int_{r/2}^{2r} + \int_{2r}^\infty.$$

Since f is bounded, each of these integrals is $O\left(\frac{1}{r}\right)$ as required. □

6.3.15 *If σ is fixed and $|t| \to \infty$, show that*

$$|\Gamma(\sigma+it)| \sim e^{-\frac{1}{2}\pi|t|}|t|^{\sigma-\frac{1}{2}}\sqrt{2\pi}.$$

This is immediate from Exercise 6.3.14.

6.3.16 *Show that* $1/\Gamma(z)$ *is of order* 1.

This is a consequence of Stirling's formula.

6.3.17 *Show that*

$$\frac{\Gamma'(z)}{\Gamma(z)} = \log z + O\left(\frac{1}{|z|}\right)$$

for $|z| \to \infty$ *in the angle* $-\pi + \delta < \arg z < \pi - \delta$ *for any fixed* $\delta > 0$.

By Exercise 6.3.13, we can differentiate the expression

$$\log \Gamma(z) = \left(z - \frac{1}{2}\right)\log z - z + \frac{1}{2}\log 2\pi + \int_0^\infty \frac{[u] - u + \frac{1}{2}}{u+z}du$$

to obtain

$$\frac{\Gamma'(z)}{\Gamma(z)} = \log z - \frac{1}{2z} - \int_0^\infty \frac{[u] - u + \frac{1}{2}}{(u+z)^2}du.$$

The integral is easily seen to be $O(1/|z|)$. □

6.3 Infinite Products for $\xi(s)$ and $\xi(s,\chi)$

6.4.1 *Show that for some constant c,*

$$|\xi(s)| < \exp(c|s|\log|s|)$$

as $|s| \to \infty$. *Conclude that* $\xi(s)$ *has order* 1.

By the functional equation,

$$\xi(s) = \xi(1-s),$$

so that it suffices to prove the result for $\sigma = \mathrm{Re}(s) \geq 1/2$.

Clearly,

$$\left|\frac{1}{2}s(s-1)\pi^{-s/2}\right| < \exp(c|s|),$$

and by Stirling's formula

$$|\Gamma(s/2)| < \exp(c|s|\log|s|),$$

which is valid in the range under consideration. We also have

$$\zeta(s) = \frac{s}{s-1} - s\int_1^\infty \frac{\{x\}}{x^{s+1}}dx,$$

valid for $\sigma > 0$. (Here $\{x\}$ denotes the fractional part of x.) The integral is bounded for $\sigma \geq 1/2$. Since

$$\lim_{|s|\to\infty} \frac{s}{s-1} = 1,$$

we see that for some constant c,

$$|\zeta(s)| < c|s|$$

as $|s| \to \infty$. Putting all this together shows that $\xi(s)$ satisfies the stated inequality. Observe also that $\zeta(s) \to 1$ as $s \to \infty$ through real values, and since $\log \Gamma(s) \sim s \log s$, we see that

$$|\xi(s)| > \exp(c_1 s \log s)$$

for such values of s. Therefore, $\xi(s)$ has order 1. □

6.4.2 *Prove that $\zeta(s)$ has infinitely many zeros in $0 \leq \text{Re}(s) \leq 1$.*

The zeros of $\zeta(s)$ in the stated region are precisely those of $\xi(s)$. If there were only finitely many zeros, $\xi(s)e^{-cs}$ would be a polynomial for some constant c. In particular,

$$|\xi(s)| \ll e^{A|s|},$$

for some constant A. This contradicts the observation(deduced from the solution to the previous exercise) that for a positive constant c_1,

$$|\xi(s)| > \exp(c_1 s \log s)$$

for real s tending to infinity. □

6.4.3 *Show that*

$$\xi(s) = e^{A+Bs} \prod_\rho \left(1 - \frac{s}{\rho}\right) e^{s/\rho},$$

where the product is over the nontrivial zeros of $\zeta(s)$ in the region $0 \leq \text{Re}(s) \leq 1$ and $A = -\log 2$, $B = -\gamma/2 - 1 + \frac{1}{2}\log 4\pi$, where γ is Euler's constant.

The existence of the factorization is clear since $\xi(s)$ has order 1. Since the trivial zeros of $\zeta(s)$ are canceled by the simple poles of $\Gamma(s/2)$, we see that the product must be over nontrivial zeros of $\zeta(s)$. Notice that

$$\begin{aligned}\xi(1) &= \lim_{s\to 1}\frac{1}{2}s(s-1)\pi^{-s/2}\Gamma\Big(\frac{s}{2}\Big)\zeta(s)\\ &= \frac{1}{2}\pi^{-1/2}\Gamma\Big(\frac{1}{2}\Big)\lim_{s\to 1}(s-1)\zeta(s)\\ &= \frac{1}{2}\pi^{-1/2}\Gamma\Big(\frac{1}{2}\Big)=\frac{1}{2}\end{aligned}$$

by Exercise 6.3.5. Therefore, $\xi(0)=\frac{1}{2}$, and consequently, $e^A=1/2$, as required. To evaluate B, we logarithmically differentiate $\xi(s)$:

$$\frac{\xi'(s)}{\xi(s)}=\frac{\zeta'(s)}{\zeta(s)}+\frac{1}{s-1}-\frac{1}{2}\log\pi+\frac{\Gamma'\Big(\frac{s}{2}+1\Big)}{2\Gamma\Big(\frac{s}{2}+1\Big)}$$

on the one hand,

$$\frac{\xi'(s)}{\xi(s)}=B+\sum_{\rho}\Big(\frac{1}{s-\rho}+\frac{1}{\rho}\Big),$$

so that

$$B=\frac{\xi'(0)}{\xi(0)}=-\frac{\xi'(1)}{\xi(1)}$$

from the functional equation. We therefore need to evaluate $\frac{\xi'(1)}{\xi(1)}$. For the Hadamard product for $1/\Gamma(s)$, we see that

$$\frac{\Gamma'\Big(\frac{s}{2}+1\Big)}{2\Gamma\Big(\frac{s}{2}+1\Big)}=\frac{\gamma}{2}+\sum_{n=1}^{\infty}\Big(\frac{1}{s+2n}-\frac{1}{2n}\Big),$$

so that

$$-\frac{\Gamma'(3/2)}{2\Gamma(3/2)}=\frac{\gamma}{2}-1+\log 2,$$

since

$$\sum_{n=1}^{\infty}\frac{(-1)^n}{n}=-\log 2.$$

Thus,

$$\frac{\xi'(1)}{\xi(1)} = \lim_{s\to 1}\left\{\frac{\zeta'(s)}{\zeta(s)} + \frac{1}{s-1}\right\} - \frac{1}{2}\log 4\pi - \frac{\gamma}{2} + 1.$$

Now,

$$\zeta(s) = \frac{s}{s-1} - sI(s),$$

where

$$I(s) = \int_1^\infty \frac{\{x\}dx}{x^{s+1}},$$

so that

$$\lim_{s\to 1}\left\{\frac{\zeta'(s)}{\zeta(s)} + \frac{1}{s-1}\right\} = 1 - I(1).$$

Now,

$$I(1) = \int_1^\infty \frac{\{x\}}{x^2}dx = \lim_{N\to\infty}\int_1^N \frac{\{x\}}{x^2}dx,$$

and the latter integral is

$$\begin{aligned}
\int_1^N \frac{x-[x]}{x^2}dx &= \log N - \sum_{n=1}^{N-1} n\int_n^{n+1}\frac{dx}{x^2} \\
&= \log N - \sum_{n=1}^{N-1} n\left(\frac{1}{n} - \frac{1}{n+1}\right) \\
&= \log N - \sum_{n=1}^{N}\frac{1}{n} + 1 \\
&= 1-\gamma.
\end{aligned}$$

Therefore,

$$\frac{\xi'(1)}{\xi(1)} = \frac{\gamma}{2} + 1 - \frac{1}{2}\log 4\pi$$

and

$$B = -\frac{\gamma}{2} - 1 + \frac{1}{2}\log 4\pi,$$

as required. □

6.4.4 *Let χ be a primitive character* (mod q). *Show that $\xi(s,\chi)$ is an entire function of order* 1.

Recall that

$$L(s,\chi) = s\int_1^\infty \frac{S(x)}{x^{s+1}}dx,$$

where $S(x) = \sum_{n\leq x}\chi(n)$. Since $|S(x)| \leq q$, the integral converges for Re$(s) > 0$. Also, by the functional equation for $\xi(s,\chi)$, it suffices to estimate it for Re$(s) \geq \frac{1}{2}$. Thus, for $\sigma =$ Re$(s) \geq 1/2$,

$$|L(s,\chi)| \leq 2q|s|,$$

so that

$$\begin{aligned} |\xi(s,\chi)| &\leq 2q^{\frac{\sigma+3}{2}}|s|\left|\Gamma\left(\frac{s+a}{2}\right)\right| \\ &\leq q^{\frac{\sigma+3}{2}}\exp(C|s|\log|s|) \end{aligned}$$

for some suitable constant C. This inequality is best possible, since for $s \to \infty$ through real values, $L(s,\chi) \to 1$, and Stirling's formula implies that the above inequality cannot be improved. □

6.4.5 *Show that $L(s,\chi)$ has infinitely many zeros in $0 \leq$ Re$(s) \leq 1$ and that*

$$\xi(s,\chi) = e^{A+Bs}\prod_\rho\left(1-\frac{s}{\rho}\right)e^{s/\rho},$$

where the product is over the nontrivial zeros of $L(s,\chi)$.

The trivial zeros of $L(s,\chi)$ are cancelled by the $\Gamma((s+a)/2)$ factor. If $L(s,\chi)$ had only finitely many zeros in the critical strip ($0 \leq \sigma \leq 1$), then it would be a polynomial and hence of order zero, which is not the case. The final product follows from the Hadamard factorization theorem. □

6.4.6 *For A and B occurring in the previous exercise, show that*

$$e^A = \xi(0,\chi)$$

and that

$$\text{Re}(B) = -\sum_\rho \text{Re}\left(\frac{1}{\rho}\right),$$

where the sum is over nontrivial zeros ρ of $L(s,\chi)$.

Setting $s = 0$ in the Hadamard factorization of $\xi(s, \chi)$ gives $e^A = \xi(0, \chi)$. (By the functional equation, we can therefore express A in terms of $L(1, \overline{\chi})$.) Logarithmic differentiation of the Hadamard product and setting $s = 0$ gives

$$B = \frac{\xi'(0, \chi)}{\xi(0, \chi)} = -\frac{\xi'(1, \overline{\chi})}{\xi(1, \overline{\chi})}$$

by the functional equation. Writing B_χ for B (since it depends on χ), we find upon logarithmic differentiation of the expression for $\xi(s, \overline{\chi})$ and setting $s = 1$ that

$$\frac{\xi'(1, \overline{\chi})}{\xi(1, \overline{\chi})} = B_{\overline{\chi}} + \sum_{\overline{\rho}} \left(\frac{1}{1 - \overline{\rho}} + \frac{1}{\overline{\rho}} \right),$$

where the sum is over nontrivial zeros ρ of $L(s, \chi)$. Thus,

$$B_\chi = -B_{\overline{\chi}} - \sum_{\overline{\rho}} \left(\frac{1}{1 - \overline{\rho}} + \frac{1}{\overline{\rho}} \right).$$

Since $B_{\overline{\chi}} = \overline{B_\chi}$, we get

$$-2 \operatorname{Re}(B_\chi) = \sum_{\overline{\rho}} \operatorname{Re} \left(\frac{1}{1 - \overline{\rho}} \right) + \operatorname{Re} \left(\frac{1}{\overline{\rho}} \right).$$

The terms in the sum are nonnegative, and we can replace $1 - \overline{\rho}$ by ρ, since by the functional equation $1 - \overline{\rho}$ is also a zero of $L(s, \chi)$ whenever ρ is. Thus,

$$2 \operatorname{Re}(B_\chi) = -\sum_{\rho} \operatorname{Re} \left(\frac{1}{\rho} + \frac{1}{\overline{\rho}} \right),$$

so that

$$\operatorname{Re}(B) = -\sum_{\rho} \operatorname{Re} \left(\frac{1}{\rho} \right)$$

as required. □

6.4 Zero-Free Regions for $\zeta(s)$ and $L(s, \chi)$

6.5.1 *Show that*

$$-3 \frac{\zeta'(\sigma)}{\zeta(\sigma)} - 4 \operatorname{Re} \left(\frac{\zeta'(\sigma + it)}{\zeta(\sigma + it)} \right) - \operatorname{Re} \left(\frac{\zeta'(\sigma + 2it)}{\zeta(\sigma + 2it)} \right) \geq 0$$

for $t \in \mathbb{R}$ and $\sigma > 1$.

Since

$$3 + 4\cos\theta + \cos 2\theta \geq 0,$$

the result is clear (See Exercise 3.2.4). □

6.5.2 *For $1 < \sigma < 2$, show that*

$$-\frac{\zeta'(\sigma)}{\zeta(\sigma)} < \frac{1}{\sigma - 1} + A$$

for some constant A.

The function $f(s) = (s-1)\zeta(s)$ is regular, and nonvanishing for $\operatorname{Re}(s) \geq 1$. Hence,

$$\frac{f'(s)}{f(s)} = \frac{1}{s-1} + \frac{\zeta'(s)}{\zeta(s)}.$$

Since the left hand side is regular in $\operatorname{Re}(s) \geq 1$,

$$\frac{f'(\sigma)}{f(\sigma)}$$

is bounded by a constant for $1 \leq \sigma \leq 2$. This proves the result. □

6.5.3 *Prove that*

$$-\operatorname{Re}\left(\frac{\zeta'(s)}{\zeta(s)}\right) < A\log|t| - \sum_{\rho} \operatorname{Re}\left(\frac{1}{s-\rho} + \frac{1}{\rho}\right)$$

for $1 \leq \sigma \leq 2$ and $|t| \geq 2$.

By Exercise 6.4.3, we know that

$$\frac{\xi'(s)}{\xi(s)} = B + \sum_{\rho}\left(\frac{1}{s-\rho} + \frac{1}{\rho}\right)$$

and

$$\frac{\xi'(s)}{\xi(s)} = \frac{\zeta'(s)}{\zeta(s)} + \frac{1}{s-1} - \frac{1}{2}\log\pi + \frac{\Gamma'\left(\frac{s}{2}+1\right)}{2\Gamma\left(\frac{s}{2}+1\right)},$$

so that

$$-\frac{\zeta'(s)}{\zeta(s)} = \frac{1}{s-1} - B - \frac{1}{2}\log\pi + \frac{\Gamma'(\frac{s}{2}+1)}{2\Gamma(\frac{s}{2}+1)} - \sum_{\rho}\left(\frac{1}{s-\rho} + \frac{1}{\rho}\right).$$

By Exercise 6.3.17, the Γ-term is $O(\log t)$ for $|t| \geq 2$ and $1 \leq \sigma \leq 2$. Thus, in this region,

$$- \operatorname{Re}\left(\frac{\zeta'(s)}{\zeta(s)}\right) < A \log |t| - \sum_{\rho} \operatorname{Re}\left(\frac{1}{s-\rho} + \frac{1}{\rho}\right),$$

since

$$\operatorname{Re}\left(\frac{1}{s-1}\right) = \operatorname{Re}\left(\frac{1}{(\sigma-1)+it}\right) = \frac{\sigma+1}{(\sigma-1)^2+t^2} = O\left(\frac{1}{t^2}\right).$$

□

6.5.4 *Show that*

$$\operatorname{Re}\left(\frac{1}{s-\rho} + \frac{1}{\rho}\right) \geq 0.$$

Deduce that

$$- \operatorname{Re}\left(\frac{\zeta'(s)}{\zeta(s)}\right) < A \log |t|$$

for $1 \leq \sigma \leq 2$, $|t| \geq 2$.

Let us write $\rho = \beta + i\gamma$. Then,

$$\operatorname{Re}\left(\frac{1}{s-\rho}\right) = \frac{\sigma-\beta}{|s-\rho|^2}$$

and

$$\operatorname{Re}\left(\frac{1}{\rho}\right) = \frac{\beta}{|\rho|^2}.$$

Thus, by Exercise 6.5.3, we get the required estimate. □

6.5.5 *Let* $\rho = \beta + i\gamma$ *be any nontrivial zero of* $\zeta(s)$. *Show that*

$$- \operatorname{Re}\left(\frac{\zeta'(\sigma+it)}{\zeta(\sigma+it)}\right) < A \log |t| - \frac{1}{\sigma-\beta}.$$

In the sum in Exercise 6.5.3, by taking one term involving β we obtain the result. □

6.5.8 *Show that*

$$- \operatorname{Re}\left(\frac{\zeta'(s)}{\zeta(s)}\right) < \operatorname{Re}\left(\frac{1}{s-1}\right) + c_1 \log(|t|+2)$$

for some constant $c_1 > 0$ *and* $\sigma > 1$.

We proceed as in Exercise 6.5.3:

$$-\frac{\zeta'(s)}{\zeta(s)} = \frac{1}{s-1} - B - \frac{1}{2}\log \pi + \frac{\Gamma'(\frac{s}{2}+1)}{2\Gamma(\frac{s}{2}+1)} - \sum_{\rho}\left(\frac{1}{s-\rho}+\frac{1}{\rho}\right).$$

The sum over the zeros is positive. The Γ-term is $O(\log(|t|+2))$. Thus

$$-\operatorname{Re}\left(\frac{\zeta'(s)}{\zeta(s)}\right) < \operatorname{Re}\left(\frac{1}{s-1}\right) + c_1\log(|t|+2).$$

□

6.5.9 *Suppose that χ is a primitive character* (mod q) *satisfying $\chi^2 \neq \chi_0$. Show that there is a constant $c > 0$ such that $L(s,\chi)$ has no zero in the region*

$$\sigma > 1 - \frac{c}{\log(q|t|+2)}.$$

We proceed as in the case of the ζ-function. We first observe that

$$-3\frac{L'(\sigma,\chi_0)}{L(\sigma,\chi_0)} - 4\operatorname{Re}\left(\frac{L'(\sigma+it,\chi)}{L(\sigma+it,\chi)}\right) - \operatorname{Re}\left(\frac{L'(\sigma+it,\chi^2)}{L(\sigma+it,\chi^2)}\right) \geq 0$$

for $t \in \mathbb{R}$ and $\sigma > 1$. (Here we are using $\chi^2 \neq \chi_0$, for otherwise, the χ^2 term above will present difficulties.)

Observe that

$$-\frac{L'(\sigma,\chi_0)}{L(\sigma,\chi_0)} = \sum_{n=1}^{\infty}\frac{\chi_0(n)\Lambda(n)}{n^\sigma} \leq -\frac{\zeta'(\sigma)}{\zeta(\sigma)} < \frac{1}{\sigma-1} + c_1$$

for $1 < \sigma < 2$ and some constant $c_1 > 0$. Also (with the notation of Exercise 6.4.5),

$$\operatorname{Re}\left(\frac{\xi'(s,\chi)}{\xi(s,\chi)}\right) = \operatorname{Re}(B) + \sum_{\rho}\operatorname{Re}\left(\frac{1}{s-\rho}+\frac{1}{\rho}\right),$$

and for $a = 0$ or 1,

$$\operatorname{Re}\left(\frac{\xi'(s,\chi)}{\xi(s,\chi)}\right) = \frac{1}{2}\log\frac{q}{\pi} + \operatorname{Re}\left(\frac{\Gamma'(\frac{s+a}{2})}{2\Gamma(\frac{s+a}{2})}\right) + \operatorname{Re}\left(\frac{L'(s,\chi)}{L(s,\chi)}\right).$$

Thus,

$$-\operatorname{Re}\left(\frac{L'(s,\chi)}{L(s,\chi)}\right)$$

$$=\frac{1}{2}\log\frac{q}{\pi}+\operatorname{Re}\left(\frac{\Gamma'(\frac{s+a}{2})}{2\Gamma(\frac{s+a}{2})}\right)-\operatorname{Re}(B)-\sum_{\rho}\operatorname{Re}\left(\frac{1}{s-\rho}+\frac{1}{\rho}\right).$$

By Exercise 6.4.6,

$$\operatorname{Re}(B)=\sum_{\rho}\operatorname{Re}\left(\frac{1}{\rho}\right),$$

and the Γ-term is $O(\log(|t|+2))$ by Exercise 6.3.17. Thus,

$$-\operatorname{Re}\left(\frac{L'(s,\chi)}{L(s,\chi)}\right)<c_2\log(q|t|+2)-\sum_{\rho}\operatorname{Re}\left(\frac{1}{s-\rho}\right).$$

This estimate holds for any primitive character χ (mod q), real or complex. Since

$$\operatorname{Re}\left(\frac{1}{s-\rho}\right)\geq 0,$$

we can omit the series or any part of it in our estimations. Thus,

$$-\operatorname{Re}\left(\frac{L'(s,\chi^2)}{L(s,\chi^2)}\right)<c_2\log(q|t|+2),$$

provided that χ^2 is a primitive character (mod q). If χ^2 is not primitive, let χ_1 be the primitive character inducing χ^2. Then

$$\left|\frac{L'(s,\chi^2)}{L(s,\chi^2)}-\frac{L'(s,\chi_1)}{L(s,\chi_1)}\right|\leq\sum_{p|q}\frac{p^{-\sigma}\log p}{1-p^{-\sigma}}\leq\sum_{p|q}\log p\leq\log q.$$

Thus, the penultimate estimate remains valid whether χ^2 is primitive or not. Hence, as before, we get (by choosing $t=\gamma$)

$$-\operatorname{Re}\left(\frac{L'(\sigma+it,\chi)}{L(\sigma+it,\chi)}\right)<c_2\log q(|t|+2)-\frac{1}{\sigma-\beta},$$

so that

$$\frac{4}{\sigma-\beta}<\frac{3}{\sigma-1}+c_3\log q(|t|+2).$$

Taking $\sigma = 1 + \delta/\log q(|t|+2)$ with δ sufficiently small gives

$$\beta < 1 - c_4/\log q(|t|+2)$$

as required. □

6.5.10 *Show that the previous result remains valid when χ is a nonreal imprimitive character.*

If χ_1 induces χ, then the zeros of $L(s,\chi)$ are the zeros of $L(s,\chi_1)$ and the zeros of a finite number of factors of the form $1 - \chi_1(p)p^{-s}$. But the additional zeros are on the line $\sigma = 0$. Thus, the result of Exercise 6.5.9 holds for all characters χ (mod q) satisfying $\chi^2 \neq \chi_0$. □

6.5 Supplementary Problems

6.6.1 *Prove that $\Gamma(s)$ has poles only at $s = 0, -1, \ldots,$ and that these are simple, with*

$$\mathrm{Res}_{s=-k}\Gamma(s) = (-1)^k/k!.$$

By Exercise 6.3.9, we know that $1/\Gamma(s)$ is entire and has simple zeros at $s = 0, -1, -2, \ldots$. By the Hadamard factorization of $1/\Gamma(s)$ (Exercise 6.3.12), these are the only zeros. Thus, the first part of the question is established. For the second part, we need to calculate

$$\lim_{s \to -k} (s+k)\Gamma(s).$$

But $s\Gamma(s) = \Gamma(s+1)$, so that

$$\Gamma(s) = \frac{\Gamma(s+1)}{s} = \frac{\Gamma(s+2)}{s(s+1)} = \cdots = \frac{\Gamma(s+k)}{s(s+1)\cdots(s+k-1)}$$

by integration. Hence

$$\begin{aligned} \lim_{s \to -k} (s+k)\Gamma(s) &= \lim_{s \to -k} \frac{(s+k)\Gamma(s+k)}{s(s+1)\cdots(s+k-1)} \\ &= \lim_{s \to -k} \frac{\Gamma(s+k+1)}{s(s+1)\cdots(s+k-1)} \\ &= (-1)^k/k!. \end{aligned}$$

□

6.6.2 *Show that*

$$e^{-1/x} = \frac{1}{2\pi i}\int_{(\sigma)} x^s\Gamma(s)ds,$$

for any $\sigma > 1$, *and* $x \geq 1$.

We first truncate the infinite line integral at R and use Stirling's formula (Exercise 6.3.15) to estimate it. Thus

$$\left|\frac{1}{2\pi i}\int_{\sigma+iR}^{\sigma+i\infty} x^s\Gamma(s)ds\right| \ll x^\sigma \int_R^\infty e^{-\frac{\pi}{2}t}t^{\sigma-\frac{1}{2}}dt,$$

and the latter integrand is clearly e^{-cR} for some constant $c > 0$. A similar analysis applies to the range from $\sigma - iR$ to $\sigma - i\infty$. Thus,

$$\frac{1}{2\pi i}\int_{(\sigma)} x^s\Gamma(s)ds = \frac{1}{2\pi i}\int_{\sigma-iR}^{\sigma+iR} x^s\Gamma(s)ds + O\left(x^\sigma e^{-cR}\right).$$

As usual, we move the line of integration to $\text{Re}(s) = -N - \frac{1}{2}$, N a positive integer. We pick up the residue at the poles of $\Gamma(s)$, namely

$$\sum_{k=0}^{N}(-1)^k x^{-k}/k!.$$

The horizontal and vertical integrals are estimated easily using Stirling's formula. Indeed, the horizontal integral

$$\frac{1}{2\pi i}\int_{\sigma+iR}^{-(N+\frac{1}{2})+iR} x^s\Gamma(s)ds$$

is bounded by $O\left(x^\sigma N e^{-\frac{\pi}{2}R}\right)$. A similar estimate holds for the other horizontal integral. The vertical integral

$$\frac{1}{2\pi i}\int_{-(N+\frac{1}{2})-iR}^{-(N+\frac{1}{2})+iR} x^s\Gamma(s)ds$$

is bounded by

$$\ll x^{-N-\frac{1}{2}}\int_{-R}^{R}\left|\Gamma(-N-\frac{1}{2}+it)\right| dt.$$

Using the functional equation $s\Gamma(s) = \Gamma(s+1)$, we find that on repeated application of this

$$\Gamma(-N - 1/2 + it) = \frac{\Gamma(\frac{1}{2} + it)}{(-N - \frac{1}{2} + it) \cdots (-\frac{1}{2} + it)},$$

so that

$$|\Gamma(-N - 1/2 + it)| \leq \frac{|\Gamma(\frac{1}{2} + it)|}{N!}.$$

By Stirling's formula, $|\Gamma(\frac{1}{2} + it)| = O\left(e^{-\frac{\pi}{2}|t|}\right)$ and we deduce

$$\left| \frac{1}{2\pi i} \int_{-(N+\frac{1}{2})-iR}^{-(N+\frac{1}{2})+iR} x^s \Gamma(s) ds \right| = O\left(\frac{1}{N! x^{N+\frac{1}{2}}} \right).$$

We now choose $R = N$ and let $N \to \infty$ through the integers to deduce

$$\frac{1}{2\pi i} \int_{(\sigma)} x^s \Gamma(s) ds = \sum_{k=0}^{\infty} \frac{(-1)^k x^{-k}}{k!} = e^{-1/x}$$

as required. This could also be derived by Mellin inversion. □

6.6.3 *Let* $f(s) = \sum_{n=1}^{\infty} a_n/n^s$ *be an absolutely convergent Dirichlet series in the half-plane* $\text{Re}(s) > 1$. *Show that*

$$\sum_{n=1}^{\infty} a_n e^{-n/x} = \frac{1}{2\pi i} \int_{(\sigma)} f(s) x^s \Gamma(s) ds$$

for any $\sigma > 1$.

We have

$$\frac{1}{2\pi i} \int_{(\sigma)} \left(\sum_{n=1}^{\infty} \frac{a_n}{n^s} \right) x^s \Gamma(s) ds = \sum_{n=1}^{\infty} a_n \frac{1}{2\pi i} \int_{(\sigma)} \left(\frac{x}{n} \right)^s \Gamma(s) ds,$$

the interchange being justified by absolute convergence of the term on the left-hand side. By Exercise 6.6.2, the integral on the right-hand side is $e^{-n/x}$, which completes the proof. □

6.6.4 *Prove that*

$$\sin z = z \prod_{n=1}^{\infty} \left(1 - \frac{z^2}{n^2 \pi^2} \right).$$

We have

$$\sin z = \frac{e^{iz} - e^{-iz}}{2i},$$

so that

$$|\sin z| \ll e^{|z|}.$$

Since $\sin z$ is entire, the above estimate shows it has order 1. By Hadamard's factorization theorem,

$$\frac{\sin z}{z} = e^{A+Bz} \prod_{\substack{n \in \mathbb{Z} \\ n \neq 0}} \left(1 - \frac{z}{\pi n}\right) e^{-z/\pi n},$$

for some constants A, B. Combining the terms corresponding to $\pm n$ in the product gives

$$\frac{\sin z}{z} = e^{A+Bz} \prod_{n=1}^{\infty} \left(1 - \frac{z^2}{\pi^2 n^2}\right).$$

Letting $z \to \infty$ gives

$$1 = e^A,$$

so that $A = 0$. Also, $\sin(-z) = -\sin z$ yields

$$e^{Bz} = e^{-Bz},$$

so that $e^{2Bz} = 0$, forcing $B = 0$. Thus,

$$\sin z = z \prod_{n=1}^{\infty} \left(1 - \frac{z^2}{\pi^2 n^2}\right),$$

as desired. □

6.6.5 *Using the previous exercise, deduce that*

$$\sum_{n=1}^{\infty} \frac{1}{n^2} = \frac{\pi^2}{6}.$$

We have

$$\frac{\sin z}{z} = 1 - \frac{z^2}{6} + \frac{z^4}{120} - \cdots = \prod_{n=1}^{\infty} \left(1 - \frac{z^2}{\pi^2 n^2}\right).$$

Expanding the product on the right-hand side and comparing the coefficient of z^2 on both sides gives

$$-\frac{1}{6} = -\sum_{n=1}^{\infty} \frac{1}{\pi^2 n^2}$$

as desired. $\square$.

7 Explicit Formulas

7.1 Counting Zeros

7.1.1 *Let L be the line joining 2 to $2+iT$ and then $\frac{1}{2}+iT$. Show that*

$$\Delta_L \arg(s-1) = \frac{\pi}{2} + O\left(\frac{1}{T}\right).$$

We have

$$\Delta_L \arg(s-1) = \arg(iT - \frac{1}{2}) = \frac{\pi}{2} + \arcsin\left(\frac{1}{\sqrt{1+4T^2}}\right).$$

Since

$$\lim_{x\to 0} \frac{\sin x}{x} = 1,$$

we have

$$\lim_{x\to 0} \frac{\arcsin x}{x} = 1.$$

Thus,

$$\arcsin\left(\frac{1}{\sqrt{1+4T^2}}\right) = O\left(\frac{1}{T}\right),$$

which proves the assertion. □

7.1.2 *With L as in the previous exercise, show that*

$$\Delta_L \arg \pi^{-s/2} = -\frac{T}{2} \log \pi.$$

We have

$$\Delta_L \arg \pi^{-s/2} = \Delta_L(-\frac{1}{2} t \log \pi) = -\frac{1}{2} T \log \pi.$$

□

7.1.3 *With L as in the previous exercise, show that*

$$\Delta_L \arg \Gamma\left(\frac{s}{2}+1\right) = \frac{T}{2} \log \frac{T}{2} - \frac{T}{2} + \frac{3}{8}\pi + O\left(\frac{1}{T}\right).$$

By Stirling's formula,

$$\Delta_L \arg \Gamma\left(\frac{s}{2}+1\right) = \operatorname{Im} \log \Gamma\left(\frac{5}{4} + \frac{iT}{2}\right)$$

$$= \operatorname{Im}\left\{\left(\frac{3}{4} + \frac{iT}{2}\right) \log\left(\frac{5}{4} + \frac{iT}{2}\right) - \frac{5}{4} - \frac{iT}{2} + \frac{1}{2}\log 2\pi + O\left(\frac{1}{T}\right)\right\}.$$

This is easily calculated to be

$$\frac{T}{2} \log \frac{T}{2} - \frac{T}{2} + \frac{3}{8}\pi + O\left(\frac{1}{T}\right),$$

as required. □

7.1.4 *Show that*

$$\sum_\rho \frac{1}{1+(T-\gamma)^2} = O(\log T),$$

where the sum is over the nontrivial zeros $\rho = \beta + i\gamma$ of $\zeta(s)$.

By Exercise 6.5.3 we know that

$$-\operatorname{Re}\left(\frac{\zeta'(s)}{\zeta(s)}\right) < A \log|t| - \sum_\rho \operatorname{Re}\left(\frac{1}{s-\rho} + \frac{1}{\rho}\right)$$

for $1 \le \sigma \le 2$ and $|t| \ge 2$ with A an absolute constant. If we take $s = 2 + iT$ in this formula, we deduce

$$\sum_\rho \operatorname{Re}\left(\frac{1}{s-\rho} + \frac{1}{\rho}\right) < A_1 \log T$$

for some constant A_1, since $|\zeta'/\zeta|$ is bounded for $\text{Re}(s) = 2$. But

$$\text{Re}\left(\frac{1}{s-\rho}\right) = \frac{2-\beta}{(2-\beta)^2 + (T-\gamma)^2} \geq \frac{1}{4 + (T-\gamma)^2}$$

and

$$\text{Re}\left(\frac{1}{\rho}\right) = \frac{\beta}{|\rho|^2}.$$

Since

$$\sum_\rho \frac{1}{|\rho|^2} < \infty,$$

we deduce

$$\sum_\rho \frac{1}{4 + (T-\gamma)^2} < A_2 \log T$$

for some constant A_2. Since

$$4 + (T-\gamma)^2 \leq 4\left(1 + (T-\gamma)^2\right),$$

the required result is now immediate. □

7.1.5 *Let $N(T)$ be the number of zeros of $\zeta(s)$ with $0 < \text{Im}(s) \leq T$. Show that*

$$N(T+1) - N(T) = O(\log T).$$

We must count zeros $\rho = \beta + i\gamma$ satisfying $T \leq \gamma \leq T + 1$. Thus, $0 \leq \gamma - T \leq 1$. From the previous exercise, the contribution of such zeros to the sum is greater than or equal to $1/2$. Hence, the estimate now follows from the previous exercise. □

7.1.6 *Let $s = \sigma + it$ with t unequal to an ordinate of a zero. Show that for large $|t|$ and $-1 \leq \sigma \leq 2$,*

$$\frac{\zeta'(s)}{\zeta(s)} = \sum_\rho{}' \frac{1}{s-\rho} + O(\log |t|),$$

where the dash on the summation is limited to those ρ for which $|t-\gamma| < 1$.

From the formula

$$-\frac{\zeta'(s)}{\zeta(s)} = \frac{1}{s-1} - B - \frac{1}{2}\log \pi + \frac{\Gamma'(s/2+1)}{2\Gamma(s/2+1)} - \sum_\rho \left(\frac{1}{s-\rho} + \frac{1}{\rho}\right)$$

evaluated first at $s = \delta + it$ and then at $2 + it$ and subtracting gives

$$\frac{\zeta'(s)}{\zeta(s)} = \sum_{\rho} \left(\frac{1}{s-\rho} - \frac{1}{2+it-\rho} \right) + O(\log |t|)$$

because of the estimate for the growth of the Γ-term (see Exercise 6.3.17). Note that

$$\left| \frac{1}{s-\rho} - \frac{1}{2+it-\rho} \right| = \frac{2-\sigma}{|(s-\rho)(2+it-\rho)|} \leq \frac{3}{|t-\gamma|^2},$$

so that the contribution of the zeros satisfying $|t-\gamma| \geq 1$ is

$$\sum_{|t-\gamma|\geq 1} \frac{3}{|t-\gamma|^2} \leq \sum_{\rho} \frac{6}{1+|t-\gamma|^2},$$

and the latter sum is $O(\log |t|)$ by Exercise 7.1.4. Finally, in the remaining terms, $|\gamma - t| < 1$, and we have

$$|2 + it - \rho| \geq 1$$

for such zeros. The number of such zeros is $O(\log |t|)$ by the previous exercise. Putting this all together gives the desired result. □

7.2 Explicit Formula for $\psi(x)$

7.2.1 *Show that if x is not a prime power and $x > 1$, then*

$$\psi(x) = \frac{1}{2\pi i} \int_{c-iR}^{c+iR} -\frac{\zeta'(s)}{\zeta(s)} \frac{x^s}{s} ds$$

$$+ O\Big(\sum_{n=1}^{\infty} \Lambda(n) \Big(\frac{x}{n}\Big)^c \min(1, R^{-1} |\log \frac{x}{n}|^{-1}) \Big).$$

Since x is not a prime power,

$$\psi(x) = \sum_{n=1}^{\infty} \Lambda(n) \delta\Big(\frac{x}{n}\Big).$$

By Theorem 4.1.4, the result is now immediate. □

7.2.2 *Prove that if x is not an integer, then*

$$\sum_{\frac{1}{2}x<n<2x} \left|\log \frac{x}{n}\right|^{-1} = O\left(\frac{x}{||x||}\log x\right),$$

where $||x||$ denotes the distance of x to the nearest integer.

Let x_1 be the largest integer less than x. Split the sum into two parts: $\frac{1}{2}x < n < x$ and $x < n < 2x$. Writing $n = x_1 - v$, we have

$$\log \frac{x}{n} \geq \log \frac{x_1}{n} = -\log\left(1 - \frac{v}{x_1}\right) > \frac{v}{x_1}.$$

Thus,

$$\sum_{\frac{1}{2}x<n<x_1} \left|\log \frac{x}{n}\right|^{-1} \leq \sum_{v=1}^{x} \frac{x_1}{v} = O(x \log x).$$

For $n = x_1$, we have

$$\log \frac{x}{x_1} = -\log \frac{x - \{x\}}{x} \geq \frac{\{x\}}{x}.$$

The analysis for the range $x < n < 2x$ is similar. Putting this all together gives the stated result. □

7.2.3 *By choosing $c = 1 + \frac{1}{\log x}$ in the penultimate exercise, deduce that*

$$\psi(x) = \frac{1}{2\pi i}\int_{c-iR}^{c+iR} -\frac{\zeta'(s)}{\zeta(s)}\frac{x^s}{s}ds + O\left(\frac{x\log^2 x}{R}\right)$$

if $x - \frac{1}{2}$ is a positive integer.

By Exercise 7.2.1, we must estimate

$$\sum_{n=1}^{\infty} \Lambda(n)\left(\frac{x}{n}\right)^c \min\left(1, R^{-1}|\log\frac{x}{n}|^{-1}\right)$$

with $c = 1 + 1/\log x$. Indeed, if $n < \frac{1}{2}x$, or $n > 2x$, $|\log\frac{x}{n}|^{-1}$ is bounded, and the contribution of such terms is

$$\ll \frac{x}{R}\sum_{n=1}^{\infty}\frac{\Lambda(n)}{n^c}.$$

By partial summation

$$\sum_{n=1}^{\infty} \frac{\Lambda(n)}{n^c} \ll c \int_1^{\infty} \frac{\psi(t)dt}{t^{c+1}} \ll \log x$$

by an application of Chebyshev's estimate for $\psi(x)$. Thus, the contribution from the terms $n < \frac{1}{2}x$ or $n > 2x$ is

$$O\left(\frac{x \log x}{R}\right).$$

For $\frac{x}{2} < n < 2x$, we apply Exercise 7.2.2 and observe that in this range x/n is bounded. Since $||x|| = \frac{1}{2}$, we find that the contribution from n in this range is

$$O\left(\frac{x \log^2 x}{R}\right),$$

where we have used $\Lambda(n) \leq \log 2x$ for $n \leq 2x$. □

7.2.4 *Let C be the rectangle with vertices $c-iR$, $c+iR$, $-U+iR$, $-U-iR$, where $c = 1 + 1/\log x$ and U is an odd positive integer. Show that*

$$\frac{1}{2\pi i} \int_C -\frac{\zeta'(s)}{\zeta(s)} \frac{x^s}{s} ds = x - \sum_{|\gamma|<R} \frac{x^\rho}{\rho} - \frac{\zeta'(0)}{\zeta(0)} + \sum_{0<2m<U} \frac{x^{-2m}}{2m},$$

where we are writing the nontrivial zeros of $\zeta(s)$ as $\rho = \beta + i\gamma$. (R is chosen so that it is not the ordinate of any zero of $\zeta(s)$.)

By Cauchy's theorem, we need to compute the residue of the integrand whenever a pole occurs. Since $\zeta(s)$ has zeros at $s = -2m$ with $m > 0$, in addition to its nontrivial zeros, we must compute the residue of the integrand there. By Exercise 6.5.3 and the partial fraction expansion for

$$-\frac{\Gamma'\left(\frac{s}{2}+1\right)}{2\Gamma\left(\frac{s}{2}+1\right)},$$

we see that $-\zeta'(s)/\zeta(s)$ has a simple pole at $s = -2m$ with residue -1. Thus, the residue of the integrand above is $x^{-2m}/2m$ when $s = -2m$. The contribution of the remaining singularities is clear. □

7.2.5 *Recall that the number of zeros $\rho = \beta + i\gamma$ satisfying $|\gamma - R| < 1$ is $O(\log R)$. Show that we can ensure $|\gamma - R| \gg (\log R)^{-1}$ by varying R by a bounded amount.*

Consider the zeros $\rho = \beta + i\gamma$ satisfying $R - 1 < \gamma < R + 1$. The number of such zeros is $O(\log R)$. We subdivide the interval $[R-1, R+1]$ into equal parts of length $c/\log R$ for some constant c. The number of parts is $O(\log R)$, and we now choose c such that the number of parts exceeds the number of zeros. By the pigeonhole principle, there is a part that contains no zero. Thus for R_i lying in such a part, we must have $|R_i - \gamma| \gg (\log R_i)^{-1}$. Since $R_i - R = O(1)$, we have proved the desired result. □

7.2.6 *Let U be a positive odd number. Prove that*

$$|\zeta'(s)/\zeta(s)| \ll (\log 2|s|)$$

for $-U \le \sigma \le -1$, provided that we exclude circles of a fixed positive radius around the trivial zeros $s = -2, -4, \ldots$ of $\zeta(s)$.

The functional equation in its asymmetric form is

$$\zeta(1-s) = 2^{1-s}\pi^{-s}\Big(\cos\frac{\pi s}{2}\Big)\Gamma(s)\zeta(s).$$

The logarithmic derivative of the right-hand side is

$$-\log 2\pi - \frac{1}{2}\pi\tan\frac{\pi s}{2} + \frac{\Gamma'(s)}{\Gamma(s)} + \frac{\zeta'(s)}{\zeta(s)}.$$

We need to estimate this for $\sigma \ge 2$. The tangent term is bounded if $|s - (2m+1)| \ge r$ for some fixed r. The second term is $O(\log|s|)$ by Stirling's formula and therefore $O(\log 2|1-s|)$ if $\sigma \ge 2$. The last term is bounded in the region. This completes the proof. □

7.2.7 *In Exercise 7.2.4, letting $U \to \infty$ along the odd numbers and $R \to \infty$ appropriately (that is, as in Exercise 7.2.5) prove that*

$$\psi(x) = x - \sum_{\rho}\frac{x^\rho}{\rho} - \frac{\zeta'(0)}{\zeta(0)} + \frac{1}{2}\log\left(1 - x^{-2}\right),$$

whenever x is half more than an integer.

By Exercise 7.2.3,

$$\psi(x) = \frac{1}{2\pi i}\int_{c-iR}^{c+iR} -\frac{\zeta'(s)}{\zeta(s)}\frac{x^s}{s}ds + O\left(\frac{x\log^2 x}{R}\right).$$

We replace the vertical line segment by the contour C and take into account the contribution of the residues:

$$\psi(x) = \sum_{|\gamma|<R} \frac{x^\rho}{\rho} - \frac{\zeta'(0)}{\zeta(0)} + \sum_{0<2m<U} \frac{x^{-2m}}{2m} - I_R - I_U + O\left(\frac{x\log^2 x}{R}\right),$$

where I_R denotes the two horizontal integrals in the contour $\mathcal{C}$ and I_U denotes the vertical integral along $\mathrm{Re}(s) = -U$. By Exercise 7.1.6 we have

$$\frac{\zeta'(s)}{\zeta(s)} = \sum_{\rho}{}' \frac{1}{s-\rho} + O(\log R),$$

where the dash on the summation means $|R-\gamma| < 1$ and $-1 \le \sigma \le 2$. With R chosen as in the previous exercise, we can arrange

$$|\gamma - R| \gg (\log R)^{-1}.$$

The number of zeros in the summation is $O(\log R)$. Thus,

$$\frac{\zeta'(s)}{\zeta(s)} = O(\log^2 R)$$

for $-1 \le \sigma \le 2$. Thus the contribution to the horizontal integral I_R for this range of σ is

$$\ll (\log^2 R) \int_{-1}^{c} \left|\frac{x^s}{s} ds\right| \ll \frac{x\log^2 R}{R\log x}.$$

In the range $\sigma \le -1$, we use Exercise 7.2.6 to get

$$I_R \ll \frac{x\log^2 R}{R\log x} + \frac{\log 2R}{R} \int_{-U}^{-1} x^\sigma \, d\sigma,$$

which is

$$O\left(\frac{x\log^2 R}{R\log x}\right).$$

The vertical integral is

$$I_U \ll \frac{\log 2U}{U} \int_{-R}^{R} x^{-U} dt \ll \frac{R\log U}{Ux^R}.$$

We first let $U \to \infty$ along odd positive integers to obtain

$$\psi(x) = x - \sum_{|\gamma|<R} \frac{x^\rho}{\rho} - \frac{\zeta'(0)}{\zeta(0)} + \frac{1}{2}\log(1-x^{-2})$$

$$+ O\left(\frac{x\log^2 x}{R} + \frac{x\log^2 R}{R\log x}\right).$$

Now let $R \to \infty$ appropriately (as in Exercise 7.2.5) to deduce the result. □

7.2.9 *Assuming the Riemann hypothesis, show that*

$$\psi(x) = x + O\left(x^{1/2}\log^2 x\right)$$

as $x \to \infty$.

Again, by Exercise 7.2.7, we have

$$\psi(x) = x - \sum_{|\gamma|<R} \frac{x^\rho}{\rho} + O\left(\frac{x\log^2 x}{R} + \frac{x\log^2 R}{R\log x}\right).$$

The Riemann hypothesis says that $\rho = \frac{1}{2} + i\gamma$. Thus, the sum over the zeros is

$$O\left(x^{1/2}\log^2 R\right).$$

Choosing $R = \sqrt{x}$ gives the desired result. □

7.2.10 *Show that if*

$$\psi(x) = x + O\left(x^{1/2}\log^2 x\right)$$

then $\zeta(s)$ *has no zeros for* $\text{Re}(s) > 1/2$.

By partial summation

$$-\frac{\zeta'(s)}{\zeta(s)} = s\int_1^\infty \frac{\psi(x)dx}{x^{s+1}}.$$

Inserting the estimate for $\psi(x)$ into the integral gives an analytic continuation of $-\zeta'(s)/\zeta(s)$ for $\text{Re}(s) > 1/2$ apart from a simple pole at $s = 1$. This means that $\zeta(s)$ has no zeros for $\text{Re}(s) > 1/2$, as required. (The same deduction can be made from the weaker estimate of $O(x^{1/2+\epsilon})$ for any $\epsilon > 0$, for the error term.) □

7.3 Supplementary Problems

7.4.1 *Using the method of Exercise 6.5.3, prove that for* $1 \le \sigma \le 2$, $|t| \ge 2$,

$$-\text{Re}\left(\frac{L'(s,\chi)}{L(s,\chi)}\right) < A_1 \log q(|t|+2) - \sum_\rho \text{Re}\left(\frac{1}{s-\rho}\right),$$

where A_1 is an absolute constant, and the summation is over all zeros ρ of $L(s,\chi)$, and χ is a primitive Dirichlet character $(\bmod\ q)$. (Of course, $s = \sigma + it$, as usual.)

This is essentially contained in the solution to Exercise 6.5.9.

7.4.2 *Let χ be a primitive Dirichlet character* $(\bmod\ q)$. *If $\rho = \beta + i\gamma$ runs through the nontrivial zeros of $L(s,\chi)$, then show that for any real t,*

$$\sum_{\rho} \frac{1}{1+(t-\gamma)^2} = O(\log q(|t|+2)).$$

We take $s = 2+it$ in the previous exercise. Since $|L'/L|$ is bounded for such s, we obtain

$$\sum_{\rho} \mathrm{Re}\left(\frac{1}{s-\rho}\right) < A_2 \log q(|t|+2).$$

Now,

$$\mathrm{Re}\left(\frac{1}{s-\rho}\right) = \frac{2-\beta}{(2-\beta)^2+(t-\gamma)^2} \geq \frac{1}{4+(t-\gamma)^2},$$

and this last quantity is greater than or equal to $\frac{1}{4}(1+(t-\gamma)^2)^{-1}$ from which the result follows. □.

7.4.3 *With χ a primitive character* $(\bmod\ q)$ *and t not coinciding with the ordinate of a zero, show that for $-3/2 \leq \sigma \leq 5/2$, $|t| \geq 2$,*

$$\frac{L'}{L}(s,\chi) = \sum_{\rho}{}' \frac{1}{s-\rho} + O(\log q(|t|+2)),$$

where the dash on the sum is over $\rho = \beta + i\gamma$ for which $|t-\gamma| < 1$.

The method is essentially the same as Exercise 7.1.6. □

7.4.4 *Let χ be a primitive Dirichlet character* $(\bmod\ q)$. *Let $N(T,\chi)$ be the number of zeros of $L(s,\chi)$ in the rectangle $0 < \sigma < 1$, $|t| < T$. Show that*

$$N(T,\chi) = \frac{T}{\pi}\log\frac{qT}{2\pi} - \frac{T}{2\pi} + O(\log qT)$$

for $T \geq 2$.

We follow the method of Theorem 7.1.7. Let R be the rectangle with vertices

$$\frac{5}{2} - iT, \quad \frac{5}{2} + iT, \quad -\frac{3}{2} + iT, \quad -\frac{3}{2} - iT.$$

(This rectangle is slightly larger than the one used for $\zeta(s)$ so as to include a possible zero at $s = -1$.)

This rectangle contains at most one trivial zero of $L(s, \chi)$, either at $s = 0$ or $s = -1$. Therefore,

$$2\pi(N(T, \chi) + 1) = \Delta_R \arg \xi(s, \chi).$$

By the functional equation (Theorem 5.4.1),

$$\arg \xi(\sigma + it, \chi) = \arg \xi(1 - \sigma - it, \overline{\chi}) + c$$

for some constant independent of s. Therefore, the contribution of the left half of the contour is equal to that of the right half. Clearly,

$$\Delta \arg \Gamma\left(\frac{s + a}{2}\right) = T \log \frac{T}{2} - T + O(1),$$

where $a = 0$ or 1 according as $\chi(-1)$ is 1 or -1, and Δ is the half contour from $\frac{1}{2} - iT$ to $\frac{5}{2} - iT$, then to $\frac{5}{2} + iT$, and then to $\frac{1}{2} + iT$. We add these two variations and then double the result. It remains to consider

$$2\pi S(t, \chi) = \Delta L(s, \chi).$$

Since $\log L(s, \chi)$ is bounded on $\text{Re}s = 5/2$, it suffices to consider the variation along the horizontal segments from $1/2 - iT$ to $5/2 - iT$, and from $5/2 + iT$ to $1/2 + iT$. By Exercise 7.4.3, this reduces to calculating $\Delta \arg(s - \rho)$ along the line segments. But this variation is at most π, and we get

$$S(t, \chi) = O(\log q(|t| + 2)).$$

This gives the desired formula for $N(T, \chi)$. □

7.4.5 *Let χ be a primitive Dirichlet character* $(\bmod q)$. *If x is not a prime power and $\chi(-1) = -1$, derive the explicit formula*

$$\begin{aligned} \psi(x, \chi) &:= \sum_{n \le x} \chi(n)\Lambda(n) \\ &= -\sum_{\rho} \frac{x^\rho}{\rho} - \frac{L'(0, \chi)}{L(0, \chi)} + \sum_{m=1}^{\infty} \frac{x^{1-2m}}{2m - 1}, \end{aligned}$$

where the first sum on the right hand side is over the nontrivial zeros of $L(s,\chi)$.

This follows easily by the method used in Exercise 7.2.7 where we replace $\zeta'(s)/\zeta(s)$ by $L'(s,\chi)/L(s,\chi)$. □

7.4.6 *Let* χ *be a primitive Dirichlet character* $(\bmod q)$. *If* x *is not a prime power and* $\chi(-1)=1$, *derive the explicit formula*

$$\psi(x,\chi) = -\sum_{\rho} \frac{x^\rho}{\rho} - \log x - b(\chi) - \frac{1}{2}\log(1-x^{-2}),$$

where $b(\chi) = \lim_{s\to 0}\left(\frac{L'(s,\chi)}{L(s,\chi)} - \frac{1}{s}\right)$, *and the sum on the right-hand side is over the nontrivial zeros of* $L(s,\chi)$.

This again follows mutatis mutandis from the method of Exercise 7.2.7. However, the only difference is that now $L(s,\chi)$ has a simple zero at $s=0$, and so

$$\frac{L'(s,\chi)}{L(s,\chi)} = \frac{1}{s} + b(\chi) + \cdots .$$

Since

$$\frac{x^s}{s} = \frac{1}{s} + \log x + \cdots ,$$

the residue of $-L'(s,\chi)x^s/sL(s,\chi)$ at $s=0$ is $-(\log x + b(\chi))$. The trivial zeros contribute

$$\sum_{m=1}^{\infty} \frac{x^{-2m}}{2m} = -\frac{1}{2}\log(1-x^{-2}).$$

□

7.4.7 *Let* χ *be a primitive Dirichlet character* $(\bmod q)$ *and set* $a=0$ *or* 1 *according as* $\chi(-1)=1$ *or* -1. *If* $x-1/2$ *is a positive integer, show that*

$$\begin{aligned}\psi(x,\chi) &= -\sum_{|\gamma|<R} \frac{x^\rho}{\rho} - (1-a)(\log x + b(\chi)) \\ &\quad + \sum_{m=1}^{\infty} \frac{x^{a-2m}}{2m-a} + O\left(\frac{x\log^2 qxR}{R}\right),\end{aligned}$$

where the first summation is over zeros $\rho=\beta+i\gamma$ *and* R *is chosen greater than or equal to* 2 *so as not to coincide with the ordinate of any zero of* $L(s,\chi)$.

We follow the method of Exercises 7.2.3, 7.2.4 and 7.2.7. The only difference is that we must use the estimate

$$\frac{L'(\sigma+iR,\chi)}{L(\sigma+iR,\chi)}=O(\log^2 qR),$$

valid for $-1\leq\sigma\leq 2$, which is easily deduced from Exercises 7.4.2 and 7.4.3. For $\sigma\leq 1$, we must use the estimate

$$\frac{L'(s,\chi)}{L(s,\chi)}=O(\log q|s|),$$

provided that we exclude circles of radius $1/2$ around the trivial zeros. The latter estimate comes from logarithmic differentiation of the functional equation in its asymmetric form:

$$L(1-s,\chi)=w(\chi)2^{1-s}\pi^{-s}q^{s-1/2}\left(\cos\frac{1}{2}\pi(s-a)\right)\Gamma(s)L(s,\bar{\chi})$$

(see Exercises 8.2.13 and 8.2.15), where $|w(\chi)|=1$. The result is now derived as in Exercise 7.2.4. □

7.4.8 *If we assume that all the nontrivial zeros of* $L(s,\chi)$ *lie on* $\mathrm{Re}(s)=1/2$ *(the generalized Riemann hypothesis), prove that*

$$\psi(x,\chi)=O\left(x^{1/2}\log^2 qx\right).$$

We choose $R=x^{1/2}$ in the previous exercise. We need to estimate

$$\sum_{|\gamma|<x^{1/2}}\frac{1}{|\rho|}$$

as well as $b(\chi)$. By partial summation and Exercise 7.4.4, we obtain

$$\sum_{|\gamma|<x^{1/2}}\frac{1}{|\rho|}=O(\log^2 qx).$$

As for $b(\chi)$, this appears only if $\chi(-1)=-1$. In that case, we have from Exercise 6.4.5 that

$$\frac{L'(s,\chi)}{L(s,\chi)}=-\frac{1}{2}\log\frac{q}{\pi}-\frac{\Gamma'(s/2)}{2\Gamma(s/2)}+B(\chi)+\sum_{\rho}\left(\frac{1}{s-\rho}+\frac{1}{\rho}\right).$$

Replacing s by 2 and subtracting gives us

$$\frac{L'(s,\chi)}{L(s,\chi)} = -\frac{\Gamma'(s/2)}{2\Gamma(s/2)} + \sum_{\rho}\left(\frac{1}{s-\rho} - \frac{1}{2-\rho}\right) + O(1),$$

so that

$$b(\chi) = -\sum_{\rho}\left(\frac{1}{\rho} + \frac{1}{2-\rho}\right) + O(1).$$

In this sum, the terms with $|\gamma| \geq 1$ are easily handled:

$$\sum_{|\gamma|\geq 1}\left|\frac{1}{\rho} + \frac{1}{2-\rho}\right| \ll \sum_{|\gamma|\geq 1}\frac{1}{|\gamma|^2} = O(\log q)$$

by Exercise 7.4.2. For $|\gamma| < 1$, we observe that $|2-\rho| \gg |2-\rho|^2$, so that

$$b(\chi) = O(\log q) - \sum_{|\gamma|<1}\frac{1}{\rho}.$$

The number of zeros in the sum is $O(\log q)$ by Exercise 7.4.4, and for each ρ we have $|\rho| \geq \frac{1}{2}$, from which the result follows. □

7.4.9 *Let*

$$\psi(x,q,a) = \sum_{\substack{n\leq x \\ n\equiv a (\mathrm{mod}\ q)}} \Lambda(n).$$

Show that the generalized Riemann hypothesis implies

$$\psi(x,q,a) = \frac{x}{\phi(q)} + O\left(x^{1/2}\log^2 qx\right)$$

when $(a,q) = 1$.

We have

$$\psi(x,q,a) = \frac{1}{\phi(q)}\sum_{\chi (\mathrm{mod}\ q)} \overline{\chi}(a)\psi(x,\chi).$$

For $\chi = \chi_0$, the trivial character, we have

$$\psi(x,\chi_0) = x + O(x^{1/2}\log^2 x)$$

by Exercise 7.2.9. For $\chi \neq \chi_0$, we have $\psi(x,\chi) = O\left(x^{1/2}\log^2 qx\right)$ by the previous exercise, from which the desired result follows. □

7.4.10 *Assuming the generalized Riemann hypothesis, show that there is always a prime $p \ll q^2 \log^4 q$ satisfying $p \equiv a \pmod{q}$ whenever $(a, q) = 1$.*

By Exercise 7.4.9, we have

$$\psi(x, q, a) = \frac{x}{\phi(q)} + O\left(x^{1/2} \log^2 qx\right).$$

Putting $x = Aq^2 \log^4 q$ for an appropriate constant A gives us the required result. □

7.4.11 *Show that if q is prime, then*

$$\frac{\varphi(q-1)}{q-1} \sum_{d|q-1} \frac{\mu(d)}{\varphi(d)} \sum_{o(\chi)=d} \chi(a) = \begin{cases} 1 & \text{if } a \text{ has order } q-1 \\ 0 & \text{otherwise.} \end{cases}$$

where the inner sum is over characters $\chi \pmod{q}$ whose order is d.

Let $f(a) = 1$ if a is a primitive root and 0 otherwise. Let g be a primitive root $\pmod{q}$ and set

$$T(g^j) = e^{2\pi i j/q-1}, \quad 1 \le j \le q-1.$$

Then T is a multiplicative character $\pmod{q}$ and all multiplicative characters $\bmod q$ can be written as T^k for some k, $1 \le k \le q-1$.

Now write

$$f(a) = \sum_{\chi} \hat{f}(\chi)\chi(a).$$

By orthogonality, we see that

$$\hat{f}(T^k) = \frac{1}{q-1} \sum_{(j,q-1)=1} e^{2\pi i jk/q-1}.$$

The right hand side is a Ramanujan sum and by Exercise 1.1.14 is easily evaluated as

$$\frac{\varphi(q-1)\mu\left(\frac{q-1}{(q-1,k)}\right)}{(q-1)\varphi\left(\frac{q-1}{(q-1,k)}\right)}.$$

If we write $d = (q-1, k)$, then $d|q-1$. Moreover, T^k has order $(q-1)/d$. As d ranges over the divisors of $q-1$, so does $(q-1)/d$, and the result is now clear. □

7.4.12 *Let q be prime and assume the generalized Riemann hypothesis. For q sufficiently large, show that there is always a prime $p < q$ such that p is a primitive root* $(\bmod q)$.

By the previous exercise, we have that

$$\frac{\varphi(q-1)}{q-1} \sum_{d|q-1} \frac{\mu(d)}{\varphi(d)} \sum_{o(\chi)=d} \psi(x,\chi)$$

is the number of prime powers p^j weighted by $\log p$ such that p^j is a primitive root $(\bmod q)$. The leading term (corresponding to $d = 1$) gives

$$\frac{\varphi(q-1)}{q-1}\psi(x).$$

For $\chi \neq \chi_0$, we use Exercise 7.4.8 to deduce that the contribution is

$$O\left(\frac{\varphi(q-1)}{q-1} d(q-1) x^{1/2} \log^2 qx\right),$$

where $d(q-1)$ is the number of divisors of $q-1$, since the number of characters of order d is $\varphi(d)$.

Since $\psi(x) = x + O\left(x^{1/2}\log^2 x\right)$, we see that for $x = q$, the main term is larger than the error term, for q sufficiently large. Moreover, if $p^j < q$ is a primitive root, so is $p < q$. □

7.4.13 *Let q be a prime. Show that the smallest primitive root* $(\bmod q)$ *is $O(2^{\nu(q-1)}q^{1/2}\log q)$, where $\nu(q-1)$ is the number of distinct prime factors of $q-1$.*

By Exercise 7.4.11, the number of primitive roots $(\bmod q)$ that are less than x is

$$\frac{\varphi(q-1)}{q-1}x + \frac{\varphi(q-1)}{q-1} \sum_{\substack{d|q-1 \\ d>1}} \frac{\mu(d)}{\varphi(d)} \sum_{o(\chi)=d} \left(\sum_{a<x} \chi(a)\right).$$

By the Pólya - Vinogradov inequality (Exercise 5.5.6) we find that the innermost sum is $O\left(q^{1/2}\log q\right)$. Thus, the number of primitive roots less than x is

$$\frac{\varphi(q-1)}{q-1}\left(x + O(2^{\nu(q-1)}q^{1/2}\log q)\right),$$

which is positive if $x \gg 2^{\nu(q-1)}q^{1/2}\log q$. This completes the proof. □

7.4.14 *Let q be a prime and assume the generalized Riemann hypothesis. Show that there is always a prime-power primitive root satisfying the bound $O\left(4^{\nu(q-1)}\log^4 q\right)$.*

We examine the solution of Exercise 7.4.12, where we showed that the number of prime-power primitive roots is

$$\frac{\varphi(q-1)}{q-1}\left(x + O(2^{\nu(q-1)}q^{1/2}\log q)\right).$$

A little reflection shows that $d(q-1)$ can be replaced by $2^{\nu(q-1)}$. Setting $x = C4^{\nu(q-1)}\log^4 q$ for a sufficiently large constant gives us the desired result. □

7.4.15 *Let q be prime and assume the generalized Riemann hypothesis. Show that the least quadratic nonresidue* $(\bmod\, q)$ *is* $O\left(\log^4 q\right)$.

Since

$$1-\left(\frac{a}{q}\right) = \begin{cases} 2 & \text{if } a \text{ is a nonresidue,} \\ 0 & \text{otherwise,} \end{cases}$$

we see that

$$\frac{1}{2}\sum_{p^j<x}\left(1-\left(\frac{p^j}{q}\right)\right)\log p$$

equals

$$\frac{\psi(x)}{2} + O(x^{1/2}\log^2 qx)$$

under the stated hypothesis. If $x = C\log^4 q$ for a sufficiently large q, the result is now clear. □

7.4.16 *Let q be prime and assume the generalized Riemann hypothesis. Show that the least prime quadratic residue* $(\bmod\, q)$ *is* $O(\log^4 q)$.

This is clear from the method of the previous exercise. □

7.4.17 *Prove that for $n > 1$,*

$$\lim_{T\to\infty}\frac{1}{T}\sum_{|\gamma|\le T} n^{\rho} = -\frac{\Lambda(n)}{\pi},$$

where the summation is over zeros $\rho = \beta + i\gamma$, $\beta \in \mathbb{R}$, of the Riemann zeta function.

Let R denote the rectangle oriented counterclockwise with vertices $3/2 - iT, 3/2 + iT, -1/2 + iT, -1/2 - iT$. Clearly,

$$\frac{1}{2\pi i}\int_R \frac{\zeta'(s)}{\zeta(s)} n^s ds = \sum_\rho n^\rho - n$$

where ρ runs over zeros of $\zeta(s)$ inside the rectangle. Let $I_1, \dots, I_4$ be the four parts of the integral relative to the sides of R starting with the vertical one in the half-plane $\mathrm{Re}(s) > 1$ and proceeding counterclockwise. Moreover, we have chosen T such that

$$-\frac{\zeta'}{\zeta}(\sigma + it) = O(\log^2 t)$$

uniformly in $-2 \leq \sigma \leq 3$, which we can do as in the solution of Exercise 7.2.7. Thus,

$$\begin{aligned} I_1 &= \frac{1}{2\pi}\int_{-T}^{T} \sum_{m=1}^{\infty} \Lambda(m)\left(\frac{n}{m}\right)^{3/2+it} dt \\ &= -\frac{T}{\pi}\Lambda(n) + O\Big(\sum_{\substack{m=1 \\ m\neq n}} \Lambda(m)\left(\frac{n}{m}\right)^{3/2} \frac{1}{|\log n/m|}\Big). \end{aligned}$$

Splitting the summation into the ranges

$$m \leq n/2, \quad n/2 < m < 2n, \quad m \geq 2n$$

and handling these sums as in Exercises 7.2.2 and 7.2.3 gives an estimate of $O\left(n^{3/2}\right)$ for the error term above. By using the estimate of $O\left(\log^2 T\right)$ for the integrand, we deduce that

$$I_2, I_4 \ll n^{3/2} \log^2 T.$$

Finally, for I_3, we use the functional equation to relate $\zeta'/\zeta(-1/2 + it)$ to $\zeta'/\zeta(3/2 - it)$. The Γ-factor gives rise to a term of the form

$$O(\log T)$$

by Stirling's formula, and after integrating we get that

$$I_3 \ll n^{3/2} \log^2 T.$$

Thus, the result is now clear. □

8

The Selberg Class

8.1 The Phragmén - Lindelöf Theorem

8.1.1 *Let $f(z)$ be an analytic function, regular in a region R and on the boundary ∂R, which we assume to be a simple closed contour. If $|f(z)| \leq M$ on ∂R, show that $|f(z)| \leq M$ for all $z \in R$.*

If $z \in R$, then by Cauchy's theorem,

$$f^n(z) = \frac{1}{2\pi i} \int_{\partial R} \frac{f^n(w)dw}{w - z},$$

so that

$$|f^n(z)| \leq KM^n,$$

where

$$K = \frac{1}{2\pi} \int_{\partial R} \left| \frac{dw}{w - z} \right|.$$

Taking nth roots and letting $n \to \infty$ gives the result. □

8.1.2 (The maximum modulus principle) *If f is as in the previous exercise, show that $|f(z)| < M$ for all interior points $z \in R$, unless f is constant.*

If z_0 is an integer point, consider the Laurent expansion of f about z_0:

$$f\left(z_0 + re^{i\theta}\right) = \sum_{n=0}^{\infty} a_n r^n e^{in\theta}.$$

Parseval's formula yields that

$$\frac{1}{2\pi}\int_0^{2\pi} \left| f\left(z_0 + re^{i\theta}\right)\right|^2 d\theta = \sum_{n=0}^{\infty} |a_n|^2 r^{2n}.$$

If z_0 is an interior point where the maximum is attained, we have $|a_0| = M$ and

$$M = |a_0|^2 \leq |a_0|^2 + |a_1|^2 r^2 + \cdots \leq |f(z_0)|^2 = 1,$$

so that we are forced to have $a_1 = a_2 = \cdots = 0$ and f is constant. □

8.1.5 *Show that for any entire function $F \in S$, we have*

$$F(s) = O\left(|t|^A\right)$$

for some $A > 0$, in the region $0 \leq \mathrm{Re}(s) \leq 1$.

This is an immediate consequence of the functional equation and Stirling's formula. Indeed, $F(s)$ is bounded on Re(s)=2. By the functional equation and Stirling's formula, it has polynomial growth on $\mathrm{Re}(s) = -1$. By the Phragmén - Lindelöf theorem, it has polynomial growth in the region $-1 \leq \mathrm{Re}(s) \leq 2$. □

8.2 Basic Properties

8.2.4 *Show that*

$$\deg F_1F_2 = \deg F_1 + \deg F_2.$$

Since

$$N_{F_1F_2}(T) = N_{F_1}(T) + N_{F_2}(T),$$

the result is immediate from Theorem 8.2.1.

□

8.2.5 *If $F \in S$ has degree 1, show that it is primitive.*

If F is not primitive, we can write $F = F_1 F_2$ with $F_1 \neq 1$, $F_2 \neq 1$. But then, $\deg F = \deg F_1 + \deg F_2$, and by Theorem 8.2.3 and Lemma 8.2.2, $\deg F_1 \geq 1$ and $\deg F_2 \geq 1$ so that $\deg F \geq 2$, a contradiction. [Observe that the proof shows that any $F \in \mathcal{S}$ of degree less than 2 is primitive.]

□

8.2.6 *Show that any $F \in \mathcal{S}$, $F \neq 1$, can be written as a product of primitive functions.*

We first show that every $F \in \mathcal{S}$ is divisible by a primitive function. If F is not primitive, we write $F = F_1 G_1$ with $F_1 \neq 1$ and $G_1 \neq 1$. Since $\deg F_1 < \deg F$, we either have F_1 primitive or not. If not, factor $F_1 = F_2 G_2$ and in this way we get $\deg F_2 < \deg F_1$. In fact, we have

$$
\begin{aligned}
0 < \deg F_1 &\leq \deg F - 1, \\
0 < \deg F_2 &\leq \deg F_1 - 1 \leq \deg F - 2,
\end{aligned}
$$

and so on. This cannot go on ad infinitum. Thus, any function $F \in \mathcal{S}$ has a primitive factor, F_1 (say). Write $F = F_1 G_1$ and now proceed to decompose G_1. Since the degree of each factor is strictly less than $\deg F$, the process terminates. □

8.2.7 *Show that the Riemann zeta function is a primitive function.*

$\zeta(s)$ has degree 1 by Theorem 5.2.2. Now apply Exercise 8.2.5. □

8.2.8 *If χ is a primitive character $\pmod q$ show that $L(s, \chi)$ is a primitive function of $\mathcal{S}$.*

By Theorem 5.4.1 and Exercise 5.4.5 we see that $L(s, \chi)$ extends to an entire function and has degree 1. □

8.2.9 *If $F \in \mathcal{S}$, show that $|a_n| \leq c(\epsilon) n^\epsilon$ implies that*

$$|b_{p^k}| \leq c(\epsilon)(2^k - 1) p^{k\epsilon}/k.$$

We have

$$F(s) = \sum_{n=1}^{\infty} \frac{a_n}{n^s} = \prod_p \exp\Big(\sum_{k=1}^{\infty} \frac{b_{p^k}}{p^{ks}}\Big)$$

so that

$$-\frac{F'(s)}{F(s)} = \sum_{p,k} \frac{kb_{p^k} \log p}{p^{ks}}.$$

We deduce that

$$a_n \log n = \sum_{p^j | n} jb_{p^j} (\log p) a_{n/p^j}.$$

Setting $n = p^k$ yields

$$kb_{p^k} \log p = ka_{p^k} \log p - \sum_{j=1}^{k-1} jb_{p^j} (\log p) a_{p^{k-j}}.$$

We now induct on k. For $k = 1$, we have $a_p = b_p$, and the result is clear.

Assume that the inequality has been proved for exponents less than or equal to $k - 1$. Then

$$\begin{aligned} k|b_{p^k}| &\leq c(\epsilon)kp^{k\epsilon} + \sum_{j=1}^{k-1} j|b_{p^j}|c(\epsilon)p^{(k-j)\epsilon} \\ &\leq c(\epsilon)p^{k\epsilon} \left\{ k + \sum_{j=1}^{k-1} (2^j - 1) \right\} \\ &\leq c(\epsilon)p^{k\epsilon} \left(2^k - 1\right), \end{aligned}$$

as desired. □

8.2.10 *Prove the asymmetric form of the functional equation for $\zeta(s)$:*

$$\zeta(1-s) = 2^{1-s}\pi^{-s} \left(\cos \frac{s\pi}{2}\right) \Gamma(s)\zeta(s).$$

We recall that $\Gamma(s)$ satisfies

$$\Gamma(s)\Gamma(1-s) = \pi / \sin \pi s,$$

by Exercise 6.3.9 and the Legendre duplication formula (Exercise 6.3.6):

$$\Gamma(2s)\sqrt{\pi} = 2^{2s-1}\Gamma(s)\Gamma\left(s + \frac{1}{2}\right).$$

Combining these two facts gives

$$\frac{\Gamma\left(\frac{s}{2}\right)}{\Gamma\left(\frac{1-s}{2}\right)} = \pi^{-1/2}2^{1-s}\left(\cos\frac{\pi s}{2}\right)\Gamma(s).$$

By the functional equation for $\zeta(s)$, we may write

$$\zeta(1-s) = \pi^{1/2-s}\frac{\Gamma(s/2)}{\Gamma((1-s)/2)}\zeta(s)$$

by Theorem 5.2.2. Putting these together gives the result. □

8.2.11 *Show that for* $k \in \mathbb{N}$,

$$|\zeta(-k)| \leq Ck!/(2\pi)^k$$

for some absolute constant C.

By the previous exercise,

$$|\zeta(-k)| = \left|2^{-k}\pi^{-k-1}\cos\left(\frac{(k+1)\pi)}{2}\right)\Gamma(k+1)\zeta(k+1)\right|.$$

Since $\lim_{k\to\infty}\zeta(k+1) = 1$, we get

$$|\zeta(-k)| \leq Ck!/(2\pi)^k$$

as required. □

8.2.12 *Show that*

$$\sum_{n=1}^{\infty} e^{-nx} = x^{-1} + \sum_{k=0}^{\infty}\frac{\zeta(-k)(-x)^k}{k!}.$$

Deduce that for $k = 2, 3, \ldots$

$$\zeta(1-k) = -B_k/k$$

and $\zeta(0) = -1/2$, *where* B_k *denotes the kth Bernoulli number.* medskip We specialize the proof of Theorem 8.2.3 to the case of the ζ-function:

$$\sum_{n=1}^{\infty} e^{-nx} = x^{-1} + \sum_{k=0}^{\infty}\frac{\zeta(-k)(-x)^k}{k!}.$$

By Exercise 8.2.11, the power series on the right-hand side converges for $|x| < 2\pi$. The left-hand side is a geometric series that is easily summed to be

$$\frac{e^{-x}}{1-e^{-x}} = \frac{1}{e^x - 1}.$$

By Exercise 2.1.7,

$$\frac{x}{e^x - 1} = \sum_{k=0}^{\infty} B_k \frac{x^k}{k!},$$

so that

$$\frac{1}{e^x - 1} = \frac{1}{x} + \sum_{k=1}^{\infty} \frac{B_k x^{k-1}}{k!}.$$

We may compare coefficients of the two power series to deduce that

$$(-1)^{k-1}\zeta(1-k) = B_k/k.$$

For k odd, $k \geq 3$, $B_k = 0$ by Exercise 2.1.8. Hence the formula is clear for k odd ≥ 3. For k even, we obtain

$$\zeta(1-k) = -B_k/k.$$

For $k = 1$, we have $\zeta(0) = B_1 = 1/2$, and we recover the result of Exercise 5.2.4. □

8.2.13 *Let χ be a primitive Dirichlet character* (mod q) *satisfying* $\chi(-1) = 1$. *Prove that*

$$L(1-s,\overline{\chi}) = \sqrt{\frac{2}{\pi}}\frac{q^{1/2}}{\tau(\chi)}\left(\frac{2\pi}{q}\right)^{1/2-s}\left(\cos\frac{\pi s}{2}\right)\Gamma(s)L(s,\chi),$$

where $\tau(\chi)$ denotes the Gauss sum.

By the functional equation (Theorem 5.4.1), we have

$$L(1-s,\overline{\chi}) = \frac{q^{1/2}}{\tau(\chi)}\left(\frac{\pi}{q}\right)^{1/2-s}\frac{\Gamma\left(\frac{s}{2}\right)}{\Gamma\left(\frac{1-s}{2}\right)}L(s,\chi).$$

As in the solution to Exercise 8.2.10, we have

$$\frac{\Gamma\left(\frac{s}{2}\right)}{\Gamma\left(\frac{1-s}{2}\right)} = \pi^{-1/2} 2^{1-s}\left(\cos\frac{\pi s}{2}\right)\Gamma(s),$$

from which the result is easily deduced. □

8.2.14 *Let χ be a primitive character* $(\bmod\, q)$ *satisfying* $\chi(-1) = 1$. *Show that for* $k \in \mathbb{N}$,

$$|L(-k,\chi)| \le Ck!(q/2\pi)^k$$

for some constant $C = O(\sqrt{q})$.

We proceed as in Exercise 8.2.11, except that we use the previous exercise instead of Exercise 8.2.10. □

8.2.15 *Let χ be a primitive Dirichlet character* $(\bmod\, q)$ *satisfying* $\chi(-1) = -1$. *Show that*

$$L(1-s,\overline{\chi}) = -(2\pi)^{-1/2}\frac{iq^{1/2}}{\tau(\chi)}\left(\frac{2\pi}{q}\right)^{1/2-s}\left(\sin\frac{\pi s}{2}\right)\Gamma(s+1)L(s,\chi).$$

This again uses the method of Exercise 8.2.10. By Exercise 5.4.5, we have

$$L(1-s,\overline{\chi}) = \frac{iq^{1/2}}{\tau(\chi)}\left(\frac{\pi}{q}\right)^{1/2-s}\frac{\Gamma\left(\frac{s+1}{2}\right)}{\Gamma\left(-\frac{s}{2}\right)}L(s,\chi).$$

By the formula

$$\frac{\Gamma\left(\frac{s}{2}\right)}{\Gamma\left(\frac{1-s}{2}\right)} = \pi^{-1/2} 2^{1-s}\left(\cos\frac{\pi s}{2}\right)\Gamma(s)$$

(derived in the solution to Exercise 8.2.10) we obtain the desired result. □

8.2.16 *Let χ be a primitive Dirichlet character* $(\bmod\, q)$ *satisfying* $\chi(-1) = -1$. *Show that for* $k \in \mathbb{N}$,

$$|L(-k,\chi)| \le C(k+1)!(q/2\pi)^k$$

for some constant $C = O(\sqrt{q})$.

We proceed as in Exercises 8.2.14 and 8.2.11, except that we use the previous exercise to derive the estimate. □

8.2.17 *Prove that*

$$\sum_{n=1}^{\infty} \chi(n)e^{-nx} = \sum_{k=0}^{\infty} \frac{L(-k,\chi)(-x)^k}{k!}.$$

Deduce that for $n \geq 1$,

$$L(1-n,\chi) = -B_{n,\chi}/n,$$

where

$$B_{n,\chi} = q^{n-1} \sum_{a=1}^{q} \chi(a) B_n\left(\frac{a}{q}\right),$$

with $B_n(x)$ *denoting the nth Bernoulli polynomial.*

From the proof of Theorem 8.2.3, the derivation of the formula

$$\sum_{n=1}^{\infty} \chi(n)e^{-nx} = \sum_{k=0}^{\infty} \frac{L(-k,\chi)(-x)^k}{k!}$$

is clear. The left-hand side can be simplified as follows.

$$\begin{aligned} \sum_{n=1}^{\infty} \chi(n)e^{-nx} &= \sum_{b \pmod q} \chi(b) \sum_{n \equiv b \pmod q} e^{-nx} \\ &= \sum_{b=1}^{q} \chi(b) \left(\sum_{r=0}^{\infty} e^{-(qr+b)x} \right) \\ &= \sum_{b=1}^{q} \chi(b) \frac{e^{-bx}}{1-e^{-qx}} \\ &= \sum_{b=1}^{q} \chi(b) \frac{e^{(q-b)x}}{e^{qx}-1}. \end{aligned}$$

Now, by Exercise 2.1.7,

$$\sum_{r=0}^{\infty} \frac{b_r(x)t^r}{r!} = \frac{te^{xt}}{e^t - 1}.$$

Thus,

$$\frac{e^{(q-b)x}}{e^{qx}-1} = \frac{e^{(1-b/q)qx}}{e^{qx}-1}$$

can be expanded as

$$\sum_{r=0}^{\infty} b_r\left(1-\frac{b}{q}\right)\frac{q^{r-1}x^{r-1}}{r!}.$$

When we insert this in the above formula, we obtain

$$\sum_{n=1}^{\infty}\chi(n)e^{-nx} = \sum_{r=0}^{\infty}\left(\sum_{b=1}^{q}\chi(b)b_r\left(1-\frac{b}{q}\right)\right)\frac{(qx)^{r-1}}{r!}$$

(notice that for $r=0$, $b_0(x)=1$), and since

$$\sum_{b=1}^{q}\chi(b)=0,$$

the polar term disappears. We deduce

$$L(1-n,\chi) = \frac{(-q)^{n-1}}{n}\sum_{b=1}^{q}\chi(b)b_n\left(1-\frac{b}{q}\right).$$

Recall that $B_n(x) = b_n(\{x\})$ and that

$$B_n(1-x) = (-1)^n B_n(x)$$

(see Exercise 2.5.22), from which the stated result follows. $\square$

8.3 Selberg's Conjectures

8.3.1 *Assuming (a) and (b), prove that any function $F \in \mathcal{S}$ can be factored uniquely as a product of primitive functions.*

Suppose

$$F = F_1^{e_1}\cdots F_r^{e_r}$$

is a factorization of F into distinct primitive functions F_i and

$$F = G_1^{f_1}\cdots G_t^{f_t}$$

is another factorization of F into distinct primitive functions G_i. Then

$$F_1^{e_1}\cdots F_r^{e_r} = G_1^{f_1}\cdots G_t^{f_t}$$

and we may suppose, without loss of generality, that no F_i is a G_j. Comparing the pth coefficient of both sides of the above equation, we deduce

$$\sum_{i=1}^{r} e_i a_p(F_i) = \sum_{j=1}^{t} f_j a_p(G_j).$$

Multiplying both sides of the equation by $\overline{a_p(F_1)}$, dividing by p, and then summing over $p \leq x$ gives us

$$e_1 \log\log x + O(1) = O(1),$$

assuming (a) and (b). Thus, $e_1 = 0$, a contradiction. This proves the unique factorization. $\square$

8.3.2 *Suppose $F, G \in \mathcal{S}$ and $a_p(F) = a_p(G)$ for all but finitely many primes p. Assuming (a) and (b), prove that $F = G$.*

Let us write

$$F = F_1^{e_1} \cdots F_r^{e_r}$$
$$G = F_1^{f_1} \cdots F_r^{f_r}$$

where $F_1, \ldots, F_r$ are distinct primitive functions and e_i, f_i are non-negative integers. We want to show that $e_i = f_i$ for all i. Without loss of generality, suppose $e_1 \neq f_1$. Then, since

$$a_p(F) = a_p(G),$$

we have

$$\sum_i e_i a_p(F_i) = \sum_i f_i a_p(F_i)$$

for all but finitely many primes p. Multiplying both sides of the equation by $\overline{a_p(F_1)}$, dividing by p, and then summing over $p \leq x$ gives

$$e_1 \sum_{p \leq x} \frac{|a_p(F_1)|^2}{p} + \sum_{i \geq 2} e_i \left(\sum_{p \leq x} \frac{a_p(F_i)\overline{a_p(F_1)}}{p} \right)$$

$$= \quad f_1 \sum_{p \leq x} \frac{|a_p(F_1)|^2}{p} + \sum_{i \geq 2} f_i \left(\sum_{p \leq x} \frac{a_p(F_i)\overline{a_p(F_1)}}{p} \right)$$

Assuming (a) and (b) gives

$$e_1 \log\log x + O(1) = f_1 \log\log x + O(1),$$

whence $e_1 = f_1$, a contradiction. Thus, $e_i = f_i$, for all i and we have $F = G$. □

8.3.3 *If $F(s) = \sum_{n=1}^{\infty} a_n n^{-s}$ and $\sigma = \mathrm{Re}(s) > \sigma_a(F)$, the abscissa of absolute convergence of F, then prove that*

$$\lim_{T\to\infty} \frac{1}{2T}\int_{-T}^{T} F(\sigma+it)y^{\sigma+it}dt = \begin{cases} a_n(F) & \text{if} \quad n = y, \\ 0 & \text{otherwise,} \end{cases}$$

for any real y.

We have

$$\frac{1}{2T}\int_{-T}^{T} F(\sigma+it)y^{\sigma+it}dt = \frac{y^\sigma}{2T}\int_{-T}^{T}\Big(\sum_{n=1}^{\infty}\frac{a_n}{n^\sigma}\Big(\frac{y}{n}\Big)^{it}\Big)dt.$$

Interchanging the summation and integration, which is justified by absolute convergence of the Dirichlet series, we obtain that the above is

$$= a_y + y^\sigma \sum_{n\neq y} \frac{a_n}{n^\sigma}\left(\frac{\sin T\log(y/n)}{T\log(y/n)}\right)$$

with the a_y term occurring only if y is a natural number. The series

$$\sum_{n\neq y} \frac{a_n}{n^\sigma}\cdot\frac{1}{\left|\log\frac{y}{n}\right|}$$

is easily seen to converge absolutely if $n > 2y$ or $n < y/2$. The intermediate range is a finite sum, and so as $T\to\infty$, the summation in the penultimate step goes to zero as required. This completes the proof. □

8.3.4 *Prove that*

$$\frac{1}{2\pi i}\int_{(c)} \frac{y^s ds}{(\alpha s+\beta)^2} = \begin{cases} \alpha^{-2}y^{-\beta/\alpha}\log y & \text{if} \quad y > 1, \\ 0 & \text{if} \quad 0 \le y \le 1, \end{cases}$$

for $c > 0$ and $\alpha, \beta > 0$.

First, suppose $y > 1$. We apply contour integration as in Exercise 4.1.6. Let ζ_R be the contour described by the line segment joining $c - iR$ to $c + iR$ and the semicircle S_R of radius R centered at c enclosing $-\beta/\alpha$. Then, by Cauchy's theorem

$$\frac{1}{2\pi i}\int_{\zeta_R}\frac{y^s ds}{(\alpha s+\beta)^2} = \text{Res}_{s=-\beta/\alpha}\frac{y^s}{(\alpha s+\beta)^2} = \alpha^{-2}y^{-\beta/\alpha}\log y.$$

Thus,

$$\frac{1}{2\pi i}\int_{c-iR}^{c+iR}\frac{y^s ds}{(\alpha s+\beta)^2} + \frac{1}{2\pi i}\int_{S_R}\frac{y^s ds}{(\alpha s+\beta)^2} = \alpha^{-2}y^{-\beta/\alpha}\log y.$$

The second integral satisfies

$$\left|\frac{1}{2\pi i}\int_{S_R}\frac{y^s ds}{(\alpha s+\beta)^2}\right| \ll \frac{y^c}{R}\int_{\pi/2}^{3\pi/2} y^{R\cos\varphi}d\varphi,$$

and the latter integral is easily seen to be bounded (see Exercise 4.1.1). Thus, as $R \to \infty$, the integral goes to zero.

If now $0 \le y \le 1$, then we choose the contour $\mathcal{D}_R$ (as in Exercise 4.1.2) described by the line segment joining $c - iR$ to $c + iR$ and the semicircle S_R to the right of the line segment of radius R, centered at c and **not** enclosing $s = -\beta/\alpha$. By Cauchy's theorem,

$$\frac{1}{2\pi i}\int_{\mathcal{D}_R}\frac{y^s ds}{(\alpha s+\beta)^2} = 0.$$

We now proceed exactly as above. □

8.3.5 *Let $f(s)$ be a meromorphic function on $\mathbb{C}$, analytic for $\text{Re}(s) \ge \frac{1}{2}$ and nonvanishing there. Suppose that $\log f(s)$ is a Dirichlet series and that $f(s)$ satisfies the functional equation*

$$H(s) = w\overline{H}(1-s),$$

where w is a complex number of absolute value 1, and

$$H(s) = A^s\frac{\prod_{i=1}^{d_1}\Gamma(\alpha_i s+\beta_i)}{\prod_{i=1}^{d_2}\Gamma(\gamma_i s+\delta_i)}f(s)$$

with certain A, α_i, $\gamma_i > 0$ and $\text{Re}(\beta_i)$, $\text{Re}(\delta_i) \ge 0$. Show that $f(s)$ is constant.

Since $f(s)$ is analytic in $\mathrm{Re}(s) \geq \frac{1}{2}$, and the Γ-function does not have any poles in $\mathrm{Re}(s) > 0$, we see immediately that $H(s)$ is analytic and nonvanishing (since f is) in the region $\mathrm{Re}(s) \geq 1/2$. By the functional equation, the same is true for $\mathrm{Re}(s) \leq 1/2$. Thus $H(s)$ is entire. By Stirling's formula and the functional equation, we see that $H(s)$ is of order 1. Since $H(s)$ has no zeros, it follows by Hadamard's theorem that $H(s) = e^{as+b}$ for some constants a and b. Hence

$$\frac{f'(s)}{f(s)} = a - \log A + \sum_{i=1}^{d_1} \frac{\Gamma'}{\Gamma}(\gamma_i s + \delta_i)\gamma_i - \sum_{i=1}^{d_2} \frac{\Gamma'}{\Gamma}(\alpha_i s + \beta_i)\alpha_i$$

is a Dirichlet series (since $\log f(s)$ is). The derivative of this is again a Dirichlet series. Since

$$\frac{d}{ds}\left(\frac{\Gamma'}{\Gamma}(s)\right) = -\sum_{m=0}^{\infty} \frac{1}{(s+m)^2},$$

then by Exercise 8.3.4 we deduce

$$\frac{1}{2\pi i}\int_{(\sigma)} \frac{d}{ds}\left(\frac{f'(s)}{f(s)}\right) y^s dy = O(1)$$

for any $y \geq 1$. By Exercise 8.3.3, this means that every coefficient of

$$\frac{d}{ds}\left(\frac{f'(s)}{f(s)}\right)$$

is zero. Since $f'(s)/f(s)$ is a Dirichlet series, this means that $f'(s)/f(s) = 0$. Hence $f(s)$ is a constant. $\square$

8.3.6 *Let $F, G \in \mathcal{S}$. Suppose $a_p(F) = a_p(G)$, $a_{p^2}(F) = a_{p^2}(G)$ for all but finitely many primes p. Show that $F = G$.*

Set

$$f(s) = \prod_p F_p(s)/G_p(s).$$

Since $\log F_p(s)$ is an absolutely convergent Dirichlet series for $\mathrm{Re}(s) > \theta$, we deduce that $F_p(s)$ is absolutely convergent for $\mathrm{Re}(s) > \theta$ and is nonvanishing there. Since $\theta < 1/2$, this holds for $\mathrm{Re}(s) \geq 1/2$. Since $a_p(F) = a_p(G)$ and $a_{p^2}(F) = a_{p^2}(G)$ for all but finitely many primes p, we can factor

$$\left(1 + \frac{a_p(F)}{p^s} + \frac{a_{p^2}(F)}{p^{2s}}\right)^{-1}$$

from the numerator and denominator of $F_p(s)/a_p(s)$ and write

$$f(s) = \prod_p f_p(s),$$

where each $f_p(s)$ is absolutely convergent for $\text{Re}(s) \geq 1/2$ and nonvanishing there. Thus, $f(s)$ satisfies the conditions of Exercise 8.3.5. Hence $f(s)$ is constant, and that constant must be 1, since

$$\lim_{s \to \infty} f_p(s) = 1$$

and consequently $\lim_{s \to \infty} f(s) = 1$. Therefore, $F = G$. $\square$

8.3.7 *Assume Selberg's conjectures (a) and (b). If $F \in \mathcal{S}$ has a pole of order m at $s = 1$, show that $F(s)/\zeta(s)^m$ is entire.*

If G is a primitive function that has a pole at $s = 1$, then

$$\sum_{p \leq x} \frac{a_p(G)}{p}$$

is unbounded as $x \to \infty$. If $G \neq \zeta$, by (b) we have

$$\sum_{p \leq x} \frac{a_p(G)}{p} = \sum_{p \leq x} \frac{a_p(G)\overline{a_p(\zeta)}}{p} = O(1),$$

a contradiction. Thus, the only primitive function with a pole at $s = 1$ is the Riemann zeta function. By Exercise 8.3.1, $\zeta(s)$ must appear in the unique factorization of F as a product of primitive functions. $\square$

8.3.8 *Assume Selberg's conjectures (a) and (b). Show that for any $F \in \mathcal{S}$, there are no zeros on* $\text{Re}(s) = 1$.

By Exercise 8.3.1, it suffices to prove this for primitive functions F. For the primitive function $\zeta(s)$, this is true by Exercise 3.2.5. So we may suppose $F \neq \zeta$. By Exercise 8.3.7, we may also suppose $F(s)$ has no pole at $s = 1$ and that it extends to an entire function. For any $t \in \mathbb{R}$, we can conclude that $G(s) = F(s + it)$ is again primitive. By conjecture (b),

$$\sum_{p \leq x} \frac{a_p(G)a_p(\zeta)}{p} = O(1)$$

as $x \to \infty$. This means that

$$\sum_{p \leq x} \frac{a_p(F)}{p^{1+it}} = O(1)$$

for all $t \in \mathbb{R}$. Hence, $F(s)$ has no zeros on $\text{Re}(s) = 1$. □

8.4 Supplementary Problems

8.4.1 *Verify that the primitive functions $\zeta(s)$, and $L(s, \chi)$, where χ is a primitive character* $(\text{mod}\, q)$, *satisfy Selberg's conjectures (a) and (b).*

To verify (a) for $\zeta(s)$, we apply Exercise 3.1.8. This also verifies (a) for all $L(s, \chi)$. To verify (b), notice that

$$\sum_{p \leq x} \frac{\chi(p)}{p} = \sum_{n \leq x} \frac{\chi(n)\Lambda(n)}{n \log n} + O(1)$$

follows easily by partial summation.

Now,

$$\sum_{n \leq x} \frac{\chi(n) \log n}{n} \to L'(1, \chi)$$

and hence is $O(1)$. On the other hand, we can write $\log n = \sum_{d|n} \Lambda(d)$, so that

$$\sum_{n \leq x} \frac{\chi(n) \log n}{n} = \sum_{d \leq x} \frac{\chi(d)\Lambda(d)}{d} \Big(\sum_{e \leq x/d} \frac{\chi(e)}{e} \Big).$$

The inner sum by Exercise 2.4.6 is

$$L(1, \chi) + O\Big(\frac{d}{x}\Big).$$

Hence

$$\sum_{n \leq x} \frac{\chi(n) \log n}{n} = L(1, \chi) \sum_{d \leq x} \frac{\chi(d)\Lambda(d)}{d} + O(1)$$

by an application of Chebyshev's theorem (Exercise 3.1.5). Therefore,

$$\sum_{d \leq x} \frac{\chi(d)\Lambda(d)}{d} = O(1),$$

since $L(1, \chi) \neq 0$ by Exercises 2.3.10 and 2.4.5. The result now follows easily by partial summation.

Finally, if χ_1 and χ_2 are distinct primitive characters $\bmod q_1$ and $\bmod q_2$ (respectively) then we may view $\chi_1\bar{\chi}_2$ as an imprimitive character $\bmod [q_1, q_2]$. Indeed, we may extend both χ_1 and χ_2 to characters $\bmod [q_1, q_2]$ in the usual way. If $q_1 = q_2$, the extended character is trivial if and only if $\chi_1 = \chi_2$. If $q_1 \neq q_2$, then $\chi_1\overline{\chi_2}$ is never trivial, and so we are done by the previous considerations. □

8.4.2 *For each F, G in S, define*

$$(F \otimes G)(s) = \prod_p H_p(s),$$

where

$$H_p(s) = \exp\left(\sum_{k=1}^{\infty} k b_{p^k}(F) b_{p^k}(G) p^{-ks}\right).$$

If $F_p(s) = \det(1 - A_p p^{-s})^{-1}$ and $G_p(s) = \det(1 - B_p p^{-s})^{-1}$ for certain nonsingular matrices A_p and B_p, show that

$$H_p(s) = \det(1 - A_p \otimes B_p p^{-s})^{-1}.$$

We use the well-known identity

$$\det(1 - At) = \exp\left(\sum_{k=1}^{\infty} \frac{\operatorname{tr}(A^k)t^k}{k}\right),$$

so that what we must show is

$$\det(1 - (A \otimes B)t) = \exp\left(\sum_{k=1}^{\infty} \frac{\operatorname{tr}(A^k)\operatorname{tr}(B^k)t^k}{k}\right).$$

Since the matrices A_p and B_p are nonsingular, the eigenvalues of the matrix $A \otimes B$ can be taken to be $\lambda_i \mu_j$ as λ_i runs through eigenvalues of A and μ_j runs through eigenvalues of B. Thus the right-hand side of the identity to be proved is $\det(1 - (A \otimes B)t)$ as required. □

8.4.3 *With notation as in the previous exercise, show that if $F, G \in S$, then $F \otimes G$ converges absolutely for* $\operatorname{Re}(s) > 1$.

This is the immediate consequence of Exercise 8.2.9.

8.4.4 *If $F \in \mathcal{S}$ and $F \otimes F$ extends to an analytic function for* $\text{Re}(s) \geq 1/2$, *except for a simple pole at $s = 1$, we will say that F is $\otimes$-simple. Prove that a $\otimes$-simple function has at most a simple pole at $s = 1$.*

Suppose F has a pole of order m at $s = 1$. Let s be real and $s \to 1^+$ Then

$$\log F(s) \sim m \log \frac{1}{s-1}.$$

But $\log F(s) = \sum_p \frac{a_p(F)}{p^s} + O(1)$. Since F is $\otimes$-simple, we have by definition

$$\sum_p \frac{|a_p(F)|^2}{p^s} \sim \log \frac{1}{s-1}$$

as $s \to 1^+$. Thus, by Cauchy's inequality

$$\left| \sum_p \frac{a_p(F)}{p^s} \right| \leq \left(\sum_p \frac{|a_p(F)|^2}{p^s} \right)^{1/2} \left(\sum_p \frac{1}{p^s} \right)^{1/2},$$

from which we deduce that $|m| \leq 1$, as required. □

8.4.5 *If $F \in \mathcal{S}$ and*

$$F = F_1^{e_1} F_2^{e_2} \cdots F_k^{e_k}$$

is a factorization of F into distinct primitive functions, show that

$$\sum_{p \leq x} \frac{|a_p(F)|^2}{p} = (e_1^2 + e_2^2 + \cdots + e_k^2) \log \log x + O(1),$$

assuming Selberg's conjectures (a) and (b).

We have clearly

$$a_p(F) = \sum_i e_i a_p(F_i),$$

from which

$$\sum_{p \leq x} \frac{|a_p(F)|^2}{p} = \sum_{i,j} e_i e_j \sum_{p \leq x} \frac{a_p(F_i)\overline{a_p(F_j)}}{p},$$

and the result is now clear. □

8.4.6 *If $F \in \mathcal{S}$, and $F \otimes \bar{F} \in \mathcal{S}$ show that F is $\otimes$-simple if and only if F is primitive, assuming Selberg's conjectures (a) and (b).*

One way is clear. If F is primitive, then F is $\otimes$-simple. Now suppose F is $\otimes$-simple. Then

$$\sum_{p\leq x}\frac{|a_p(F)|^2}{p} = \log\log x + O(1).$$

If $F = F_1^{e_1}F_2^{e_2}\cdots F_k^{e_k}$ is the factorization of F into distinct primitive functions, then by Exercise 8.4.5, we get

$$1 = e_1^2 + e_2^2 + \cdots + e_k^2,$$

from which we deduce that F is primitive. □

8.4.7 *If $F \in \mathcal{S}$ is $\otimes$-simple and entire, prove that $F(1+it) \neq 0$ for all $t \in \mathbb{R}$.*

Suppose F has a zero on $\mathrm{Re}(s) = 1$. By translating, we may suppose F has a zero at $s = 1$. Consider

$$G(s) = \zeta(s)F(s)\bar{F}(s)(F\otimes\bar{F})(s).$$

Then $G(s)$ is a Dirichlet series that is analytic for $\mathrm{Re}(s) \geq 1/2$. Also, $\log G(s)$ is a Dirichlet series with nonnegative coefficients. By Exercise 3.2.11, $G(1+it) \neq 0$ for all $t \in \mathbb{R}$. By Landau's theorem (Exercise 2.5.14) the abscissa of convergence is a real singularity σ_0 (say). Thus $\log G(\sigma) \geq 0$ for $\sigma > \sigma_0$. Hence

$$|G(\sigma)| \geq 1$$

for $\sigma > \sigma_0$. By continuity, $|G(\sigma_0)| \geq 1$. However, σ_0 is a singularity of $\log G(s)$, which must come from a zero of $G(s)$. Thus $G(s_0) = 0$, which is a contradiction. Hence, $F(1) \neq 0$. □

8.4.8 *Let $F \in \mathcal{S}$ and write*

$$-\frac{F'}{F}(s) = \sum_{n=1}^{\infty}\Lambda_F(n)n^{-s}.$$

For $T > 1$ and $n \in \mathbb{N}$, $n > 1$, show that

$$\sum_{|\gamma|\leq T} n^{\rho} = -\frac{T}{\pi}\Lambda_F(n) + O\left(n^{3/2}\log^2 T\right)$$

where $\rho = \beta + i\gamma$, $\beta > 0$ runs over the non-trivial zeros of $F(s)$.

This is a generalization of Exercise 7.4.17 and the proof is similar. (The result shows how to reconstruct $F(s)$ from a knowledge of its zeros.) □

8.4.9 *Suppose $F, G \in \mathcal{S}$. Let*

$$Z_F(T) = \{\rho = \beta + i\gamma, \beta > 0, F(\rho) = 0 \, and \, |\gamma| \leq T\}.$$

Suppose that as $T \to \infty$,

$$|Z_F(T) \Delta Z_G(T)| = o(T),$$

where Δ denotes the symmetric difference $A \Delta B = (A \setminus B) \cup (B \setminus A)$. Show that $F = G$.

By the previous exercise,

$$-\Lambda_F(n) = \lim_{T \to \infty} \frac{\pi}{T} \sum_{|\gamma| \leq T} n^\rho$$

where the summation runs over zeros of $F(s)$ with imaginary part γ satisfying $|\gamma| \leq T$. Since the zeros of $G(s)$ are the same apart from $o(T)$ of them, we find that the above limit is $-\Lambda_G(n)$. Thus, $F = G$, as required. □

9

Sieve Methods

9.1 The Sieve of Eratosthenes

9.1.2 *Prove that there is a constant c such that*

$$\prod_{p\le z}\left(1-\frac{1}{p}\right)=\frac{e^{-c}}{\log z}\left(1+O\left(\frac{1}{\log z}\right)\right).$$

Let $V(z)=\prod_{p\le z}\left(1-\frac{1}{p}\right)$. Then

$$-\log V(z)=\sum_{p\le z}\frac{1}{p}+\sum_{\substack{k\ge 2\\ p\le z}}\frac{1}{kp^k}.$$

The second sum satisfies

$$\sum_{\substack{k\ge 2\\ p\le z}}\frac{1}{kp^k}\le\sum_{p\le z}\sum_{k\ge 2}\frac{1}{p^k}=\sum_{p\le z}\frac{1}{p(p-1)},$$

so that

$$-\log V(z)=\sum_{p\le z}\frac{1}{p}+c_0+O\left(\frac{1}{z}\right),$$

with

$$c_0=-\sum_{p}\left\{\log\left(1-\frac{1}{p}\right)+\frac{1}{p}\right\}.$$

On the other hand, we have

$$R(z) := \sum_{p \le z} \frac{\log p}{p} = \log z + O(1)$$

by Exercise 3.1.7, so that by partial summation

$$\begin{aligned} \sum_{p \le z} \frac{1}{p} &= \frac{R(z)}{\log z} + \int_2^z \frac{R(t)dt}{t \log^2 t} \\ &= \log \log z + c_1 + O\left(\frac{1}{\log z}\right) \end{aligned}$$

for some constant c_1. Thus,

$$-\log V(z) = \log \log z + (c_0 + c_1) + O\left(\frac{1}{\log z}\right),$$

so that with $c = c_0 + c_1$,

$$\prod_{p \le z} \left(1 - \frac{1}{p}\right) = \frac{e^{-c}}{\log z}\left(1 + O\left(\frac{1}{\log z}\right)\right),$$

as required. □

9.1.4 *For $z \le \log x$, prove that*

$$\pi(x, z) = (1 + o(1)) \frac{x e^{-\gamma}}{\log z}$$

whenever $z = z(x) \to \infty$ as $x \to \infty$.

By Exercise 9.1.2,

$$\pi(x, z) = x \prod_{p \le z} \left(1 - \frac{1}{p}\right) + O(2^z).$$

For $z \le \log x$, the error term is $O(x^\theta)$ with $\theta < 1$. The result now follows by applying Mertens's theorem. □

9.1.5 (Rankin's trick) *Prove that*

$$\Phi(x, z) \le x^\delta \prod_{p \le z} \left(1 - \frac{1}{p^\delta}\right)^{-1}$$

for any $\delta > 0$.

For any $\delta > 0$, we have

$$\begin{aligned}\Phi(x,z) &\le \sum_{\substack{n\le x\\ p|n\Rightarrow p\le z}} 1 \le \sum_{\substack{n\le x\\ p|n\Rightarrow p\le z}} \left(\frac{x}{n}\right)^{\delta}\\ &\le x^{\delta}\prod_{p\le z}\left(1-\frac{1}{p^{\delta}}\right)^{-1}.\end{aligned}$$

□

9.1.6 *Choose $\delta = 1 - \frac{1}{\log z}$ in the previous exercise to deduce that*

$$\Phi(x,z) \ll x(\log z)\exp\left(-\frac{\log x}{\log z}\right).$$

Choosing $\delta = 1 - \eta$ with $\eta \to 0$ as $z \to \infty$, we see that

$$\Phi(x,z) \ll x^{\delta}\prod_{p\le z}\left(1+\frac{1}{p^{\delta}}\right).$$

Applying the elementary inequality $1 + x \le e^{x}$, we obtain

$$\Phi(x,z) \ll \exp\left(\delta\log x + \sum_{p\le z}\frac{1}{p^{\delta}}\right).$$

Writing $p^{-\delta} = p^{-1}p^{\eta} = p^{-1}e^{\eta\log p}$, and using the inequality $e^{x} \le 1 + xe^{x}$, we deduce

$$\sum_{p\le z}\frac{1}{p^{\delta}} \le \sum_{p\le z}\frac{1}{p}\left(1+(\eta\log p)z^{\eta}\right),$$

since $p \le z$. Now choosing $\eta = \frac{1}{\log z}$ gives the desired result. □

9.1.7 *Prove that*

$$\pi(x,z) = x\sum_{\substack{d|P_z\\ d\le x}}\frac{\mu(d)}{d} + O\left(x(\log z)\exp\left(-\frac{\log x}{\log z}\right)\right)$$

for $z = z(x) \to \infty$ as $x \to \infty$.

Observe that

$$\pi(x,z) = x\sum_{\substack{d|P_z\\ d\le x}}\frac{\mu(d)}{d} + O(\Phi(x,z)),$$

since $[x/d] = 0$ unless $d \leq x$. Now use Exercise 9.1.6. □

9.1.8 *Prove that*

$$\sum_{\substack{d|P_z \\ d\leq x}} \frac{\mu(d)}{d} = \prod_{p\leq z}\left(1-\frac{1}{p}\right) + O\Big((\log z)^2 \exp\Big(-\frac{\log x}{\log z}\Big)\Big),$$

with $z = z(x) \to \infty$ *as* $x \to \infty$.

We have

$$\sum_{\substack{d|P_z \\ d\leq x}} \frac{\mu(d)}{d} = \prod_{p\leq z}\left(1-\frac{1}{p}\right) - \sum_{\substack{d|P_z \\ d>x}} \frac{\mu(d)}{d}.$$

The last sum is dominated by

$$\sum_{\substack{d|P_z \\ d>x}} \frac{1}{d} \leq -\frac{\Phi(x,z)}{x} + \int_x^\infty \frac{\Phi(t,z)dt}{t^2},$$

on using partial summation. Using the estimate derived for $\Phi(t,z)$ in Exercise 9.1.6, we get that the integral is bounded by

$$\begin{aligned}(\log z)\int_x^\infty \exp\Big(-\frac{\log t}{\log z}\Big)\frac{dt}{t} &= (\log z)\int_x^\infty \frac{dt}{t^{1+1/\log z}} \\ &\ll (\log z)^2 \exp\Big(-\frac{\log x}{\log z}\Big).\end{aligned}$$

This completes the proof. □

9.1.9 *Prove that*

$$\pi(x,z) = xV(z) + O\left(x(\log z)^2 \exp\left(-\frac{\log x}{\log z}\right)\right),$$

where

$$V(z) = \prod_{p\leq z}\left(1-\frac{1}{p}\right)$$

and $z = z(x) \to \infty$ *as* $x \to \infty$.

This essentially follows from Exercises 9.1.7 and 9.1.8. □

9.1.10 *Prove that*

$$\pi(x) \ll \frac{x}{\log x}\log\log x$$

by setting $\log z = \epsilon \log x / \log\log x$ *for some sufficiently small* ϵ *in the previous exercise.*

We have

$$\pi(x) \leq \pi(x, z) + \pi(z).$$

Choosing z as stated shows that

$$\pi(x, z) \ll \frac{x}{\log x} \log\log x$$

from Mertens's theorem and Exercise 9.1.9. Here, the implied constant depends on ϵ. □

9.1.11 *For any* $A > 0$, *show that*

$$\pi\left(x, (\log x)^A\right) \sim \frac{xe^{-\gamma}}{A \log\log x}$$

as $x \to \infty$.

Apply Exercise 9.1.9 with $z = (\log x)^A$. □

9.1.12 *Suppose that*

$$\sum_{\substack{p \leq z \\ p \in \mathcal{P}}} \frac{\omega(p) \log p}{p} \leq \kappa \log z + O(1).$$

Show that

$$F_\omega(t, z) := \sum_{\substack{d \leq t \\ d | P(z)}} \omega(d)$$

is bounded by

$$O\Big(t(\log z)^\kappa \exp\Big(-\frac{\log t}{\log z}\Big)\Big).$$

We apply Rankin's trick for any $\delta > 0$,

$$F_\omega(t, z) \leq \sum_{d | P(z)} \omega(d)(t/d)^\delta.$$

Since ω is multiplicative (by definition), we see that

$$F_\omega(t, z) \leq \exp\left(\delta \log t + \sum_{\substack{p \leq z \\ p \in \mathcal{P}}} \frac{\omega(p)}{p^\delta}\right),$$

on applying the elementary inequality $1 + x \le e^x$. Setting $\delta = 1 - \eta$ and using the inequality $e^x \le 1 + xe^x$, which is valid for $x \ge 0$, we obtain

$$F_\omega(t, z) \le t \exp\left(-\eta \log t + \sum_{\substack{p \le z \\ p \in \mathcal{P}}} \frac{\omega(p)}{p} + \eta z^\eta \sum_{\substack{p \le z \\ p \in \mathcal{P}}} \frac{\omega(p) \log p}{p} \right).$$

The hypothesis gives by partial summation that

$$\sum_{\substack{p \le z \\ p \in \mathcal{P}}} \frac{\omega(p)}{p} \le \kappa \log\log z + O(1),$$

so that

$$F_\omega(t, z) \ll t \exp(-\eta \log t + \kappa \log\log z + \kappa\eta(\log z) z^\eta).$$

Choosing $\eta = 1/\log z$ gives the result. □

9.1.13 *Let C be a constant. With the same hypothesis as in the previous exercise, show that*

$$\sum_{\substack{d | P(z) \\ d > Cx}} \frac{\omega(d)}{d} = O\Big((\log z)^{\kappa+1} \exp\Big(-\frac{\log x}{\log z} \Big) \Big).$$

With the notation of Exercise 9.1.12, we have

$$\sum_{\substack{d | P(z) \\ d > Cx}} \frac{\omega(d)}{d} \ll \int_{Cx}^\infty \frac{F_\omega(t, z) dt}{t^2},$$

and the previous exercise immediately gives the result. □

9.1.14 (Sieve of Eratosthenes) *Suppose there is a constant $C > 0$ such that $|A_d| = 0$ for $d > Cx$. Then*

$$S(A, \mathcal{P}, z) = XW(z) + O\Big(x (\log z)^{\kappa+1} \exp\Big(-\frac{\log x}{\log z} \Big) \Big).$$

By the inclusion - exclusion principle, we have

$$\begin{aligned} S(A, \mathcal{P}, z) &= \sum_{d|\mathcal{P}(z)} \mu(d)|A_d| \\ &= \sum_{\substack{d|\mathcal{P}(z) \\ d\leq Cx}} \mu(d)\frac{X\omega(d)}{d} + O(F_\omega(Cx, z)), \end{aligned}$$

in the notation of the previous exercise. Then, the first sum can be rewritten

$$X\left(\sum_{d|\mathcal{P}(z)} \frac{\mu(d)\omega(d)}{d} - \sum_{\substack{d|\mathcal{P}(z) \\ d>Cx}} \frac{\mu(d)\omega(d)}{d}\right),$$

so that we can use the estimate of Exercise 9.1.13 on the second sum. Exercise 9.1.12 gives an estimate for $F_\omega(Cx, z)$. This completes the proof. □

9.2 Brun's Elementary Sieve

9.2.1 *Show that for r even,*

$$\pi(x, z) \leq x\sum_{d|P_z} \frac{\mu_r(d)}{d} + O(z^r).$$

Recall that

$$\begin{aligned} \pi(x, z) &= \sum_{n\leq x} \sum_{d|(n,P_z)} \mu(d) \\ &\leq \sum_{n\leq x} \sum_{d|(n,P_z)} \mu_r(d) \\ &\leq \sum_{d|P_z} \mu_r(d)\left[\frac{x}{d}\right] \\ &\leq x\sum_{d|P_z} \frac{\mu_r(d)}{d} + \sum_{d|P_z} |\mu_r(d)|. \end{aligned}$$

The last term is easily seen to be $O(z^r)$, as required. □

9.2.3 *Show that*

$$\sum_{d|P(z)} \frac{\mu_r(d)\omega(d)}{d} = \prod_{\substack{p\le z\\ p\in\mathcal{P}}}\left(1-\frac{\omega(p)}{p}\right)\sum_{\delta|P(z)}\frac{\psi_r(\delta)\omega(\delta)}{\Omega(\delta)},$$

where $\Omega(\delta) = \prod_{p|\delta}(p-\omega(p))$.

By Möbius inversion, we have

$$\mu_r(d) = \sum_{\delta|d}\mu(d/\delta)\psi_r(\delta),$$

so that

$$\begin{aligned}\sum_{d|P(z)} \frac{\mu_r(d)\omega(d)}{d} &= \sum_{d|P(z)}\frac{\omega(d)}{d}\sum_{\delta|d}\mu(d/\delta)\psi_r(\delta)\\ &= \sum_{\delta|P(z)}\frac{\psi_r(\delta)\omega(\delta)}{\delta}\sum_{d|P(z)/\delta}\frac{\mu(d)\omega(d)}{d}\\ &= \prod_{\substack{p\le z\\ p\in\mathcal{P}}}\left(1-\frac{\omega(p)}{p}\right)\cdot\sum_{\delta|P(z)}\frac{\psi_r(\delta)\omega(\delta)}{\Omega(\delta)},\end{aligned}$$

where $\Omega(\delta) = \prod_{p|\delta}(p-\omega(p))$. □

9.2.4 *Suppose that* $\omega(p)\le c$, *and that* $\sum_{\substack{p\le z\\ p\in\mathcal{P}}}\frac{\omega(p)}{p}\le c_1\log\log z + c_2$ *for some constants* c, c_1, *and* c_2. *Show that there are constants* c_3, c_4 *and* c_5 *such that*

$$\sum_{\substack{\delta|P(z)\\ \delta>1}}\frac{\psi_r(\delta)\omega(\delta)}{\Omega(\delta)} \ll \frac{1}{r!}(c_3\log\log z + c_4)^r(\log z)^{c_5}.$$

Recall that

$$\psi_r(\delta)\le\binom{\nu(\delta)-1}{r}$$

so that the sum under consideration is

$$\begin{aligned}&\le \sum_{\substack{\delta|P(z)\\ \delta>1}}\binom{\nu(\delta)-1}{r}\frac{\omega(\delta)}{\Omega(\delta)}\\ &\le \sum_{r\le m\le\pi(z)}\binom{m}{r}\frac{1}{m!}\Big(\sum_{\substack{p\le z\\ p\in\mathcal{P}}}\frac{\omega(p)}{p-c}\Big)^m\\ &\le \frac{1}{r!}(c_3\log\log z + c_4)^r\exp(c_3\log\log z + c_4),\end{aligned}$$

which gives the result. □

9.2.6 *Show that the number of primes $p \leq x$ such that $p+2$ is also prime is $\ll x(\log\log x)^2/(\log x)^2$.*

Let $A = \{n : n \leq x\}$; $P = \{p : 2 < p \leq z\}$, the set of odd primes less than or equal to z. For each odd prime p, we distinguish the residue classes 0 and $-2(\text{mod } p)$, so that $\omega(p) = 2$. Then, $\omega(d) = 2^{\nu(d)}$, where $\nu(d)$ is the number of prime factors of d, and $A_d = \cap_{p|d} A_p$. By the Chinese remainder theorem,

$$|A_d| = \frac{x\omega(d)}{d} + R_d$$

with $|R_d| = O(2^{\nu(d)})$. Applying (9.1), we get

$$S(A, P, z) = xW(z) + O\Big(x(\exp\Big(-\frac{c_3 \log x}{\log z}\Big)\Big).$$

We choose $\log z = \log x / A \log\log x$ for an appropriate constant A This gives the result, since

$$W(z) = \prod_{3 \leq p \leq z}\left(1 - \frac{2}{p}\right) \leq \prod_{3 \leq p \leq z}\left(1 - \frac{1}{p}\right)^2,$$

so that an application of Mertens's theorem completes the proof. □

9.2.7 (Brun, 1915) *Show that*

$$\sum{}' \frac{1}{p} < \infty,$$

where p is such that $p+2$ is prime.

Let $\pi_2(x)$ be the number of twin primes less than or equal to x. By partial summation, the sum is

$$\ll \int_3^\infty \frac{\pi_2(t)dt}{t^2} \ll \int_3^\infty \frac{(\log\log t)^2 dt}{t(\log t)^2} < \infty.$$

□

9.3 Selberg's Sieve

9.3.1 *Let $P_z = \prod_{p \le z} p$ be the product of the primes $p \le z$. Show that*

$$\pi(x, z) \le \sum_{n \le x} \Big(\sum_{d | (n, P_z)} \lambda_d \Big)^2,$$

for any sequence λ_d of real numbers satisfying $\lambda_1 = 1$.

This is clear from $\lambda_1 = 1$. The quantity on the right-hand side is always nonnegative and is equal to 1 when $(n, P_z) = 1$. □

9.3.2 *Show that if $|\lambda_d| \le 1$, then*

$$\pi(x, z) \le \sum_{d_1, d_2 \le z} \frac{\lambda_{d_1} \lambda_{d_2}}{[d_1, d_2]} x + O(z^2),$$

where $[d_1, d_2]$ is the least common multiple of d_1 and d_2.

In Exercise 9.3.1, we expand the sequence,

$$\begin{aligned} \pi(x, z) &\le \sum_{n \le x} \sum_{d_1, d_2 | (n, P_z)} \lambda_{d_1} \lambda_{d_2} \\ &\le \sum_{d_1, d_2 \le z} \lambda_{d_1} \lambda_{d_2} \Big(\sum_{\substack{n \le x \\ d_1, d_2 | n}} 1 \Big), \end{aligned}$$

since $\lambda_d = 0$ for $d > z$. Since

$$\sum_{\substack{n \le x \\ d_1, d_2 | n}} 1 = \frac{x}{[d_1, d_2]} + O(1)$$

and $|\lambda_d| \le 1$, the estimate is clear. □

9.3.3 *Prove that*

$$[d_1, d_2](d_1, d_2) = d_1 d_2,$$

where (d_1, d_2) is the greatest common divisor of d_1 and d_2.

This is clear from unique factorization. Write

$$d_1 = \prod_p p^{\alpha_p}, \quad d_2 = \prod_p p^{\beta_p}.$$

Then

$$[d_1, d_2] = \prod_p p^{\max(\alpha_p, \beta_p)}$$

and

$$(d_1, d_2) = \prod_p p^{\min(\alpha_p, \beta_p)}.$$

□

9.3.4 *Show that*

$$\sum_{d_1, d_2 \leq z} \frac{\lambda_{d_1} \lambda_{d_2}}{[d_1, d_2]} = \sum_{\delta \leq z} \phi(\delta) \Big(\sum_{\substack{\delta | d \\ d \leq z}} \frac{\lambda_d}{d} \Big)^2.$$

By the previous exercise, we can write the left-hand side as

$$\begin{aligned} \sum_{d_1, d_2 \leq z} \frac{\lambda_{d_1} \lambda_{d_2}}{d_1 d_2} (d_1, d_2) &= \sum_{d_1, d_2 \leq z} \frac{\lambda_{d_1} \lambda_{d_2}}{d_1 d_2} \sum_{\delta | (d_1, d_2)} \phi(\delta) \\ &= \sum_{\delta \leq z} \phi(\delta) \Big(\sum_{\substack{\delta | d \\ d \leq z}} \frac{\lambda_d}{d} \Big)^2, \end{aligned}$$

as required (notice that this is a "diagonalization" of the quadratic form). □

9.3.5 *If*

$$u_\delta = \sum_{\substack{\delta | d \\ d \leq z}} \frac{\lambda_d}{d},$$

show that

$$\frac{\lambda_\delta}{\delta} = \sum_{\delta | d} \mu(d/\delta) u_d.$$

(Note that $u_\delta = 0$ for $\delta > z$, since $\lambda_d = 0$ for $d > z$.)

This is an application of the dual Möbius inversion formula (Exercise 1.5.16). □

9.3.6 *Show that if $\lambda_1 = 1$, then*

$$\sum_{d_1, d_2 \leq z} \frac{\lambda_{d_1} \lambda_{d_2}}{[d_1, d_2]}$$

attains the minimum value $1/V(z)$, *where*

$$V(z) = \sum_{d \leq z} \frac{\mu^2(d)}{\phi(d)}.$$

By Exercise 9.3.4, we must minimize

$$\sum_{\delta \leq z} \phi(\delta) \Big(\sum_{\substack{\delta | d \\ d \leq z}} \frac{\lambda_d}{d} \Big)^2$$

subject to the constraint $\lambda_1 = 1$. By Exercise 9.3.5 we must minimize

$$\sum_{\delta \leq z} \phi(\delta) u_\delta^2$$

subject to

$$\sum_{d \leq z} \mu(d) u_d = 1.$$

By the Lagrange multiplier method, this minimum is attained when

$$2\phi(\delta) u_\delta = \lambda \mu(\delta)$$

for some scalar λ. Thus,

$$u_\delta = \frac{\lambda \mu(\delta)}{2\phi(\delta)},$$

so that

$$\frac{\lambda}{2} \sum_{\delta \leq z} \frac{\mu^2(\delta)}{\phi(\delta)} = 1$$

and the minimum is

$$\sum_{\delta \leq z} \phi(\delta) \cdot \frac{\lambda^2 \mu^2(\delta)}{4\phi^2(\delta)} = \frac{\lambda^2}{4} \sum_{\delta \leq z} \frac{\mu^2(\delta)}{\phi(\delta)} = \frac{1}{V(z)}$$

as desired. □

9.3.7 *Show that for the choice of*

$$u_\delta = \mu(\delta)/(\phi(\delta)V(z)),$$

we have $|\lambda_d| \leq 1$.

We have, by Exercise 9.3.5,

$$\begin{aligned} \frac{\lambda_d}{d} &= \sum_t \mu(t) u_{dt} \\ &= \frac{\mu(d)}{\phi(d)} \sum_{\substack{(d,t)=1 \\ t \leq z/d}} \frac{\mu^2(t)}{\phi(t)V(z)}. \end{aligned}$$

Hence

$$\begin{aligned} V(z)\lambda_d &= \mu(d) \prod_{p|d} \left(1 + \frac{1}{p-1}\right) \sum_{\substack{t \leq z/d \\ (t,d)=1}} \frac{\mu^2(t)}{\phi(t)} \\ &= \mu(d) \sum_{\delta|d} \frac{\mu^2(\delta)}{\phi(\delta)} \sum_{\substack{t \leq z/d \\ (t,d)=1}} \frac{\mu^2(t)}{\phi(t)}. \end{aligned}$$

Thus,

$$|\lambda_d V(z)| \leq \sum_{t \leq z} \frac{\mu^2(t)}{\phi(t)} = V(z),$$

so that $|\lambda_d| \leq 1$, as required. □

9.3.8 *Show that*

$$\pi(x, z) \leq \frac{x}{V(z)} + O(z^2).$$

Deduce that $\pi(x) = O\left(\frac{x}{\log x}\right)$ *by setting* $z = x^{1/2-\epsilon}$.

We have

$$V(z) = \sum_{\delta \leq z} \frac{\mu^2(\delta)}{\phi(\delta)} \gg \log z$$

by the following elementary argument. We have

$$\sum_{\delta \leq z} \frac{\mu^2(\delta)}{\phi(\delta)} \geq \sum_{\delta \leq z} \frac{\mu^2(\delta)}{\delta}.$$

Now,

$$\sum_{\delta \leq z} \frac{1}{\delta} = \log z + O(1),$$

and it is clear that

$$\sum_{\delta \leq z} \frac{\mu^2(\delta)}{\delta} = \sum_{\delta \leq z} \frac{1}{\delta} - \sum_{m \leq z}{}' \frac{1}{m},$$

where the dash on the sum means that m has a squared prime factor. Clearly,

$$\sum_{m \leq z}{}' \frac{1}{m} \leq \frac{1}{4} \sum_{\delta \leq z/4} \frac{1}{\delta} \leq \frac{1}{4}(\log z + O(1)).$$

Thus,

$$\sum_{\delta \leq z} \frac{\mu^2(\delta)}{\phi(\delta)} \gg \log z.$$

Now choose $z = x^{1/2-\epsilon}$ to obtain the desired result. □

9.3.9 *Let f be a multiplicative function. Show that*

$$f([d_1, d_2])f((d_1, d_2)) = f(d_1)f(d_2).$$

We can write

$$[d_1, d_2] = (d_1, d_2)e_1 e_2,$$

where $e_1(d_1, d_2) = d_1$, $e_2(d_1, d_2) = d_2$. Thus e_1, e_2, (d_1, d_2) are mutually coprime. Therefore,

$$f([d_1, d_2]) = f((d_1, d_2))f(e_1)f(e_2).$$

Multiplying both sides by $f(d_1, d_2)$ gives

$$f([d_1, d_2])f((d_1, d_2)) = f(d_1)f(d_2)$$

as desired, since e_1 and (d_1, d_2) are coprime, as well as e_2 and (d_1, d_2). □

9.3.11 *Show that*

$$U(z) \geq \sum_{\delta \leq z} \frac{1}{\tilde{f}(\delta)},$$

where $\tilde{f}(n)$ is the completely multiplicative function defined by $\tilde{f}(p) = f(p)$.

We have

$$U(z) = \sum_{d \le z} \frac{\mu^2(d)}{f_1(d)}.$$

Now, for square-free n,

$$\begin{aligned}\frac{f(n)}{f_1(n)} &= \prod_{p|n} \frac{f(p)}{f(p)-1} = \prod_{p|n}\left(1 - \frac{1}{f(p)}\right)^{-1} \\ &= \sum_d{}' \frac{1}{\tilde{f}(d)},\end{aligned}$$

where the dash on the summation means that d ranges over elements of the monoid generated by the prime divisors of n. Hence, for square-free n,

$$\frac{1}{f_1(n)} = \frac{1}{\tilde{f}(n)} \sum_d{}' \frac{1}{\tilde{f}(d)},$$

so that

$$U(z) = \sum_{d \le z} \frac{\mu^2(d)}{f_1(d)} \ge \sum_{\delta \le z} \frac{1}{\tilde{f}(\delta)},$$

as required. □

9.3.12 *Let $\pi_2(x)$ denote the number of twin primes $p \le x$. Using Selberg's sieve, show that*

$$\pi_2(x) = O\left(\frac{x}{\log^2 x}\right).$$

We consider the sequence $a_n = n(n+2)$ and count the number of elements coprime to P_z. The number of $n \le x$ such that $d|a_n$ is clearly

$$\frac{x 2^{\nu(d)}}{d} + O\left(2^{\nu(d)}\right)$$

by an application of the Chinese remainder theorem. Thus, $f(d) = d/2^{\nu(d)}$ in the notation of Selberg's sieve, and we have

$$\begin{aligned}N(x,z) &\le \frac{x}{U(z)} + O\Big(\sum_{d_1, d_2 \le z} 2^{\nu[d_1,d_2]}\Big) \\ &\le \frac{x}{U(z)} + O\Big(\sum_{d \le z} 2^{\nu(d)}\Big)^2.\end{aligned}$$

By Exercise 1.4.1, the error term is easily seen to be $O\left(z^2\log^2 z\right)$. By Exercise 9.3.11,

$$U(z) \geq \sum_{d\leq z} \frac{2^{\omega(d)}}{d},$$

where $\omega(d)$ is the number of prime factors of d counted with multiplicity. By partial summation (using the result of Exercise 4.4.18) we deduce

$$\sum_{d\leq z} \frac{2^{\omega(d)}}{d} \sim c(\log z)^2$$

for some nonzero constant c. Thus,

$$N(x,z) \ll \frac{x}{(\log z)^2} + O\left(z^2\log^2 z\right).$$

The number of twin primes is clearly less than or equal to $z+N(x,z)$ for any value of z. Choosing $z = x^{1/4}$ (say) gives us the required result. □

9.3.13 (The Brun - Titchmarsh theorem) *For $(a,k) = 1$, and $k \leq x$, show that*

$$\pi(x,k,a) \leq \frac{(2+\epsilon)x}{\varphi(k)\log(2x/k)}$$

for $x > x_0(\epsilon)$, where $\pi(x,k,a)$ denotes the number of primes less than x which are congruent to $a \pmod k$.

We consider the set of numbers $n \leq x$, $n \equiv a \pmod k$ that are not divisible by primes p such that $p \leq z$ and $(p,k) = 1$. Clearly, the primes counted by

$$\pi(x,k,a) - \pi(z,k,a)$$

are contained in this set. In the notation of the Selberg sieve, we obtain

$$N(d) = \frac{x}{kd} + O(1),$$

and the upper bound becomes

$$\frac{x}{kU(z)} + O(z^2).$$

By Exercise 9.3.11,

$$U(z) \geq \sum_{\substack{d\leq z\\(d,k)=1}} \frac{1}{d}.$$

Now,

$$\frac{k}{\phi(k)}U(z) \geq \prod_{p|k}\left(1-\frac{1}{p}\right)^{-1} \sum_{\substack{d\leq z\\(d,k)=1}} \frac{1}{d} \geq \sum_{m\leq z} \frac{1}{m},$$

and the latter quantity is asymptotic to $\log z$. This gives a final estimate of

$$\pi(x,k,a) \leq \frac{x}{\phi(k)\log z} + O(z^2),$$

and choosing $z = (2x/k)^{1/2-\epsilon}$ gives the final result. □

9.3.14 (Titchmarsh divisor problem) *Show that* $\sum_{p\leq x} d(p-1) = O(x)$, *where the sum is over primes and* $d(n)$ *denotes the divisor function.*

We have, trivially,

$$d(n) \leq 2 \sum_{\substack{d|n\\d\leq\sqrt{n}}} 1,$$

so that

$$\sum_{p\leq x} d(p-1) \leq 2 \sum_{d\leq\sqrt{x}} \pi(x,d,1).$$

By an application of the Brun - Titchmarsh theorem we get

$$\sum_{p\leq x} d(p-1) \ll \frac{x}{\log x} \sum_{\delta\leq\sqrt{x}} \frac{1}{\phi(\delta)}.$$

By Exercise 4.4.14 (or the weaker 4.4.13) we are done. □

9.4 Supplementary Problems

9.4.1 *Show that*

$$\sum_{\substack{p\leq x\\p\equiv 1\ (\mathrm{mod}\ k)}} \frac{1}{p} \ll \frac{\log\log x + \log k}{\varphi(k)},$$

where the implied constant is absolute.

By partial summation, we have

$$\sum_{\substack{p\leq x\\p\equiv 1 (\mathrm{mod}\ k)}} \frac{1}{p} = \frac{\pi(x,k,1)}{x} + \int_2^x \frac{\pi(t,k,1)dt}{t^2}.$$

We have the trivial estimate $\pi(x,k,1) \leq x/k$, so the first term is negligible.

For the integral, we break the interval of integration into two parts: $[2,k^2]$ and $[k^2,x]$. On the first interval we use the trivial estimate to get an estimate of $O((\log k)/k)$. On the second interval, we use the Brun - Titchmarsh theorem (Exercise 9.3.13) to obtain the final result. □

9.4.2 *Suppose that P is a set of primes such that*

$$\sum_{p\in P}\frac{1}{p} = +\infty.$$

Show that the number of $n \leq x$ not divisible by any prime $p \in P$ is $o(x)$ as $x \to \infty$.

We apply the sieve of Eratosthenes. The number is clearly bounded by

$$x\prod_{\substack{p\leq z\\ p\in P}}\left(1-\frac{1}{p}\right)+O(2^z)$$

for any value of z. Now, for $0<x<1$,

$$e^x < (1-x)^{-1}$$

so that $1-x<e^{-x}$. Hence the bound in question is

$$\leq x\exp\Big(-\sum_{\substack{p\leq z\\ p\in P}}\frac{1}{p}\Big)+O(2^z).$$

Since

$$\sum_{p\in P}\frac{1}{p} = +\infty,$$

the result follows upon choosing $z=\log x$. □

9.4.3 *Show that the number of solutions of $[d_1,d_2]\leq z$ is $O(z(\log z)^3)$.*

The number in question is clearly

$$\begin{aligned}
&\leq \sum_{d_1,d_2\leq z}\frac{z}{[d_1,d_2]}\sum_{\delta|d_1,d_2}\phi(\delta)\\
&\leq z\sum_{\delta\leq z}\phi(\delta)\Big(\sum_{\substack{\delta|d\\ d\leq z}}\frac{1}{d}\Big)^2 \ll z(\log z)^2\sum_{\delta\leq z}\frac{\phi(\delta)}{\delta^2}\\
&\ll z(\log z)^3,
\end{aligned}$$

as required. □

9.4.4 *Prove that*

$$\sum_{p \leq x/2} \frac{1}{p \log(x/p)} = O\left(\frac{\log \log x}{\log x}\right),$$

where the summation is over prime numbers.

We subdivide the interval $[1, x/2]$ into subintervals of the form $I_j = \left[e^j, e^{j+1}\right]$. We estimate

$$\sum_{p \in I_j} \frac{1}{p \log(x/p)} \leq \frac{1}{\log(x/e^j)} \sum_{p \in I_j} \frac{1}{p}.$$

By Chebyshev's theorem,

$$\sum_{p \in I_j} \frac{1}{p} \ll e^{-j}\left(\frac{e^j}{j}\right) \ll \frac{1}{j}.$$

We need to estimate

$$\sum_{j=1}^{\log(x/2)} \frac{1}{j \log(x/e^j)} = O\left(\frac{\log \log x}{\log x}\right)$$

by an easy partial summation. □

9.4.5 *Let $\pi_k(x)$ denote the number of $n \leq x$ with k prime factors (not necessarily distinct). Using the sieve of Eratosthenes, show that*

$$\pi_k(x) \leq \frac{x(A \log \log x + B)^k}{k! \log x}$$

for some constants A and B.

We prove it by induction on k. For $k = 1$, this is Exercise 1.5.12. Clearly,

$$\pi_k(x) \leq \frac{1}{k} \sum_{p \leq x/2} \pi_{k-1}(x/p),$$

since a number $p_1 \cdots p_k \leq x$ is counted k times in the summation

$$\sum_{p \leq x/2} \pi_{k-1}(x/p).$$

(Also, we may suppose that each $p_i \leq x/2$, since $k \geq 2$.) By the induction hypothesis and Exercise 9.4.4, we are done. □

9.4.6 *Let a be an even integer. Show that the number of primes $p \leq x$ such that $p + a$ is also prime is*

$$\ll \frac{x}{(\log x)^2} \prod_{p|a} \left(1 + \frac{1}{p}\right),$$

where the implied constant is absolute.

We let $a_n = n(n + a)$ and apply the Selberg sieve. For P, we take the set of all primes, and in the notation of Theorem 9.3.10 we take $2 < z \leq \sqrt{x}$. If $n > \sqrt{x}$, then $a_n \equiv 0 \pmod{p}$ implies that either n or $n + a$ is composite. Thus, the number to be estimated is less than or equal to

$$\sqrt{x} + N(x, z).$$

Let us write each square-free d as

$$d = p_1 \cdots p_k q_1 \cdots q_t,$$

where the p_i's divide a and the q_i's are coprime to a. By the Chinese remainder theorem it is easily seen that for square-free d,

$$N(d) = \frac{x}{f(d)} + R_d,$$

where $R_d \leq 2^{\omega(d)}$ and $f(d)$ is the completely multiplicative function defined by

$$f(p) = \begin{cases} p/2 & \text{if } (p, a) = 1, \\ p & \text{if } p|a. \end{cases}$$

Thus, by Exercise 9.3.11 and Theorem 9.3.10, we obtain that the number of primes in question is

$$\leq \sqrt{x} + \frac{x}{U(z)} + O\Big(\sum_{d<z^2} 2^{\omega(d)}\Big).$$

The error term is easily seen to be $O(z^2 \log z)$. As for the other term, we have

$$U(z) \geq \sum_{m \leq z} \frac{1}{f(m)},$$

where

$$1/f(m) = \frac{2^{\omega_a(m)}}{m}$$

with $w_a(m)$ equal to the total number (including multiplicity) of prime factors of m that are coprime to a. If we let $d_a(m)$ be the number of divisors of m coprime to a, then we see that $1/f(m) \geq d_a(m)/m$. Hence

$$\left(\sum_{m \leq z} \frac{d_a(m)}{m}\right) \prod_{p|a} \left(1 - \frac{1}{p}\right)^{-1} \geq \sum_{m \leq z} \frac{d_a(m)}{m} \sum_{\gamma(n)|a} \frac{1}{n},$$

where $\gamma(n)$ is the product of the distinct prime divisors of a. Rearranging the sums, we find that the above sum is

$$\geq \sum_{t=1}^{\infty} \frac{1}{t} \sum_{\substack{m \leq z \\ m|t, \gamma(t/m)|a}} d_a(m) \geq \sum_{t \leq z} \frac{1}{t} \sum_{\substack{m|t \\ \gamma(t/m)|a}} d_a(m).$$

The inner sum is clearly greater than or equal to $d(t)$. Thus

$$U(z) \geq \prod_{p|a} \left(1 - \frac{1}{p}\right) \sum_{t \leq z} \frac{d(t)}{t}.$$

By Exercise 2.5.9, this gives

$$U(z) \geq \prod_{p|a} \left(1 - \frac{1}{p}\right) (\log z)^2.$$

Choosing $z = x^{1/4}$ and observing that

$$\begin{aligned} \prod_{p|a} \left(1 - \frac{1}{p}\right)^{-1} &= \prod_{p|a} \left(1 - \frac{1}{p^2}\right)^{-1} \left(1 + \frac{1}{p}\right) \\ &\ll \prod_{p|a} \left(1 + \frac{1}{p}\right) \end{aligned}$$

gives the final result. □

9.4.7 *Let k be a positive even integer greater than 1. Show that the number of primes $p \leq x$ such that $kp + 1$ is also prime is*

$$\ll \frac{x}{(\log x)^2} \prod_{p|k} \left(1 + \frac{1}{p}\right).$$

We proceed as in Exercise 9.4.6 and take the sequence $a_n = n(kn+1)$. As before, we obtain

$$N(d) = \frac{x}{f(d)} + R_d$$

with $|R_d| \leq 2^{\omega(d)}$, and $f(d)$ as in Exercise 9.4.6. We proceed as in the previous exercise to deduce the result. □

9.4.8 *Let k be even and satisfy $2 \leq k < x$. The number of primes $p \leq x$ such that $p - 1 = kq$ with q prime is*

$$\ll \frac{x}{\varphi(k) \log^2(x/k)}.$$

We substitute x/k for x in the previous exercise and observe that we have actually proved

$$\prod_{p|k} \left(1 - \frac{1}{p}\right)^{-1} \frac{x}{\log^2 x}$$

as the upper bound. Since the product is $k/\varphi(k)$, the result follows. □

9.4.9 *Let n be a natural number. Show that the number of solutions of the equation $[a, b] = n$ is $d(n^2)$, where $d(n)$ is the number of divisors of n.*

Clearly, a and b can only have prime factors dividing n. Writing

$$n = \prod_{p|n} p^{\nu_p(n)}, \quad a = \prod_{p|a} p^{\nu_p(a)}, \quad b = \prod_{p|b} p^{\nu_p(b)}$$

we must have $\nu_p(n) = \max(\nu_p(a), \nu_p(b))$. The number of solutions for this latter equation is enumerated as follows. We can set $\nu_p(a) = \nu_p(n)$ and vary $\nu_p(b)$ between 0 and $\nu_p(n)$ or the other way around.

But we have counted the pair $(\nu_p(a), \nu_p(b)) = (\nu_p(n), \upsilon_p(n))$ twice. Thus the number of solutions is

$$\prod_{p|n}(2\nu_p(n)+1) = d(n^2).$$

□

9.4.10 *Show that the error term in Theorem 9.3.10 can be replaced by*

$$O\Big(\sum_{a<z^2} d(a^2)|R_a|\Big).$$

By the previous exercise, the number of solutions of $[d_1, d_2] = a$ is $d(a^2)$, and we are done. □

9.4.11 *Show that*

$$\sum_{p\le x} \frac{p-1}{\varphi(p-1)} = O\Big(\frac{x}{\log x}\Big),$$

where the summation is over prime numbers.

Observe that

$$\frac{n}{\varphi(n)} = \prod_{p|n}\Big(1-\frac{1}{p}\Big)^{-1} \ll \prod_{p|n}\Big(1+\frac{1}{p}\Big),$$

so that

$$\frac{n}{\varphi(n)} \ll \sum_{d|n} \frac{1}{n}.$$

Therefore,

$$\sum_{p\le x} \frac{p-1}{\varphi(p-1)} \ll \sum_{d\le x} \frac{\pi(x,d,1)}{d}.$$

The latter sum is split into two parts: $d \le \sqrt{x}$ and $d > \sqrt{x}$. On the second part we use the trivial estimate $\pi(x,d,1) \le x/d$, and on the first part, we use the Brun - Titchmarsh theorem to deduce the desired estimate. □

9.4.12 *Prove that*

$$\prod_{r<p\le x}\Big(1-\frac{r}{p}\Big) \ll \frac{1}{(\log x)^r}.$$

We have the inequality

$$1 - x \leq e^{-x},$$

easily verified to be valid for $x \geq 0$. Indeed, let $f(x) = e^{-x} + x - 1$. Then $f'(x) = -e^{-x} + 1$, which is nonnegative for $x \geq 0$. Hence, $f(x)$ is increasing, so that $f(x) \geq f(0) = 0$, for $x \geq 0$. This fact, combined with the elementary fact

$$\sum_{p \leq x} \frac{1}{p} = \log\log x + O(1),$$

gives the desired result. □

9.4.13 *Prove that for some constant $c > 0$, we have*

$$\sum_{n \leq x} \frac{d(n^2)}{\varphi(n)} = c(\log x)^3 + O\left(\log^2 x\right).$$

We consider the Dirichlet series

$$f(s) = \sum_{n=1}^{\infty} \frac{d(n^2)}{\varphi(n) n^s} = \prod_p \left(1 + \frac{3}{p^s(p-1)} + \cdots\right).$$

We see that $g(s) = f(s-1)$ has a pole of order 3 at $s = 1$. Moreover, we can write $g(s) = \zeta^3(s) h(s)$, where $h(s)$ is regular for $\mathrm{Re}(s) > 1/2$. Hence by the methods of Chapter 4, we deduce that

$$\sum_{n \leq x} d(n^2) n / \varphi(n) \sim c_1 x (\log x)^2.$$

The result now follows by partial summation. □

9.4.14 *Let $d(n)$ denote the number of divisors of n. Show that*

$$\sum_{p \leq x} d^2(p-1) = O(x \log^2 x \log\log x),$$

where the summation is over prime numbers.

The sum in question is

$$\sum_{[d_1, d_2] \leq x} \pi(x, [d_1, d_2], 1) = \sum_{n \leq x} \pi(x, n, 1) d(n^2)$$

by Exercise 9.4.4. By the Brun - Titchmarsh theorem, this latter sum is bounded by

$$\ll \sum_{n<x} \frac{xd(n^2)}{\varphi(n)\log\frac{x}{n}}.$$

We split the summation over dyadic intervals of the form $[2^k, 2^{k+1}] = I_k$ (say). The sum is

$$\ll \sum_{k=1}^{N-1} \frac{x}{N-k}\left(\sum_{n\in I_k} \frac{d(n^2)}{\varphi(n)}\right),$$

where $N = [\log x/\log 2]$. The inner sum by the previous exercise is $O(k^2)$, and we must estimate

$$\begin{aligned} x\sum_{k=1}^{N-1} \frac{k^2}{N-k} &= x\sum_{j=1}^{N-1} \frac{1}{j}(N-j)^2 \\ &\ll xN^2 \log N = O(x\log^2 x \log\log x), \end{aligned}$$

as desired. □

9.4.15 *Show that the result in the previous exercise can be improved to $O(x\log^2 x)$ by noting that $d^2(n) \le d_4(n)$, where $d_4(n)$ is the number of ways of writing n as a product of four natural numbers.*

If we write $n = d_1d_2d_3d_4 \le x$, then we must have some $d_i \ge n^{1/4}$. It is than not difficult to see that

$$\begin{aligned} \sum_{p\le x} d^2(p-1) &\le \sum_{p\le x} d_4(p-1) \\ &\ll \sum_{d_1d_2d_3\le x^{3/4}} \pi(x, d_1d_2d_3, 1). \end{aligned}$$

Now apply the Brun - Titchmarsh theorem (Exercise 9.3.13) to get the desired result. □

10

p-adic Methods

10.1 Ostrowski's Theorem

10.1.1 *If F is a field with norm $||\cdot||$, show that $d(x,y) = ||x-y||$ defines a metric on F.*

We may suppose $0 \neq 1$ in F, in which case $||1|| = ||1||^2$ implies $||1|| = 1$. Hence $||-1||^2 = 1$ gives $||-1|| = 1$. Now, $d(x,y) = 0$ $\Leftrightarrow ||x-y|| = 0 \Leftrightarrow x = y$; also, $d(x,y) = d(y,x)$, since $||-1|| = 1$. Finally, $d(x,y) = ||x-y|| \leq ||x-z|| + ||z-x|| = d(x,z) + d(z,x)$, which is the triangle inequality. $\square$

10.1.2 *Show that $|\cdot|_p$ is a norm on $\mathbb{Q}$.*

Clearly $|x|_p = 0$ if and only if $x = 0$. Also, we can write $x = p^{\nu_p(x)}x_1$, $y = p^{\nu_p(y)}y_1$ with x_1, y_1 coprime to p. Then, it is clear that $|xy|_p = |x|_p|y|_p$. To prove the triangle inequality, suppose first that $\nu_p(x) \neq \nu_p(y)$ and without loss of generality $\nu_p(x) < \nu_p(y)$. Then $x + y = p^{\nu_p(x)}x_1 + p^{\nu_p(y)}y_1 = p^{\nu_p(x)}(x_1 + p^{\nu_p(y)-\nu_p(x)}y_1)$, so that $|x + y|_p \leq |x|_p = \max(|x|_p, |y|_p)$ in this case. If $\nu_p(x) = \nu_p(y)$, the number $x_1 + y_1$ when written in lowest terms has denominator coprime to p. Thus,

$$|x+y|_p \leq \max(|x|_p, |y|_p)$$

in this case also. Thus, we have the triangle inequality satisfied in a sharper form. □

10.1.3 *Show that the usual absolute value on $\mathbb{Q}$ is archimedean.*

We must show that $|x+y| \leq \max(|x|, |y|)$ is not satisfied for some pair of rational numbers x, y. If $x > y > 0$, we have $|x+y| = x+y > x = |x|$. □

10.1.4 *If $0 < c < 1$ and p is prime, define*

$$||x|| = \begin{cases} c^{\nu_p(x)} & \text{if } x \neq 0 \\ 0 & \text{if } x = 0, \end{cases}$$

for all rational numbers x. Show that $|| \cdot ||$ is equivalent to $| \cdot |_p$ on $\mathbb{Q}$.

Since $\nu_p(x+y) \leq \min(\nu_p(x), \nu_p(y))$ the result is clear. □

10.1.6 *Let F be a field with norm $|| \cdot ||$, satisfying*

$$||x+y|| \leq \max(||x||, ||y||).$$

If $a \in F$ and $r > 0$, let $B(a, r)$ be the open disk $\{x \in F : ||x-a|| < r\}$. Show that $B(a, r) = B(b, r)$ for any $b \in B(a, r)$. (This result says that every point of the disk is the 'center' of the disk.)

If $x \in B(a, r)$, then $||x-a|| < r$, so that $||x-b|| = ||(x-a)+(a-b)|| \leq \max(||x-a||, ||a-b||) < r$, so that $x \in B(b, r)$. The converse is also clear. □

10.1.7 *Let F be a field with $|| \cdot ||$. Let R be the set of all Cauchy sequences $\{a_n\}_{n=1}^{\infty}$. Define addition and multiplication of sequences pointwise: that is,*

$$\{a_n\}_{n=1}^{\infty} + \{b_n\}_{n=1}^{\infty} = \{a_n + b_n\}_{n=1}^{\infty},$$
$$\{a_n\}_{n=1}^{\infty} \times \{b_n\}_{n=1}^{\infty} = \{a_n b_n\}_{n=1}^{\infty}.$$

Show that $(R, +, \times)$ is a commutative ring. Show further that the subset R consisting of null Cauchy sequences (namely those satisfying $||a_n|| \to 0$ as $n \to \infty$) forms a maximal ideal m.

We must first show that the sum and product of two Cauchy sequences is again Cauchy. Let $\epsilon > 0$. Choose N_1 such that $||a_n - a_m|| < \epsilon/2$ for $n, m \geq N_1$. Choose N_2 such that $||b_n - b_m|| < \epsilon/2$ for $n, m \geq N_2$. Then, for $N = \max(N_1, N_2)$, we have

$$\begin{aligned} ||(a_n + b_n) - (a_m + b_m)|| &\leq ||a_n - a_m|| + ||b_n - b_m|| \\ &< \epsilon/2 + \epsilon/2 = \epsilon, \end{aligned}$$

for $n, m \geq N$. Thus, the sum of two Cauchy sequences is again Cauchy. Now let K be such that $||a_n|| \leq K$, $||b_n|| \leq K$ for all n (this is clear from the Cauchy property). Then given $\epsilon > 0$, choose M_1 such that for $n, m \geq M_1$, we have $||a_n - a_m|| < \epsilon/2K$. Let M_2 be such that $||b_n - b_m|| < \epsilon/2K$ for $n, m \geq M_2$. For $M = \max(M_1, M_2)$ and $n, m \geq M$, we have

$$\begin{aligned} ||a_n b_n - a_m b_m|| &\leq ||a_n|| ||b_n - b_m|| + ||b_m|| ||a_m - a_n|| \\ &< \epsilon/2 + \epsilon/2 = \epsilon. \end{aligned}$$

Thus, the product of two Cauchy sequences is again Cauchy. Therefore R is closed under taking sums and products. The other ring axioms are easily verified. Clearly, the sum and product of two null sequences is again a null sequence. It is also clear that given a null sequence $\{a_n\}_{n=1}^{\infty}$ and a Cauchy sequence $\{b_n\}_{n=1}^{\infty} \in R$, $\{a_n b_n\}_{n=1}^{\infty}$ is again a null sequence. Therefore, the null sequences form an ideal m of R. To show that m is a maximal ideal, it suffices to show that R/m is a field. To do this, we must show that any nonzero element has an inverse. Thus, given $\{a_n\}_{n=1}^{\infty} \notin m$, we know that there is an $\epsilon_1 > 0$ such that $|a_n|_p > \epsilon_1$ for all n sufficiently large. By adjusting a few of the initial elements (if necessary) we may suppose that $a_n \neq 0$ for all n, because the adjusted sequence would still be in the same equivalence class $(\bmod\, m)$. It is now clear that $\{1/a_n\}_{n=1}^{\infty}$ is a Cauchy sequence and is inverse to the given sequence. Thus, R/m is a field and m is a maximal ideal. □.

10.1.9 *Show that*

$$\mathbb{Z}_p = \{x \in \mathbb{Q}_p : |x|_p \leq 1\}$$

is a ring. (This ring is called the ring of p-adic integers.)

Each $x \in \mathbb{Q}_p$ is a Cauchy sequence, say $\{a_n\}_{n=1}^{\infty}$. We have defined $|x|_p = \lim_{n\to\infty} |a_n|_p$. Thus, $|a_n|_p \leq 1$ for n sufficiently large, since the values taken on by $|a_n|_p$ are integral powers of p. If $x, y \in \mathbb{Q}_p$ are such that $|x_p| \leq 1$ and $|y_p| \leq 1$, then writing $y = \{b_n\}_{n=1}^{\infty}$, we see that $|a_n + b_n|_p \leq \max(|a_n|_p, |b_n|_p) \leq 1$ for n sufficiently large. The same is true for $|a_n b_n|_p = |a_n||b_n|$. Thus, it is clear that $\mathbb{Z}_p$ is a ring. This completes the proof. □

10.1.10 *Given $x \in \mathbb{Q}$ satisfying $|x|_p \leq 1$, and any natural number i, show that $|x - a_i|_p \leq p^{-i}$. Moreover, we can choose a_i satisfying $0 \leq a_i < p^i$.*

Let $x = a/b$, with $(a, b) = 1$. Since $|x|_p \le 1$, p does not divide b, so that p^i and b are coprime. We can therefore find integers u and v such that $ub + vp^i = 1$. Let $a_i = ua$. Then

$$\begin{aligned} |a_i - x|_p &= \left|ua - \frac{a}{b}\right|_p = \left|\frac{a}{b}\right|_p |ub - 1|_p \\ &\le p^{-i}, \end{aligned}$$

so that a_i does the job. By translating a_i by a multiple of p^i we can ensure $0 \le a_i < p^i$, and the above inequality is not altered. □

10.1.12 *Show that the p-adic series*

$$\sum_{n=1}^{\infty} c_n, \quad c_n \in \mathbb{Q}_p$$

converges if and only if $|c_n|_p \to 0$.

It is clear that if the series converges, then $|c_n|_p \to 0$. Now suppose $|c_n|_p \to 0$. Let $s_N = \sum_{n=1}^{N} c_n$. Since $\mathbb{Q}_p$ is complete, it suffices to show that $\{s_N\}_{N=1}^{\infty}$ is Cauchy. We have for $M > N$,

$$\begin{aligned} |s_M - s_N|_p &= |c_{N+1} + c_{N+2} + \cdots + c_M|_p \\ &\le \max\left(|c_{N+1}|_p, |c_{N+2}|_p, \ldots, |c_M|_p\right), \end{aligned}$$

which goes to 0 as $N \to \infty$. □

10.1.13 *Show that*

$$\sum_{n=1}^{\infty} n!$$

converges in $\mathbb{Q}_p$.

Clearly, $|n!|_p \to 0$ as $n \to \infty$, and we are done by Exercise 10.1.12. □

10.1.14 *Show that*

$$\sum_{n=1}^{\infty} n \cdot n! = -1$$

in $\mathbb{Q}_p$.

We have

$$s_N = \sum_{n=1}^{N} n \cdot n! = (N+1)! - 1,$$

as an easy induction argument shows.

Indeed, $s_1 = 2! - 1 = 1$ and

$$s_{N+1} = s_N + (N+1)(N+1)! = (N+2)! - 1$$

by the induction hypothesis. Thus, $\lim_{N \to \infty} s_N = -1$. □

10.1.15 *Show that the power series*

$$\sum_{n=0}^{\infty} \frac{x^n}{n!}$$

converges in the disk $|x|_p < p^{-\frac{1}{p-1}}$.

The power of p dividing $n!$ is

$$\sum_{i=1}^{\infty} \left[\frac{n}{p^i}\right] < \sum_{i=1}^{\infty} \frac{n}{p^i} = \frac{n}{p-1}.$$

Therefore,

$$|n!|_p > p^{-n/(p-1)},$$

so that

$$|x^n/n!|_p < |x|_p^n p^{n/(p-1)},$$

which goes to 0 as $n \to \infty$. □

10.1.16 (Product formula) *Prove that for* $x \in \mathbb{Q}$,

$$\prod_p |x|_p = 1,$$

where the product is taken over all primes p including ∞.

This is just a restatement of unique factorization. □

10.1.17 *Prove that for any natural number n and a finite prime p,*

$$|n|_p \geq \frac{1}{|n|_\infty}.$$

This also is clear from

$$|n|_p = p^{-\nu_p(n)}$$

and

$$|n|_\infty = p^{\nu_p(n)}(n/p^{\nu_p(n)}).$$

□

10.2 Hensel's Lemma

10.2.1 *Show that $x^2 = 7$ has no solution in $\mathbb{Q}_5$.*

If it did, then we could write x as a 5-adic number

$$x = \sum_{n=-N}^{\infty} a_n 5^n.$$

The 5-adic expansion of 7 is $2 + 1 \cdot 5$, so that $N = 0$. Thus

$$\left(\sum_{n=0}^{\infty} a_n 5^n \right)^2 = 2 + 1 \cdot 5.$$

Reducing $\pmod 5$ shows that $a_0^2 \equiv 2 \pmod 5$ has a solution, which is not the case. □

10.2.4 *Let $f(x) \in \mathbb{Z}_p[x]$. Suppose that for some N and $a_0 \in \mathbb{Z}_p$ we have $f(a_0) \equiv 0 \left(\operatorname{mod} p^{2N+1}\right)$, $f'(a_0) \equiv 0 \left(\operatorname{mod} p^{N}\right)$ but $f'(a_0) \not\equiv 0 \left(\operatorname{mod} p^{N+1}\right)$. Show that there is a unique $a \in \mathbb{Z}_p$ such that $f(a) = 0$ and $a \equiv a_0 \pmod{p^{N+1}}$.*

We proceed as in the proof of Theorem 10.2.3. Write $f(x) = \sum_i c_i x^i$. We will solve inductively

$$f(a_n) \equiv 0 \left(\operatorname{mod} p^{2N+n+1}\right)$$

satisfying $a_{n+1} \equiv a_n \left(\operatorname{mod} p^{N+n+1}\right)$, $f'(a_n) \equiv 0 \left(\operatorname{mod} p^{N}\right)$ and $f'(a_n) \not\equiv 0 \left(\operatorname{mod} p^{N+1}\right)$. Writing $a_{n+1} = a_n + tp^{N+n+1}$, we need to solve

$$f(a_n + tp^{N+n+1}) \equiv 0 \left(\operatorname{mod} p^{2N+n+2}\right),$$

which reduces (as before) to

$$f(a_n) + p^{N+n+1} t f'(a_n) \equiv 0 \left(\operatorname{mod} p^{2N+n+2}\right).$$

We can divide through by p^{2N+n+1}, since

$$f'(a_n) \equiv f'(a_0) \equiv 0 \left(\operatorname{mod} p^{N}\right),$$

which gives a congruence $(\operatorname{mod} p)$ since $f'(a_n)/p^N$ is coprime to p Thus, we can solve for t. The sequence $\{a_n\}_{n=1}^{\infty}$ is Cauchy, and its limit satisfies the required conditions. □

10.2.5 *For any prime p and any positive integer m coprime to p, show that there exists a primitive mth root of unity in $\mathbb{Q}_p$ if and only if $m|(p-1)$.*

First suppose $m|(p-1)$. The polynomial $f(x) = x^m - 1$ has m distinct roots $(\bmod\, p)$ because $(\mathbb{Z}/p\mathbb{Z})^*$ is a cyclic group of order $(p-1)$. Moreover, each of these roots lifts to $\mathbb{Z}_p$ by Hensel's lemma. Among the roots $(\bmod\, p)$, precisely $\varphi(m)$, where $\varphi(m)$ denotes Euler's function, have order exactly m. For the converse, notice that if $\alpha \in \mathbb{Q}_p$ such that α has order m then, since $f(x)$ is monic, $\alpha \in \mathbb{Z}_p$ and α is an element of order m $(\bmod\, p)$. Thus, $m|(p-1)$. $\square$

10.2.6 *Show that the set of $(p-1)$st roots of unity in $\mathbb{Q}_p$ is a cyclic group of order $(p-1)$.*

This is again a consequence of Hensel's lemma. Each of the residue classes $\bmod\, p$ lifts to a unique $(p-1)$st root of unity in $\mathbb{Z}_p$. It is clear that the set of such roots of unity is a group. The cyclicity follows from the fact that there is an element of order $(p-1)$ established in the previous exercise. $\square$

10.2.7 (Polynomial form of Hensel's lemma) *Suppose $f(x) \in \mathbb{Z}_p[x]$ and that there exist $g_1, h_1 \in (\mathbb{Z}/p\mathbb{Z})[x]$ such that $f(x) \equiv g_1(x)\, h_1(x)$ $(\bmod\, p)$, with $(g_1, h_1) = 1$, $g_1(x)$ monic. Then there exist polynomials $g(x), h(x) \in \mathbb{Z}_p[x]$ such that $g(x)$ is monic, $f(x) = g(x)h(x)$, and $g(x) \equiv g_1(x)$ $(\bmod\, p)$, $h(x) \equiv h_1(x)$ $(\bmod\, p)$.*

The idea is to construct two sequences of polynomials g_n and h_n such that

$$g_{n+1} \equiv g_n (\bmod\, p^n), \quad h_{n+1} \equiv h_n (\bmod\, p^n),$$

and $f(x) \equiv g_n(x)h_n(x)(\bmod\, p^n)$, with each g_n monic and of degree equal to deg g_1 and then take the limit. The idea is as in Hensel's lemma. We do this first for $n = 2$. Write $g_2(x) = g_1(x) + pr_1(x)$, for some polynomial $r_1(x) \in \mathbb{Z}_p[x]$. Similarly, $h_2(x) = h_1(x) + ps_1(x)$. We want

$$f(x) \equiv g_2(x)h_2(x) \ (\bmod\, p^2).$$

That is,

$$f(x) \equiv g_1(x)h_1(x) + pr_1(x)h_1(x) + ps_1(x)g_1(x) \ (\bmod\, p^2).$$

Since $f(x) \equiv g_1(x)h_1(x)$ $(\bmod\, p)$, we can write $f(x) - g_1(x)h_1(x) = pk_1(x)$ for some $k_1(x) \in \mathbb{Z}_p[x]$. Therefore, we get

$$k_1(x) \equiv r_1(x)h_1(x) + s_1(x)g_1(x) \ (\bmod\, p).$$

Since $(g_1, h_1) = 1$, we can find polynomials $a(x)$, $b(x)$ such that

$$a(x)g_1(x) + b(x)h_1(x) \equiv 1 \pmod{p}.$$

If we set $\tilde{r}_1(x) = b(x)k_1(x)$, $\tilde{s}_1(x) = a(x)k_1(x)$, these polynomials almost work for r_1, s_1. We have to ensure that $\deg g_2 = \deg g_1$ and that g_2 is monic. By the Euclidean algorithm for $(\mathbb{Z}/p\mathbb{Z})[x]$,

$$\tilde{r}_1(x) = g_1(x)q(x) + r_1(x)$$

with $\deg r_1 < \deg g_1$. Set $s_1(x) = \tilde{s}_1(x) + h_1(x)q(x)$; then

$$r_1(x)h_1(x) + s_1(x)g_1(x) \equiv k_1(x) \pmod{p}$$

as required. Also, since $\deg r_1 < \deg g_1$, we have g_2 monic and $\deg g_2 = \deg g_1$. We now continue in this way for $g_3, g_4, \ldots$ and take the limit. □.

10.2.9 *Show that for* $p \neq 2$, *the only solution to* $x^2 \equiv 1 \pmod{p^n}$ *is* $x = \pm 1$, *for every* $n \geq 1$.

For $n = 1$, this is clear. Since the polynomial $f(x) = x^2 - 1$ satisfies $f'(x) = 2x$ and $f'(\pm 1) \not\equiv 0 \pmod{p}$ (since $p \neq 2$), we can apply Hensel's lemma to obtain that both $x \equiv 1 \pmod{p}$ and $x \equiv -1 \pmod{p}$ extend to p-adic solutions. These are clearly $x = \pm 1$. □

10.3 p-adic Interpolation

10.3.1 *Show that there is no continuous function* $f : \mathbb{Z}_p \to \mathbb{Q}_p$ *such that* $f(n) = n!$

Let $x \in \mathbb{Z}_p \setminus \mathbb{Z}$. We want $n! \to f(x)$ as $n \to x$. But $n!$ is getting p-adically closer to 0 as $n \to x$ (since n gets large in the usual sense as $n \to x$). Therefore, $\lim_{n \to x} n! = 0$, so that there is no continuous p-adic function interpolating the factorials. □

10.3.2 *Let* $p \neq 2$ *be prime. Prove that for any natural numbers* n, s *we have*

$$\prod_{\substack{j=1 \\ (n+j,p)=1}}^{p^s-1} (n+j) \equiv -1 \pmod{p^s}.$$

The numbers $n, n+1, \ldots, n+p^s-1$ form a complete set of residues $\bmod p^s$. The product therefore is congruent to the product of all the coprime residue classes $\bmod p^s$. Now, in any abelian group A,

$$\prod_{g \in A} g = \prod_{\substack{g \in A \\ g^2=1}} g,$$

since we can pair g and g^{-1} in the left-hand product. By Exercise 10.2.9,

$$x^2 \equiv 1 \pmod{p^s}$$

has only 2 solutions, namely $x = \pm 1$. Thus,

$$\prod_{\substack{j=1 \\ (n+j,p)=1}}^{p^s-1} (n+j) \equiv -1 \pmod{p^s}.$$

(Notice that for $s = 1, n = 0$, this is just Wilson's theorem.) □

10.3.3 *Show that if* $p \neq 2$,

$$a_k = \prod_{\substack{j \leq k \\ (j,p)=1}} j \quad ,$$

then $a_{k+p^s} \equiv -a_k \pmod{p^s}$.

We have

$$\begin{aligned}\prod_{\substack{j\le k+p^s\\(j,p)=1}} j &= \prod_{\substack{j\le p^s\\(j,p)=1}} j \prod_{\substack{p^s<j\le k+p^s\\(j,p)=1}} j\\ &\equiv -\prod_{\substack{j\le k\\(j,p)=1}} j \pmod{p^s}\end{aligned}$$

by application of Exercise 10.3.2. Therefore,

$$a_{k+p^s} \equiv -a_k \pmod{p^s}.$$

□

10.3.4 *Prove that for* $p \neq 2$,

$$\Gamma_p(k+p^s) \equiv \Gamma_p(k) \pmod{p^s}.$$

We have $\Gamma_p(n) = (-1)^n a_{n-1}$, in the notation of the previous exercise. Thus,

$$\Gamma_p(k+p^s) = (-1)^{k+p^s} a_{k+p^s-1} \equiv (-1)^k a_{k-1} \pmod{p^s},$$

which gives the result. (Note that p is odd.) □

10.3.5 *Let* n, k *be natural numbers and write*

$$n = a_0 + a_1 p + a_2 p^2 + \cdots,$$

$$k = b_0 + b_1 p + b_2 p^2 + \cdots,$$

for the p-adic expansions of n and k, respectively. Show that

$$\binom{n}{k} \equiv \binom{a_0}{b_0}\binom{a_1}{b_1}\binom{a_2}{b_2}\cdots \pmod{p}.$$

We have

$$\begin{aligned}(1+x)^n &= (1+x)^{a_0}(1+x)^{a_1 p}(1+x)^{a_2 p^2}\cdots\\ &\equiv (1+x)^{a_0}(1+x^p)^{a_1}(1+x^{p^2})^{a_2}\cdots \pmod{p}.\end{aligned}$$

Now compare coefficients of x^k on both sides. Since $k = b_0 + b_1 p + \cdots$ is the unique p-adic expansion, the result is now evident. □

10.3.6 *If p is prime, show that*

$$\binom{p^n}{k} \equiv 0 \pmod{p}$$

for $1 \leq k \leq p^n - 1$ and all n.

The p-adic expansion of p^n is just p^n, so that $a_0 = a_1 = \cdots = a_{n-1} = 0$ from which the result now follows. □

10.3.7 (Binomial inversion formula) *Suppose for all n,*

$$b_n = \sum_{k=0}^{n} \binom{n}{k} a_k.$$

Show that

$$a_n = \sum_{k=0}^{n} \binom{n}{k} (-1)^{n-k} b_k,$$

and conversely.

Consider the multiplication of formal power series:

$$\left(\sum_{n=0}^{\infty} \frac{a_n x^n}{n!}\right)\left(\sum_{n=0}^{\infty} \frac{c_n x^n}{n!}\right) = \sum_{n=0}^{\infty} \frac{b_n x^n}{n!}.$$

It is easily seen that

$$b_n = \sum_{k=0}^{n} \binom{n}{k} a_k c_{n-k}.$$

Thus, the given relation for b_n implies

$$\sum_{n=0}^{\infty} \frac{b_n x^n}{n!} = e^x \sum_{n=0}^{\infty} \frac{a_n x^n}{n!}$$

from which the result is clear. □

10.3.8 *Prove that*

$$\sum_{k=0}^{n} \binom{n}{k} (-1)^k \binom{k}{m} = \begin{cases} (-1)^m & \text{if} \quad n = m, \\ 0 & \text{otherwise.} \end{cases}$$

Put

$$a_k = \begin{cases} (-1)^m & \text{if} \quad k = m, \\ 0 & \text{otherwise.} \end{cases}$$

In the notation of the previous exercise,

$$b_n = \binom{n}{m}(-1)^m,$$

so that

$$a_n = \sum_{k=0}^{n} \binom{n}{k}(-1)^{n-k}\binom{k}{m}(-1)^m = \begin{cases} (-1)^m & \text{if} \quad n = m, \\ 0 & \text{otherwise,} \end{cases}$$

as desired. □

10.3.9 *Define*

$$\Delta^n f(x) = \sum_{k=0}^{n} \binom{n}{k}(-1)^{n-k} f(x+k).$$

Show that

$$\Delta^n f(x) = \sum_{j=0}^{m} \binom{m}{j} \Delta^{n+j} f(x-m).$$

It suffices to show that

$$f(x) = \sum_{j=0}^{m} \binom{m}{j} \Delta^j f(x-m),$$

for the result follows by applying the operator Δ^n to both sides of the equation. But then

$$\begin{aligned} \sum_{j=0}^{m} \binom{m}{j} \Delta^j f(x-m) &= \sum_{j=0}^{m} \binom{m}{j} \sum_{k=0}^{j} \binom{j}{k} (-1)^{j-k} f(x-m+k) \\ &= \sum_{k=0}^{m} (-1)^k f(x-m+k) \sum_{j=0}^{m} \binom{m}{j}\binom{j}{k}(-1)^j, \end{aligned}$$

and the inner sum is 0 unless $k = m$, in which case it is $(-1)^m$, by Exercise 10.3.8. Thus, the result is immediate. □

10.3.10 *Prove that*

$$\sum_{j=0}^{m} \binom{m}{j} a_{n+j}(f) = \sum_{k=0}^{n} (-1)^{n-k} \binom{n}{k} f(k+m)$$

with $a_n(f)$ defined by

$$a_n(f) = \sum_{k=0}^{n} (-1)^{n-k} \binom{n}{k} f(k).$$

For $m = 0$, the formula is clear. By the previous exercise,

$$\Delta^n f(m) = \sum_{j=0}^{m} \binom{m}{j} \Delta^{n+j} f(0).$$

Now,

$$\Delta^n f(m) = \sum_{k=0}^{n} \binom{n}{k} (-1)^{n-k} f(k+m),$$

and we need only observe that $\Delta^n f(0) = a_n(f)$ to complete the proof. □

10.3.11 *Show that the polynomial*

$$\binom{x}{n} = \begin{cases} \frac{x(x-1)\cdots(x-n+1)}{n!} & \text{if } \; n \geq 1 \\ 1 & \text{if } \; n = 0. \end{cases}$$

takes integer values for $x \in \mathbb{Z}$. Deduce that

$$\left| \binom{x}{n} \right|_p \leq 1$$

for all $x \in \mathbb{Z}_p$.

For x a natural number, this is clear. If $x = -m$ $(m \in \mathbb{N})$ then

$$\binom{-m}{n} = (-1)^n \binom{m+n-1}{n} \in \mathbb{Z}.$$

The polynomial $\binom{x}{n}$ is continuous. Since $\mathbb{Z}$ is dense in $\mathbb{Z}_p$, it follows that for all $x \in \mathbb{Z}_p$

$$\left|\binom{x}{n}\right|_p \leq 1. \qquad \square$$

10.3.13 *If $f(x) \in \mathbb{C}[x]$ is a polynomial taking integral values at integral arguments, show that*

$$f(x) = \sum_k c_k \binom{x}{k}$$

for certain integers c_k.

This is purely formal, and a consequence of Exercise 10.3.7. Indeed, set

$$a_n(f) = \sum_{k=0}^{n} \binom{n}{k} (-1)^{n-k} f(k),$$

which gives a sequence of integers, since the $f(k)$ are all integers. By the binomial inversion formula,

$$f(n) = \sum_{k=0}^{n} \binom{n}{k} a_k(f).$$

Let D be the degree of f. Set

$$f^*(x) = \sum_{k=0}^{D} \binom{x}{k} a_k(f).$$

Now, for $0 \leq n \leq D$,

$$f^*(n) = \sum_{k=0}^{D} \binom{n}{k} a_k(f) = \sum_{k=0}^{n} \binom{n}{k} a_k(f) = f(n).$$

Since the polynomials $f(x)$ and $f^*(x)$ have the same degree and agree on $D+1$ points, we must have $f(x) = f^*(x)$. This completes the proof. $\square$

10.3.14 *If $n \equiv 1 \pmod{p}$, prove that $n^{p^m} \equiv 1 \pmod{p^{m+1}}$. Deduce that the sequence $a_k = n^k$ can be p-adically interpolated.*

We prove the congruence by induction. For $m = 1$, we may write $n = 1 + tp$, for some t, so that $n^p = (1 + tp)^p \equiv 1 \pmod{p^2}$. Assume that the result has been shown for $m \leq n$. Then, we must show that that $n^{p^n} \equiv 1 \pmod{p^{n+2}}$. By induction, we have $n^{p^n} = 1 + jp^{n+1}$ for some j. Hence, $n^{p^{n+1}} = (1 + jp^{n+1})^p \equiv 1 \pmod{p^{n+2}}$ as required. To prove that the sequence of a_k's can be p-adically interpolated, it suffices to show that if $k \equiv k' \pmod{p^m}$, then $a_k \equiv a_{k'} \pmod{p^{m+1}}$. Indeed, we have

$$n^{k-k'} \equiv 1 \pmod{p^{m+1}}$$

by what we have just shown. □

10.3.15 *Let* $(n, p) = 1$. *If* $k \equiv k' \pmod{(p-1)p^N}$, *then show that*

$$n^k \equiv n^{k'} \left(\operatorname{mod} p^{N+1}\right).$$

We have to prove

$$n^{k-k'} \equiv 1 \pmod{p^{N+1}}.$$

But this follows from Euler's theorem. □

10.3.16 *Fix* $s_0 \in \{0, 1, 2, \ldots, p-2\}$ *and let* A_{s_0} *be the set of integers congruent to* $s_0 \pmod{p-1}$. *Show that* A_{s_0} *is a dense subset of* $\mathbb{Z}_p$.

This is an application of the Chinese reminder theorem. Given $m \in \mathbb{Z}_p$, we must find an integer n such that $n \equiv m \pmod{p^N}$ and $n \equiv s_0 \pmod{p-1}$, which we can do since p and $p-1$ are coprime. □

10.3.17 *If* $(n, p) = 1$, *show that* $f(k) = n^k$ *can be extended to a continuous function on* A_{s_0}.

For $s \in A_{s_0}$, we write $s = s_0 + (p-1)s_1$, and hence $f(s) = n^{s_0}(n^{p-1})^{s_1}$. Since $n^{p-1} \equiv 1 \pmod{p}$, the function $(n^{p-1})^{s_1}$ can be p-adically interpolated for all $s_1 \in \mathbb{Z}_p$ by Exercise 10.3.14. Thus, f extends to a continuous function on A_{s_0}. □

10.4 The p-adic ζ-Function

10.4.1 *Verify that* μ_k *extends to a distribution on* $\mathbb{Z}_p$.

We must verify that

$$\mu_k(a + p^n\mathbb{Z}_p) = \sum_{b=0}^{p-1} \mu_k(a + bp^n + p^{n+1}\mathbb{Z}_p).$$

The right-hand side equals

$$p^{(n+1)(k-1)} \sum_{b=0}^{p-1} b_k \left(\frac{a + bp^n}{p^{n+1}} \right).$$

After multiplying both sides by $p^{-n(k-1)}$, the identity to be proved reduces to

$$b_k(px) = p^{k-1} \sum_{b=0}^{p-1} b_k \left(x + \frac{b}{p} \right),$$

and this is easily deduced from the power series generating function for the Bernoulli polynomials □

10.4.3 *Show that $\mu_{1,\alpha}$ is a measure.*

We have

$$\begin{aligned} \mu_{1,\alpha}(a + p^N \mathbb{Z}_p) &= \frac{a}{p^N} - \frac{1}{2} - \frac{1}{\alpha} \left(\frac{(\alpha a)_N}{p^N} - \frac{1}{2} \right) \\ &= \frac{(1/\alpha) - 1}{2} + \frac{a}{p^N} - \frac{1}{\alpha} \left(\frac{\alpha a}{p^N} - \left[\frac{\alpha a}{p^N} \right] \right), \end{aligned}$$

where $[\cdot]$ denotes the greatest integer function. Thus,

$$\mu_{1,\alpha}(a + p^N \mathbb{Z}_p) = \frac{1}{\alpha} \left[\frac{\alpha a}{p^N} \right] + \frac{(1/\alpha) - 1}{2}.$$

Since $\alpha \in \mathbb{Z}_p^*$, $1/\alpha \in \mathbb{Z}_p$ and $((1/\alpha) - 1)/2 \in \mathbb{Z}_p$ if $p \neq 2$. If $p = 2$, then $\alpha^{-1} \equiv 1 \pmod 2$ and $(\alpha^{-1} - 1)/2 \in \mathbb{Z}_p$ in this case also. Thus,

$$\mu_{1,\alpha}(a + p^N \mathbb{Z}_p) \in \mathbb{Z}_p,$$

and hence

$$\left| \mu_{1,\alpha} \left(a + p^N \mathbb{Z}_p \right) \right| \leq 1.$$

Since every compact-open set U is a finite disjoint union of intervals of the form $a + p^N \mathbb{Z}_p$, the result immediately follows from the nonarchimedean property of the p-adic norm. □

10.4.4 *Let d_k be the least common multiple of the denominators of coefficients of $b_k(x)$. Show that*

$$d_k \mu_{k,\alpha}(a + p^N \mathbb{Z}_p) \equiv d_k k a^{k-1} \mu_{1,\alpha} \left(a + p^N \mathbb{Z}_p \right) \pmod{p^N}.$$

The Bernoulli polynomial begins as

$$x^k - \frac{k}{2}x^{k-1} + \cdots,$$

as is easily checked. Now,

$$d_k\mu_{k,\alpha}(a + p^N\mathbb{Z}_p) = d_k p^{N(k-1)}\left(b_k\left(\frac{a}{p^N}\right) - \alpha^{-k}b_k\left(\frac{(\alpha a)_N}{p^N}\right)\right).$$

The polynomial $d_kB_k(x)$ has integral coefficients, and its first two terms are $d_kx^k - k(d_k/2)x^{k-1}$. Since $x = a/p^N$ has denominator p^N, and we are multiplying by $p^{N(k-1)}$, the terms after x^{k-2} will be divisible by p^N for $x = a/p^N$. Thus,

$$\begin{aligned}
d_k\mu_k(a + p^N\mathbb{Z}_p) &\equiv d_k p^{N(k-1)}\Big(\frac{a^k}{p^{Nk}} - \alpha^{-k}\Big(\frac{(\alpha a)_N}{p^N}\Big)^k \\
&\quad -\frac{k}{2}\Big(\frac{a^{k-1}}{p^{N(k-1)}} - \alpha^{-k}\Big(\frac{(\alpha a)_N}{p^N}\Big)^{k-1}\Big)\Big) \pmod{p^N} \\
&\equiv d_k\Big(\frac{a^k}{p^N} - \alpha^{-k}p^{N(k-1)}\Big(\frac{\alpha a}{p^N} - \Big[\frac{\alpha a}{p^N}\Big]\Big)^k \\
&\quad -\frac{k}{2}\Big(a^{k-1} - \alpha^{-k}p^{N(k-1)}\Big(\frac{\alpha a}{p^N} - \Big[\frac{\alpha a}{p^N}\Big]\Big)^{k-1}\Big)\Big) \\
&\equiv d_k\Big(\frac{a^k}{p^N} - \alpha^{-k}\Big(\frac{\alpha^k a^k}{p^N} - k\alpha^{k-1}a^{k-1}\Big[\frac{\alpha a}{p^N}\Big]\Big) \\
&\quad -\frac{k}{2}(a^{k-1} - \alpha^{-k}(\alpha^{k-1}a^{k-1}))\Big) \pmod{p^N} \\
&\equiv d_k k a^{k-1}\Big(\frac{1}{\alpha}\Big[\frac{\alpha a}{p^N}\Big] + \frac{\alpha^{-1} - 1}{2}\Big) \pmod{p^N} \\
&\equiv d_k k a^{k-1}\mu_{1,\alpha}(a + p^N\mathbb{Z}_p) \pmod{p^N}.
\end{aligned}$$

10.4.5 *Show that*

$$\int_{\mathbb{Z}_p} d\mu_{k,\alpha} = k\int_{\mathbb{Z}_p} x^{k-1}d\mu_{1,\alpha}.$$

In the notation of the previous exercise we see that

$$\begin{aligned} d_k \int_{\mathbb{Z}_p} d\mu_{k,\alpha} &\equiv \sum_{0\le a\le p^N-1} \mu_{k,\alpha}\left(a+p^N\mathbb{Z}_p\right) \pmod{p^N} \\ &\equiv d_k k \sum_{0\le a\le p^N-1} a^{k-1}\mu_{1,\alpha}\left(a+p^N\mathbb{Z}_p\right) \pmod{p^N}, \end{aligned}$$

from which the result now follows. □

10.4.6 *If $\mathbb{Z}_p^*$ is the group of units of $\mathbb{Z}_p$, show that*

$$\mu_{k,\alpha}(\mathbb{Z}_p^*) = (1-\alpha^{-k})(1-p^{k-1})B_k,$$

where B_k is the kth Bernoulli number.

Clearly,

$$\begin{aligned} \mu_{k,\alpha}(\mathbb{Z}_p^*) &= \mu_{k,\alpha}(\mathbb{Z}_p) - \mu_{k,\alpha}(p\mathbb{Z}_p) \\ &= \mu_k(\mathbb{Z}_p) - \alpha^{-k}\mu_k(\alpha\mathbb{Z}_p) - \mu_k(p\mathbb{Z}_p) + \alpha^{-k}\mu_k(\alpha\mathbb{Z}_p). \end{aligned}$$

Now, $\mu_k(\mathbb{Z}_p) = B_k$, and $\mu_k(p\mathbb{Z}_p) = p^{k-1}B_k$. Also, since α is an integer coprime to p, $\alpha\mathbb{Z}_p = \mathbb{Z}_p$, so that $\mu_k(\alpha\mathbb{Z}_p) = B_k$ and $\mu_k(\alpha p\mathbb{Z}_p) = p^{k-1}B_k$. The result now follows. □

10.4.8 (Kummer congruences) *If $(p-1)\nmid i$ and $i \equiv j \pmod{p^n}$ show that*

$$(1-p^{i-1})B_i/i \equiv (1-p^{j-1})B_j/j \pmod{p^{n+1}}.$$

Let α be a primitive root $\pmod p$. Since $(p-1)\nmid i$, we have $\alpha^i \not\equiv 1 \pmod p$, so that $\alpha^{-i}-1 \in \mathbb{Z}_p^*$. By Theorem 10.4.7, it suffices to prove $\alpha^{-i}-1 \equiv \alpha^{-j}-1 \pmod{p^{n+1}}$ and

$$\int_{\mathbb{Z}_p^*} x^{i-1}d\mu_{1,\alpha} \equiv \int_{\mathbb{Z}_p^*} x^{j-1}d\mu_{1,\alpha} \pmod{p^{n+1}}.$$

The former congruence follows from Euler's theorem. The latter follows from $x^{i-1} \equiv x^{j-1} \pmod{p^{n+1}}$, by the same theorem. □

10.4.9 (Kummer) *If $(p-1)\nmid i$, show that $|B_i/i|_p \le 1$.*

As in Exercise 10.4.8,

$$|B_i/i|_p = |\alpha^{-i}-1|_p^{-1}|1-p^{j-1}|_p^{-1}\left|\int_{\mathbb{Z}_p^*} x^{i-1}d\mu_{1,\alpha}\right|_p.$$

Since $(p-1) \not| i$, $\alpha^i - 1$ is coprime to p. Thus,

$$\begin{aligned} |B_i/i|_p &= \left| \int_{\mathbb{Z}_p^*} x^{i-1} d\mu_{1,\alpha} \right|_p \\ &\leq 1, \end{aligned}$$

because $|\mu_{1,\alpha}(U)|_p \leq 1$ for all compact-open sets U. □

10.4.10 (Clausen and von Staudt) *If* $(p-1)|i$ *and* i *is even, then*

$$pB_i \equiv -1 \pmod{p}.$$

By Exercise 2.5.23,

$$(m+1)s_m(p) = \sum_{k=0}^{m} \binom{m+1}{k} B_k p^{m+1-k},$$

where

$$s_m(p) = 1^m + 2^m + \cdots + (p-1)^m.$$

Therefore,

$$pB_m = s_m(p) - \sum_{k=0}^{m-1} \frac{1}{m+1} \binom{m+1}{k} B_k p^{m+1-k}$$

which is equal to

$$s_m(p) - \frac{p^{m+1}}{m+1} - \sum_{k=1}^{m-1} \binom{m}{k-1} \frac{B_k}{k} p^{m+1-k}.$$

By Exercise 10.4.9, $|B_k/k|_p \leq 1$ if $(p-1) \nmid k$. We now write $m = (p-1)t$ and induct on t. For $t = 1$,

$$pB_{p-1} = s_{p-1}(p) - p^{p-1} - \sum_{k=1}^{p-2} \binom{p-1}{k-1} \frac{B_k}{k} p^{p-k} \equiv -1 \pmod{p}$$

by Fermat's little theorem. The result is now deduced by an easy induction argument. □

10.5 Supplementary Problems

10.5.1 *Let $1 \leq a \leq p-1$, and set $\phi(a) = (a^{p-1}-1)/p$. Prove that $\phi(ab) \equiv \phi(a) + \phi(b) \pmod{p}$.*

We have

$$\begin{aligned}(ab)^{p-1} &= a^{p-1}b^{p-1} = (1+p\phi(a))(1+p\phi(b)) \\ &\equiv 1 + p(\phi(a)+\phi(b)) \pmod{p^2},\end{aligned}$$

and the result is now clear. □

10.5.2 *With ϕ as in the previous exercise, show that*

$$\phi(a+pt) \equiv \phi(a) - \bar{a}t \pmod{p},$$

where $a\bar{a} \equiv 1 \pmod{p}$.

We have

$$\begin{aligned}(a+pt)^{p-1} &\equiv a^{p-1} + p(p-1)ta^{p-2} \pmod{p^2} \\ &\equiv 1 + p\phi(a) - pta^{p-1}\bar{a} \pmod{p^2} \\ &\equiv 1 + p\phi(a) - pt(1+p\phi(a))\bar{a} \pmod{p^2} \\ &\equiv 1 + p\phi(a) - pt\bar{a} \pmod{p^2},\end{aligned}$$

from which the congruence follows. □

10.5.3 *Let $[x]$ denote the greatest integer less than or equal to x. For $1 \leq a \leq p-1$, show that*

$$\frac{a^p - a}{p} \equiv \sum_{j=1}^{p-1} \frac{1}{j}\left[\frac{aj}{p}\right] \pmod{p}.$$

We have

$$\begin{aligned}\sum_{j=1}^{p-1} \phi(aj) &\equiv \sum_{j=1}^{p-1} \phi(a) + \sum_{j=1}^{p-1} \phi(j) \pmod{p} \\ &\equiv (p-1)\phi(a) + \sum_{j=1}^{p-1} \phi(j) \pmod{p}.\end{aligned}$$

Thus

$$\phi(a) \equiv \sum_{j=1}^{p-1} \phi(j) - \sum_{j=1}^{p-1} \phi(aj) \pmod{p}.$$

Write $aj = r_j + pq_j$, where $1 \leq r_j \leq p-1$. Then by Exercise 10.5.2,

$$\phi(aj) = \phi(r_j + pq_j) \equiv \phi(r_j) - \frac{q_j}{r_j} \pmod{p},$$

so that

$$\sum_{j=1}^{p-1} \phi(aj) = \sum_{j=1}^{p-1} \phi(r_j) - \sum_{j=1}^{p-1} \frac{q_j}{r_j}; \pmod{p}.$$

Clearly, as j runs through 1 to $p-1$, so does r_j. Hence

$$\phi(a) \equiv \sum_{j=1}^{p-1} \frac{q_j}{r_j} \pmod{p}.$$

Now, $aj \equiv r_j \pmod{p}$ and $q_j = [aj/p]$, so that

$$a\phi(a) \equiv \sum_{j=1}^{p-1} \frac{1}{j} \left[\frac{aj}{p}\right] \pmod{p}$$

as desired. □

10.5.4 *Prove the following generalization of Wilson's theorem:*

$$(p-k)!(k-1)! \equiv (-1)^k \pmod{p}$$

for $1 \leq k \leq p-1$.

Write

$$\begin{aligned} -1 &\equiv (p-1)! \equiv (p-1)(p-2)\cdots(p-(k-1))(p-k)! \pmod{p} \\ &\equiv (-1)^{k-1}(k-1)!(p-k)! \pmod{p}, \end{aligned}$$

from which the result follows. □

10.5.5 *Prove that for an odd prime* p,

$$\frac{2^{p-1}-1}{p} \equiv \sum_{j=1}^{p-1} \frac{(-1)^{j+1}}{2j} \pmod{p}.$$

Deduce that $2^{p-1} \equiv 1 \pmod{p^2}$ *if and only if the numerator of*

$$1 - \frac{1}{2} + \frac{1}{3} - \cdots - \frac{1}{p-1}$$

is divisible by p.

We have,

$$\begin{aligned} \frac{2^{p-1}-1}{p} &= \frac{(1+1)^p - 2}{2p} = \frac{1}{2p}\sum_{j=1}^{p-1}\binom{p}{j} \\ &= \frac{1}{2}\sum_{j=1}^{p-1}\frac{(p-1)!}{(p-j)!j!}. \end{aligned}$$

By Wilson's theorem the numerator of each summand is congruent to $-1 \pmod{p}$. By Exercise 10.5.4, the denominator is congruent to $(-1)^j j \pmod{p}$. Thus

$$\frac{2^{p-1}-1}{p} \equiv \sum_{j=1}^{p-1}\frac{(-1)^{j+1}}{2j} \pmod{p},$$

as desired. □

10.5.6 *Let* p *be an odd prime. Show that for all* $x \in \mathbb{Z}_p$, $\Gamma_p(x+1) = h_p(x)\Gamma_p(x)$, *where*

$$h_p(x) = \begin{cases} -x & \text{if} \quad |x|_p = 1, \\ -1 & \text{if} \quad |x|_p < 1. \end{cases}$$

From the definition, we have

$$\Gamma_p(n+1) = \begin{cases} -n\Gamma_p(n) & \text{if} \quad (n,p) = 1, \\ -\Gamma_p(n) & \text{if} \quad (n,p) \neq 1. \end{cases}$$

The result now follows by continuity. □

10.5.7 *For* $s \geq 2$, *show that the only solutions of* $x^2 \equiv 1 \pmod{2^s}$ *are* $x \equiv 1, -1, 2^{s-1} - 1$, *and* $2^{s-1} + 1$.

We have $2^s|(x^2-1)$. Since $x^2-1=(x-1)(x+1)$, exactly one of $(x-1)$ or $(x+1)$ is divisible by 4. Either $2||(x-1)$ or $2||(x+1)$. In the former case, $x \equiv -1 \pmod{2^{s-1}}$, so that

$$x = 2^{s-1}t - 1$$

for some t. If t is even, we get $x \equiv -1 \pmod{2^s}$. If t is odd, we get $x \equiv 2^{s-1} - 1 \pmod{2^s}$.) In the latter case, $x \equiv 1 \pmod{2^{s-1}}$, and if t is odd, we get $x \equiv 2^{s-1} + 1 \pmod{2^s}$. □

10.5.8 (The 2-adic Γ-function) *Show that the sequence defined by*

$$\Gamma_2(n) = (-1)^n \prod_{\substack{1 \le j < n \\ (j,2)=1}} j$$

can be extended to a continuous function on $\mathbb{Z}_2$.

We have

$$\Gamma_2(n+2^s) = \Gamma_2(n) \prod_{\substack{0 \le j < 2^s \\ (n+j,2)=1}} (n+j).$$

As we remarked earlier, the product of all the elements in an abelian group is equal to the product of the elements of order 2. We must therefore solve

$$x^2 \equiv 1 \pmod{2^s}.$$

By Exercise 10.5.7, these are precisely $1, -1, 2^{s-1}+1$, and $2^{s-1}-1$. Therefore,

$$\Gamma_2(n+2^s) \equiv \Gamma_2(n) \pmod{2^s},$$

from which the result follows by an application of Mahler's theorem. This completes the proof. □

10.5.9 *Prove that for all natural numbers* n,

$$\Gamma_p(-n)\Gamma_p(n+1) = (-1)^{[n/p]+n+1}.$$

By Exercise 10.5.6, we have

$$1 = \Gamma_p(0) = \Gamma_p(-1)h_p(-1) = \Gamma_p(-2)h_p(-2)h_p(-1),$$

and so on. Thus,

$$\Gamma_p(-n)^{-1} = \prod_{j=1}^{n} h_p(-j)$$

for any natural number n. Again by Exercise 10.5.6, we know that $h_p(-j) = -1$ if $p|j$, and j otherwise. Thus,

$$\begin{aligned}\Gamma_p(-n)^{-1} &= (-1)^{[n/p]} \prod_{\substack{1\le j\le n\\(j,p)=1}} j \\ &= (-1)^{[n/p]+n+1}\Gamma_p(n+1),\end{aligned}$$

as desired. □

10.5.10 *If p is an odd prime, prove that for $x \in \mathbb{Z}_p$,*

$$\Gamma_p(x)\Gamma_p(1-x) = (-1)^{\ell(x)},$$

where $\ell(x)$ is defined as the element of $\{1, 2, \ldots, p\}$ satisfying $\ell(x) \equiv x \pmod{p}$. (This is the p-adic analogue of Exercise 6.3.4.)

From Exercise 10.5.9, we have

$$\Gamma_p(n+1)\Gamma_p(-n) = (-1)^{n+1+[n/p]}.$$

Write $n-1$ instead of n:

$$\Gamma_p(n)\Gamma_p(1-n) = (-1)^{n+[(n-1)/p]}.$$

If $n = a_0 + a_1p + a_2p^2 + \cdots$ is the p-adic expansion of n, then

$$[(n-1)/p] = [((a_0 - 1) + a_1p + \cdots)/p].$$

First suppose $a_0 \neq 0$. Then

$$[(n-1)/p] = a_1 + a_2p + \cdots,$$

so that $n - p[(n-1)/p] = a_0 = \ell(n)$. Clearly,

$$(-1)^{n+[(n-1)/p]} = (-1)^{n-p[(n-1)/p]} = (-1)^{\ell(n)},$$

and the formula is proved in this case. If $a_0 = 0$, then

$$n - 1 = (p-1) + b_1p + \cdots$$

and

$$[(n-1)/p] = b_1 + b_2 p + \cdots,$$

which gives

$$n - p[(n-1)/p] = p = \ell(n),$$

and again the formula is proved. □

10.5.11 *Show that*

$$\Gamma_p(1/2)^2 = \begin{cases} 1 & \textit{if} \quad p \equiv 3 \pmod 4, \\ -1 & \textit{if} \quad p \equiv 1 \pmod 4. \end{cases}$$

By Exercise 10.5.10,

$$\Gamma_p(1/2)^2 = (-1)^{\ell(1/2)}.$$

Now, $\ell(1/2) = \ell((p+1)/2) = (p+1)/2$, so the result follows. □

11
Equidistribution

11.1 Uniform distribution modulo 1

11.1.1 *Let us write the sequence of non-zero rational numbers in* $[0,1]$ *as follows:*

$$1, \frac{1}{2}, \frac{1}{3}, \frac{2}{3}, \frac{1}{4}, \frac{3}{4}, \frac{1}{5}, \frac{2}{5}, \frac{3}{5}, \frac{4}{5}, \frac{1}{6}, \frac{5}{6}, \ldots$$

where we successively write all the fractions with denominator b *for* $b = 1, 2, 3, \ldots$. *Show that this sequence is u.d. mod 1.*

We denote by x_n the sequence thus formed. Our goal is to show that the number of $n \leq M$ with $x_n \leq x$ is asymptotically Mx. Let us first consider all the fractions with denominator at most N. The number of such fractions is

$$V_N = \sum_{b=1}^{N} \phi(b).$$

For each x, we count the number of fractions $\leq x$. This number is

$$\sum_{b=1}^{N} \sum_{a \leq bx} \sum_{d|a, d|b} \mu(d),$$

since the inner sum is zero unless $(a, b) = 1$. We find easily that this is

$$\sum_{b=1}^{N} \sum_{d|b} \mu(d) \sum_{a \leq bx, d|a} 1.$$

The innermost sum is $[bx/d]$ and so the sum in question is

$$x \sum_{b=1}^{N} \phi(b) + O\left(\sum_{b \leq N} d(b) \right),$$

where $d(b)$ denotes the number of divisors of b. By Exercise 1.4.1, the error term is $O(N \log N)$. By Exercise 1.4.2, the main term is asymptotic to cxN^2 for some constant c. In other words, $V_N \sim cN^2$. Now let M be an arbitrary integer. As the sequence V_N is strictly increasing, there is an N such that

$$V_N \leq M < V_{N+1}.$$

The number $n \leq M$ with $x_n \leq x$ is equal to

$$xV_N + O(N \log N) + O(\phi(N + 1)).$$

Since $M = V_N + O(\phi(N + 1))$, this completes the proof. □

11.1.2 *If a sequence of real numbers* $\{x_n\}_{n=1}^{\infty}$ *is u.d., show that for any* a *with* $0 \leq a < 1$, *we have*

$$\#\{n \leq N : (x_n) = a\} = o(N).$$

For any $\epsilon > 0$, we take $b = a + \epsilon$ so that from the definition of u.d, we have

$$\#\{n \leq N : (x_n) \in [a, a + \epsilon]\} \leq 2\epsilon N,$$

for $N \geq N_0(\epsilon)$. Since the quantity in question is bounded by the above, we are done. □

11.1.3 *If the sequence* $\{x_n\}_{n=1}^{\infty}$ *is u.d. and* $f : [0, 1] \to \mathbb{C}$ *is a continuous function, show that*

$$\lim_{N \to \infty} \frac{1}{N} \sum_{n=1}^{N} f(x_n) \to \int_0^1 f(x) dx,$$

and conversely.

It suffices to establish the result for real-valued functions. For any characteristic function of an interval, we have the result by virtue of the definition of uniform distribution. Let $\epsilon > 0$ be fixed. By the theory of the Riemann integral, we know that there are step functions (that is, finite $\mathbb{R}$-linear combinations of characteristic functions) f_1, f_2 such that

$$f_1(x) \leq f(x) \leq f_2(x),$$

$$\int_0^1 f_1(x)dx \leq \int_0^1 f(x)dx \leq \int_0^1 f_2(x)dx,$$

and

$$0 \leq \int_0^1 (f_2(x) - f_1(x))dx \leq \epsilon.$$

The result is now immediate from our initial remark. For the converse, we observe that given any $\epsilon > 0$, the characteristic function $\chi_{[a,b]}$ of the interval $[a, b]$ can be approximated by continuous functions f_1, f_2 such that

$$f_1(x) \leq \chi_{[a,b]}(x) \leq f_2(x),$$

and

$$\int_0^1 (f_2(x) - f_1(x))dx \leq \epsilon.$$

Indeed, we may take

$$f_1(x) = \begin{cases} 0 & \text{if } x \leq a \\ (x-a)/\epsilon & \text{if } a \leq x \leq a+\epsilon \\ 1 & \text{if } a+\epsilon \leq x \leq b-\epsilon \\ (b-x)/\epsilon & \text{if } b-\epsilon \leq x \leq b \\ 0 & \text{if } b \leq x \end{cases}$$

with $f_2(x)$ analogously defined. Note that

$$\int_0^1 (f_2(x) - f_1(x))dx \leq 2\epsilon.$$

This completes the proof. □

11.1.4 *If $\{x_n\}_{n=1}^{\infty}$ is u.d. then*

$$\lim_{N\to\infty}\frac{1}{N}\sum_{n=1}^{N} f(x_n) = \int_0^1 f(x)dx,$$

for any piecewise C^1-function $f : [0,1] \to \mathbb{C}$.

This is clear from the previous exercise. □

11.1.6 *Show that Weyl's criterion need only be checked for positive integers m.*

This is immediate upon taking complex conjugation in Weyl's criterion. □

11.1.7 *Show that the sequence $\{x_n\}_{n=1}^{\infty}$ is u.d. mod 1 if and only if*

$$\lim_{N\to\infty}\frac{1}{N}\sum_{n=1}^{N} f(x_n) = \int_0^1 f(x)dx,$$

for any family of functions f which is dense in $C[0,1]$. Here, $C[0,1]$ is the metric space of continuous functions on $[0,1]$ with the sup norm.

The proof of this result follows the method of Theorem 11.1.5. We simply replace trigonometric polynomials by finite linear combinations of functions in our family. □

11.1.8 *Let θ be an irrational number. Show that the sequence $x_n = n\theta$ is u.d.*

By Weyl's criterion, it suffices to check

$$\sum_{n=1}^{N} e^{2\pi i m n\theta} = o(N)$$

for $m = 1, 2, \ldots$ Indeed, the sum on the left hand side is the sum of a geometric progression and equals

$$\frac{e^{2\pi i m(N+1)\theta} - 1}{e^{2\pi i m\theta} - 1}.$$

This is bounded by $2/|e^{2\pi i m\theta} - 1|$, where the denominator is nonzero since θ is irrational. Thus, the sum in question is clearly $o(N)$. □

11.1.9 *If θ is rational, show that the sequence $x_n = n\theta$ is not u.d.*

Let $\theta = a/b$ with a, b coprime integers. Then, Weyl's criterion fails with $m = b$ since

$$\sum_{n=1}^{N} e^{2\pi i b(na/b)} = N.$$

□

11.1.10 *Show that the sequence $x_n = \log n$ is not u.d. but is dense mod 1.*

By Weyl's criterion, we need to consider the sum

$$\sum_{n=1}^{N} e^{2\pi im \log n} = \sum_{n=1}^{N} n^{2\pi im}.$$

We may apply the Euler-Maclaurin summation formula to the right hand side to deduce that it is

$$\int_1^N t^{2\pi im} dt + \frac{1}{2}(N^{2\pi im} - 1) + \int_1^N B_1(t)(2\pi im)t^{2\pi im-1} dt.$$

This is easily seen to be

$$\frac{N^{2\pi im+1} - 1}{2\pi im + 1} + O(\log N).$$

Dividing by N and letting N tend to infinity shows that the first term does not converge. For example, for $m = 1$, we have

$$N^{2\pi i} = \cos(2\pi \log N) + i \sin(2\pi \log N).$$

If $N = 2^r$, we get

$$N^{2\pi i} = \cos(2\pi r \log 2) + i \sin(2\pi r \log 2).$$

Since $\log 2$ is irrational, the sequence $r \log 2$ is u.d. and we can make $r \log 2$ (mod 1) to be close to any number for infinitely many choices of r. Thus, the limit does not exist. To show the sequence is dense mod 1, we need only note that $m \log 2$ is u.d. mod 1 since $\log 2$ is irrational. □

11.1.11 *Let $0 \leq x_n < 1$. Show that the sequence $\{x_n\}_{n=1}^{\infty}$ is u.d. mod 1 if and only if*

$$\lim_{N\to\infty} \frac{1}{N} \sum_{n=1}^{N} x_n^r = \frac{1}{r+1},$$

for every natural number r.

If the sequence is u.d., then the value of the limit follows from Exercise 11.1.3. The converse is immediate upon applying the Weierstrass approximation theorem that states that every continuous function can be approximated by a polynomial. □

11.1.12 *If $\{x_n\}_{n=1}^{\infty}$ is u.d. mod 1, then show that $\{mx_n\}_{n=1}^{\infty}$ is u.d. mod 1 for m a non-zero integer.*

This is an immediate application of Weyl's criterion. □

11.1.13 *If $\{x_n\}_{n=1}^{\infty}$ is u.d. mod 1, and c is a constant, show that $\{x_n + c\}_{n=1}^{\infty}$ is u.d. mod 1.*

This is again an immediate consequence of Weyl's criterion. □

11.1.14 *If $\{x_n\}_{n=1}^{\infty}$ is u.d. mod 1 and $y_n \to c$ as $n \to \infty$, show that $\{x_n + y_n\}_{n=1}^{\infty}$ is u.d. mod 1.*

By the previous exercise, we may suppose that $c = 0$. We must show that for any interval $[a, b]$,

$$\#\{n \leq N : (x_n + y_n) \in [a, b]\} = (b - a)N + o(N).$$

To this end, let $\epsilon > 0$ be such that $2\epsilon < b - a$ and $|y_n| < \epsilon$ for $n \geq N_0$. Then,

$$\#\{n \leq N : (x_n) \in [a+\epsilon, b-\epsilon]\} - N_0 \leq \#\{n \leq N : (x_n + y_n) \in [a, b]\}$$

and

$$\#\{n \leq N : (x_n + y_n) \in [a, b]\} \leq \#\{n \leq N : (x_n) \in [a-\epsilon, b+\epsilon]\} + N_0$$

Using the known u.d. of the sequence of x_n's, we now deduce the desired result. □

11.1.15 *Let F_n denote the nth Fibonacci number defined by the recursion $F_0 = 1, F_1 = 1, F_{n+1} = F_n + F_{n-1}$. Show that $\log F_n$ is u.d. mod 1.*

It is easy to deduce (by induction, for example) that

$$F_n = \frac{\alpha^{n+1} - \beta^{n+1}}{\alpha - \beta}$$

where $\alpha = (1 + \sqrt{5})/2$ and $\beta = (1 - \sqrt{5})/2$. Thus, it suffices to show $\log(\alpha^{n+1} - \beta^{n+1})$ is u.d. mod 1. Since $|\beta/\alpha| < 1$, and we must study the sequence

$$(n + 1) \log \alpha + \log(1 - (\beta/\alpha)^{n+1}),$$

it suffices to show that $(n + 1) \log \alpha$ is u.d. since the second term tends to zero as n tends to infinity. By a classical theorem of Hermite, $\log \alpha$ is irrational and thus the sequence $(n + 1) \log \alpha$ is u.d. mod 1. □

11.1.18 *Let $y_1, ..., y_N$ be complex numbers. Let $\mathcal{H}$ be a subset of $[0, H]$ with $1 \leq H \leq N$. Show that*

$$\left|\sum_{n=1}^{N} y_n\right|^2 \leq \frac{N+H}{|\mathcal{H}|}\sum_{n=1}^{N}|y_n|^2 + \frac{2(N+H)}{|\mathcal{H}|^2}\sum_{r=1}^{H} N_r \left|\sum_{n=1}^{N-r} y_{n+r}\overline{y_n}\right|,$$

where N_r is the number of solutions of $h - k = r$ with $h > k$ and $h, k \in \mathcal{H}$.

We proceed as in the proof of Theorem 11.1.16. We have

$$|\mathcal{H}|^2\left|\sum_n y_n\right|^2 = \left|\sum_{h\in\mathcal{H}}\sum_n y_{n+h}\right|^2 = \left|\sum_n\sum_{h\in\mathcal{H}} y_{n+h}\right|^2.$$

As before, we note that the inner sum is zero if $n \notin [-H+1, N]$. Applying the Cauchy-Schwarz inequality, we get a bound of

$$\leq (N+H)\sum_n\left|\sum_{h\in\mathcal{H}} y_{n+h}\right|^2.$$

Expanding the sum, we obtain

$$\sum_n\sum_{h,k\in\mathcal{H}} y_{n+h}\overline{y}_{n+k} = |\mathcal{H}|\sum_n|y_n|^2 + \sum_n\sum_{h\neq k, h,k\in\mathcal{H}} y_{n+h}\overline{y}_{n+k}.$$

In the second sum, we combine the terms corresponding to (h, k) and (k, h) to get

$$2\,\mathrm{Re}\left(\sum_n\sum_{h\in\mathcal{H}}\sum_{k<h,k\in\mathcal{H}} y_{n+h}\overline{y}_{n+k}\right).$$

As before, writing $m = n + k$, we can re-write the above as

$$2\,\mathrm{Re}\left(\sum_m\sum_{r=1}^{H} y_{m+r}\overline{y}_m N_r\right)$$

from which the result is now easily deduced. □

11.1.19 *Let θ be an irrational number. Show that the sequence $\{n^2\theta\}_{n=1}^{\infty}$ is u.d. mod 1.*

For each fixed h, the sequence $(n+h)^2\theta - n^2\theta = 2hn\theta + h^2\theta$ is u.d. and so by Corollary 11.1.17, we are done. □

11.1.20 *Show that the sequence $\{an^2 + bn\}_{n=1}^{\infty}$ is u.d. provided that one of a or b is irrational.*

Suppose first that a is irrational. An application of Corollay 11.1.17 immediately gives the result. If a is rational, then b must be irrational by hypothesis. Writing $a = A/B$ with $B > 0$, we can re-write the corresponding Weyl sum as:

$$\sum_{d=0}^{B-1} \sum_{k=1}^{[N/B]} e^{2\pi i m(A(Bk+d)^2/B + b(Bk+d))}.$$

This simplifies to

$$\sum_{d=0}^{B-1} e^{2\pi i m(Ad^2/B+bd)} \sum_{k=1}^{[N/B]-1} e^{2\pi i m Bbk} + O(B).$$

Since b is irrational, the inner sum is $o(N/B)$ from which the result follows. □

11.1.21 *Let $P(n) = a_d n^d + a_{d-1}n^{d-1} + \cdots + a_1 + a_0$ be a polynomial with real coefficients with at least one coefficient a_i with $i \geq 1$ irrational. Show that the sequence of fractional parts of $P(n)$ is u.d. mod 1.*

We proceed by induction on the degree of P. If the degree of P is 1 or 2, the result follows from the previous exercises. Suppose first that a_1 is irrational and $a_2, \ldots, a_d$ are rational. Letting B be the least common multiple of all the denominators of $a_2, \ldots, a_d$, we have as in the previous exercise,

$$\sum_{n=1}^{N} e^{2\pi i m P(n)} = \sum_{d=0}^{B-1} \sum_{k=1}^{[N/B]-1} e^{2\pi i m P(Bk+d)} + O(B).$$

Since a_1 is irrational, the sequence $a_1 Bk$ is u.d. and we find the inner sum is

$$\sum_{k=1}^{[N/B]-1} e^{2\pi i m(Bka_1 + a_1 d + a_0)},$$

which is $o(N/B)$. The result now follows in this case. Now suppose that the highest index for which a_i is irrational is t. If $t = 1$ we are done by the argument just given. For fixed h, consider $P_h(n) = P(n+h) - P(n)$, which is a polynomial of degree $d-1$ and whose corresponding highest index irrational coefficient is the coefficient of n^{t-1}. By induction, this sequence is u.d. By Corollary 11.1.17, we are done. □

11.2 Normal numbers

11.2.1 *Show that a normal number is irrational.*

A rational number has a b-adic expansion which is eventually periodic and thus cannot be normal since we can find a block B_k which does not occur at all in the expansion. □

11.2.3 *If x is normal to the base b, show that mx is normal to the base b for any non-zero integer m.*

By Theorem 11.2.2, we need to check that for every $h \neq 0$,

$$\sum_{n=1}^{N} e^{2\pi i h m x b^n} = o(N).$$

But this is clear since x is normal. □

11.2.4 *Let $\{v_n\}_{n=1}^{\infty}$ be a sequence of distinct integers and set for a non-zero integer h,*

$$S(N, x) = \frac{1}{N} \sum_{n=1}^{N} e^{2\pi i v_n x h}.$$

Show that

$$\int_0^1 |S(N, x)|^2 dx = \frac{1}{N},$$

and

$$\sum_{N=1}^{\infty} \int_0^1 |S(N^2, x)|^2 dx < \infty.$$

We have

$$\int_0^1 |S(N, x)|^2 dx = \frac{1}{N^2} \sum_{n,m=1}^{N} \int_0^1 e^{2\pi i (v_n - v_m) x} dx.$$

The integral on the right hand side is zero if $v_n \neq v_m$ and 1 otherwise. As the v_n's are distinct, this means the integral is 1 if $n = m$ and zero otherwise. The result is now immediate. □

11.2.6 *Show that the sequence $n!e$ is not u.d. mod 1.*

We have

$$n!e = \left(\frac{n!}{1!} + \frac{n!}{2!} + \cdots + \frac{n!}{n!} \right) + \left(\frac{1}{n+1} + \frac{1}{(n+1)(n+2)} + \cdots \right).$$

The term in the first set of brackets is a positive integer and the term in the second set is bounded by

$$\frac{1}{n+1} + \frac{1}{(n+1)^2} + \cdots = \frac{1}{n},$$

which is less than 1 if $n \geq 2$. Thus, the fractional part tends to zero as $n \to \infty$ and hence cannot be u.d. mod 1. □

11.2.7 *If x is normal to the base b, show that it is simply normal to the base b^m for every natural number m.*

If x has b-adic expansion,

$$x = \sum_{n=1}^{\infty} \frac{a_n}{b^n},$$

then its expansion in base b^m is

$$\sum_{r=1}^{\infty} \frac{A_r(m)}{b^{mr}}$$

where

$$A_r(m) = \sum_{k=1}^{m} a_{(r-1)m+k} b^{m-k}.$$

The result now follows from the definition of normality. □

11.3 Asymptotic distribution functions mod 1

11.3.1 *A sequence $\{x_n\}_{n=1}^{\infty}$ has a.d.f. $g(x)$ if and only if for every piecewise continuous function f on $[0,1]$, we have*

$$\lim_{N\to\infty} \frac{1}{N} \sum_{n=1}^{N} f(x_n) = \int_0^1 f(x)dg(x).$$

This is immediate from the solution to Exercise 11.1.3 where the theory of the Riemann integral is replaced by the theory of the Riemann-Stieltjes integral. □

11.3.2 *A sequence $\{x_n\}_{n=1}^{\infty}$ has a.d.f $g(x)$ if and only if*

$$\lim_{N\to\infty} \frac{1}{N} \sum_{n=1}^{N} e^{2\pi i m x_n} = \int_0^1 e^{2\pi i m x} dg(x),$$

for all integers m.

The necessity follows from Exercise 11.3.1. For the sufficiency, we modify the proof of Theorem 11.1.5 by using the Riemann-Stieltjes integral instead of the Riemann integral. □

11.3.4 *Suppose that* $\{x_n\}_{n=1}^{\infty}$ *is a sequence such that for all integers* m, *the limits*

$$a_m := \lim_{N\to\infty} \frac{1}{N}\sum_{n=1}^{N} e^{2\pi i m x_n},$$

exist and

$$\sum_{m=-\infty}^{\infty} |a_m|^2 < \infty.$$

Put

$$g_1(x) = \sum_{m=-\infty}^{\infty} a_m e^{2\pi i m x}.$$

Show that

$$\lim_{N\to\infty} \frac{\#\{n \le N : x_n \in [\alpha,\beta]\}}{N} = \int_{\alpha}^{\beta} g_1(x)dx,$$

for any interval $[\alpha, \beta]$ *contained in* $[0, 1]$.

By the Wiener-Schoenberg theorem, the sequence has a continuous a.d.f. Letting f be the characteristic function of the interval and using Exercise 11.3.1, the result follows. □

11.4 Discrepancy

11.4.1 *Show that the sequence* $\{x_n\}_{n=1}^{\infty}$ *is u.d. mod 1 if and only if* $D_N \to 0$ *as* $N \to \infty$.

The sufficiency is clear. To show necessity, let m be an integer ≥ 2 and for $0 \le k \le m-1$, let $I_k = [k/m, (k+1)/m]$. As the sequence is u.d. mod 1, there is an $N_0 = N_0(m)$ so that for $N \ge N_0$, and for every $k = 0, 1, ..., m-1$, we have

$$\frac{1}{m} - \frac{1}{m^2} \le \frac{\#\{n \le N : (x_n) \in I_k\}}{N} \le \frac{1}{m} + \frac{1}{m^2}.$$

Now consider $J = [a, b]$. We "approximate" J by intervals of the type I_k. Indeed, there exist intervals J_1, J_2 which are a finite union of intervals I_k so that

$$J_1 \subseteq J \subseteq J_2,$$

and
$$|J_2| - \frac{2}{m} \le |J| \le |J_1| + \frac{2}{m}.$$

Clearly,

$$\#\{n \le N : (x_n) \in J_1\} \le \#\{n \le N : (x_n) \in J\} \le \#\{n \le N : (x_n) \in J_2\},$$

so that
$$\left| \frac{\#\{n \le N : (x_n) \in J\}}{N} - |J| \right| \le \frac{3}{m} + \frac{2}{m^2}.$$

Therefore $D_N \le 3/m + 2/m^2$ for $N \ge N_0$. Since m can be taken to be arbitrarily large, we deduce that $D_N \to 0$ as $N \to \infty$. □

11.4.2 *Show that*

$$\left(\frac{\sin \pi z}{\pi} \right)^2 \sum_{n=-\infty}^{\infty} \frac{1}{(z-n)^2} = 1, \quad z \notin \mathbb{Z}$$

By exercise 6.3.4, we get upon logarithmic differentiation that

$$z \cot z = 1 + 2 \sum_{n=1}^{\infty} \frac{z^2}{z^2 - n^2 \pi^2}.$$

We can rewrite this as a conditionally convergent series:

$$\pi \cot \pi z = \sum_{n \in \mathbb{Z}} \frac{1}{(z-n)}.$$

Differentiating this once more, we obtain the desired result. □

11.4.5 *For any $\delta > 0$, and any interval $I = [a, b]$, show that there are continuous functions $H_+(x), H_-(x) \in L_1(\mathbb{R})$ such that*

$$H_-(x) \le \chi_I(x) \le H_+(x),$$

with $\hat{H}_\pm(t) = 0$ for $|t| \ge \delta$ and

$$\int_{-\infty}^{\infty} (\chi_I(x) - H_-(x)) dx = \int_{-\infty}^{\infty} (H_+(x) - \chi_I(x)) dx = \frac{1}{\delta}.$$

Choose $S_\pm(x)$ for the interval $I = [\delta a, \delta b]$ as in Theorem 11.4.4. Put $H_\pm(x) = S_\pm(\delta x)$. These functions have the stated property. □

11.4.6 *Let $f \in L^1(\mathbb{R})$. Show that the series*

$$F(x) = \sum_{n \in \mathbb{Z}} f(n+x),$$

is absolutely convergent for almost all x, has period 1 and satisfies $\hat{F}(k) = \hat{f}(k)$.

The fact that $F(x)$ is absolutely convergent for almost all x follows from

$$\int_0^1 \sum_{n \in \mathbb{Z}} |f(n+x)| dx = \int_{-\infty}^{\infty} |f(t)| dt < \infty.$$

The periodicity is clear. Finally,

$$\hat{F}(k) = \int_0^1 F(x) e(-kx) dx = \sum_{n \in \mathbb{Z}} \int_0^1 f(n+x) e(-kx) dx = \int_{-\infty}^{\infty} f(t) e(-kt) dt.$$

□

11.4.9 *Let $x_1, \ldots, x_N$ be N points in $(0,1)$. For $0 \le x \le 1$, let*

$$R_N(x) = \#\{m \le N : 0 \le x_m \le x\} - Nx.$$

Show that

$$\int_0^1 R_N^2(x) dx = \left(\sum_{n=1}^{N} (x_n - 1/2) \right)^2 + \frac{1}{2\pi^2} \sum_{h=1}^{\infty} \frac{1}{h^2} \left| \sum_{n=1}^{N} e^{2\pi i h x_n} \right|^2.$$

$R_N(x)$ is a piecewise linear function of x with discontinuities only at $x_1, \ldots, x_N$. Also, $R_N(0) = R_N(1)$. Thus, we may expand $R_N(x)$ as a Fourier series which represents $R_N(x)$ apart from a finite set of values of x. Writing

$$R_N(x) = \sum_{h=-\infty}^{\infty} a_h e^{2\pi i h x},$$

we have that

$$a_h = \int_0^1 R_N(x) e^{-2\pi i h x}.$$

For $1 \le n \le N$, let c_n be the characteristic function of the interval $[x_n, 1]$. Then,

$$\sum_{n=1}^{N} c_n(x) = \#\{n \le N : 0 \le x_n \le x\}$$

so that

$$a_h = \int_0^1 \left(\sum_{n=1}^{N} c_n(x) - Nx \right) e^{-2\pi i h x} dx.$$

In particular,

$$a_0 = \sum_{n=1}^{N} \int_0^1 c_n(x) dx - \frac{N}{2} = -\sum_{n=1}^{N} (x_n - 1/2).$$

For $h \ne 0$,

$$a_h = \sum_{n=1}^{N} \int_{x_n}^1 e^{-2\pi i h x} dx + \frac{N}{2\pi i h} = \frac{1}{2\pi i h} \sum_{n=1}^{N} e^{-2\pi i h x_n}.$$

The result now follows using Parseval's identity. □

11.4.10 *Let α be irrational. Let $||x||$ denote the distance of x from the nearest integer. Show that the discrepancy D_N of the sequence $n\alpha$ satisfies*

$$D_N \ll \frac{1}{M} + \frac{1}{N} \sum_{m=1}^{M} \frac{1}{m||m\alpha||},$$

for any natural number M.

This is immediate from the Erdös-Turán inequality and the observation that

$$\left| \sum_{n=1}^{N} e^{2\pi i h n \alpha} \right| \le \frac{1}{|\sin \pi h \alpha|} \le \frac{1}{2||h\alpha||}.$$

□

11.5 Equidistribution and L-functions

11.5.1 *Show that $L(s, \rho)$ defines an analytic function in the region* $\mathrm{Re}(s) > 1$.

This is clear since the representations are unitary and finite dimensional. □ **11.5.3** (Serre) *Suppose that for each irreducible representation $\rho \neq 1$, we have that $L(s,\rho)$ extends to an analytic function for* $\mathrm{Re}(s) \geq 1$ *and does not vanish there. Prove that the sequence x_v is μ-equidistributed in the space of conjugacy classes, with respect to the image of the normalized Haar measure μ of G.*

By logarithmic differentiation and an application of the Tauberian theorem, we deduce that for $\chi = \mathrm{tr}\,\rho$,

$$\sum_{Nv \leq x} \chi(x_v) = o(\pi_K(x)).$$

The result now follows from Theorem 11.5.2. □

11.5.4 *Let G be the additive group of residue classes mod k. Show that a sequence of natural numbers $\{x_n\}_{n=1}^{\infty}$ is equidistributed in G if and only if*

$$\sum_{n=1}^{N} e^{2\pi i a x_n/k} = o(N),$$

for $a = 1, 2, ..., k-1$.

This is immediate upon applying Weyl's criterion (Theorem 11.5.2) to the group $\mathbb{Z}/k\mathbb{Z}$ and noting that its irreducible characters are given by $x \mapsto e^{2\pi i a x/k}$. □

11.5.5 *Let p_n denote the n-th prime. Show that the sequence $\{\log p_n\}_{n=1}^{\infty}$ is not u.d. mod 1.*

Let $a_n = 1$ if n is prime and 0 otherwise. By Weyl's criterion, we must examine

$$\sum_{n \leq N} a_n n^{2\pi i m \log n}.$$

Using the prime number theorem and partial summation, we easily find that this sum is

$$\frac{N^{2\pi i m+1}}{(2\pi i m+1)\log N} + O\left(\frac{N}{\log^2 N}\right).$$

Reasoning as in Exercise 11.1.10, we deduce that upon dividing by $N/\log N$ and letting N tend to infinity, the first term does not converge. □

11.5.6 *Let $v_1, v_2, ...$ be a sequence of vectors in $\mathbb{R}^k/\mathbb{Z}^k$. Show that the sequence is equidistributed in $\mathbb{R}^k/\mathbb{Z}^k$ if and only if*

$$\sum_{n=1}^{N} e^{2\pi i b \cdot v_n} = o(N),$$

for every $b \in \mathbb{Z}^k$ with b unequal to the zero vector.

This follows upon writing down Weyl's criterion (Theorem 11.5.2) for the group $\mathbb{R}^k/\mathbb{Z}^k$ and observing that all its characters are given by

$$v \mapsto e^{2\pi i b \cdot v},$$

as b ranges over all the vectors of $\mathbb{Z}^k$. □

11.5.7 *Let $1, \alpha_1, \alpha_2, ..., \alpha_k$ be linearly independent over $\mathbb{Q}$. Show that the vectors $v_n = (n\alpha_1, ..., n\alpha_k)$ are equidistributed in $\mathbb{R}^k/\mathbb{Z}^k$.*

We apply the previous exercise and consider the Weyl sums:

$$\sum_{n \leq N} e^{2\pi i n(b_1\alpha_1 + \cdots + b_k\alpha_k)}.$$

Since 1 and the α_i's are linearly independent over $\mathbb{Q}$, the term $b_1\alpha_1 + \cdots + b_k\alpha_k$ is irrational and the geometric sum is easily estimated to be $O(1)$ when the b_i's are not all zero. □

11.5.8 *Let a be a squarefree number and for primes p coprime to a, consider the map*

$$p \mapsto x_p := \left(\frac{a}{p}\right),$$

where (a/p) denotes the Legendre symbol. Show that the sequence of x_p's is equidistributed in the group of order 2 consisting of $\{\pm 1\}$.

We use Serre's theorem. The L-series to consider is

$$\prod_{p,(p,a)=1} \left(1 - \left(\frac{a}{p}\right) p^{-s}\right)^{-1},$$

and this converges for $\text{Re}(s) > 1$ and extends to an entire function since it is a Dirichlet series attached to a quadratic character mod a, by quadratic reciprocity. Thus, by Dirichlet's theorem, it is nonvanishing on $\text{Re}(s) = 1$ and so the equidistribution result follows. □

11.6 Supplementary Problems

11.6.1 *Show that Exercise 1.1.2 cannot be extended to Lebesgue integrable functions f.*

Let $f(x) = 1$ if $x = x_n$ for some n and zero otherwise. Then,

$$\frac{1}{N}\sum_{n=1}^{N} f(x_n) = 1,$$

whereas

$$\int_0^1 f(x)dx = 0.$$

□

11.6.2 (Féjer) *Let f be a real valued differentiable function, with $f'(x) > 0$ and monotonic. If $f(x) = o(x)$ and $xf'(x) \to \infty$ when $x \to \infty$, show that the sequence $\{f(n)\}_{n=1}^{\infty}$ is u.d. mod 1.*

We apply Weyl's criterion to show that

$$\sum_{n=1}^{N} e^{2\pi i m f(n)} = o(N),$$

for every non-zero integer m. By the Euler summation formula (Theorem 2.1.9), we have

$$\sum_{n=1}^{N} e^{2\pi i m f(n)}$$

$$= \int_1^N e^{2\pi i m f(x)}dx + \int_1^N B_1(x)2\pi i m f'(x)e^{2\pi i m f(x)}dx + O(1).$$

Since $f'(x) > 0$, the second integral is bounded by

$$2\pi|m|\int_1^N f'(x)dx \leq 2\pi|m|(f(N) - f(1)) = o(N),$$

by the first hypothesis. To estimate the first integral, let $u(x) = \cos 2\pi m x$ and $v(x) = \sin 2\pi m x$ and let us consider the integrals

$$\int_1^N u(f(x))dx, \quad \text{and} \quad \int_1^N v(f(x))dx.$$

We estimate the first one, the estimation of the second one being similar. By the second mean value theorem for integrals, there is a ξ with $1 \leq \xi \leq N$ such that

$$\int_1^N u(f(x))dx = \int_1^N \frac{dv(f(x))}{2\pi m f'(x)}$$

$$= \frac{1}{2\pi m f'(1)} \int_1^{\xi} dv(f(x)) + \frac{1}{2\pi m f'(N)} \int_{\xi}^{N} dv(f(x)).$$

The two integrals on the right are easily seen to be bounded so that the integral is $O(1/f'(N))$. The result is now immediate. □

11.6.3 *For any $c \in (0,1)$, and $\alpha \neq 0$, show that the sequence αn^c is u.d. mod 1.*

Let $f(x) = \alpha x^c$ and apply the previous exercise. □

11.6.4 *For any $c > 1$, show that the sequence $(\log n)^c$ is u.d. mod 1.*

Let $f(x) = (\log x)^c$ and apply the previous exercise. □

11.6.5 *Let f be real-valued and have a monotone derivative f' in $[a,b]$ with $f'(x) \geq \lambda > 0$. Show that*

$$\left| \int_a^b e^{2\pi i f(x)} dx \right| \leq \frac{2}{\pi \lambda}.$$

The integral in question is

$$\frac{1}{2\pi i} \int_a^b \frac{de^{2\pi i f(x)}}{f'(x)}.$$

Using the hypotheses satisfied by f', we may apply the second mean value theorem to see that the integral is equal to

$$\frac{1}{2\pi i} \left(\frac{1}{f'(a)} \int_a^c de^{2\pi i f(x)} + \frac{1}{f'(b)} \int_c^b de^{2\pi i f(x)} \right),$$

for some c with $a \leq c \leq b$. The final estimate is now easily deduced from this. □

11.6.6 *Let f be as in the previous exercise but now assume that $f'(x) \leq -\lambda < 0$. Show that the integral estimate is still valid.*

This is immediate from the previous exercise by replacing f by $-f$ and this does not change the absolute value of the integral. □

11.6.7 *Let f be real-valued and twice differentiable on $[a,b]$ with $f''(x) \geq \delta > 0$. Prove that*

$$\left| \int_a^b e^{2\pi i f(x)} dx \right| \leq \frac{4}{\sqrt{\delta}}.$$

Clearly $f'(x)$ is increasing. Suppose first that $f'(x) \geq 0$ in $[a,b]$. By the mean value theorem of differential calculus, we have for $a < c < b$,

$$\frac{f'(c) - f'(a)}{c - a} = f''(\xi),$$

for some $\xi \in [a, c]$. Thus,

$$f'(x) \geq f'(c) \geq (c-a)\delta + f'(a) \geq (c-a)\delta,$$

for $x \in [c, b]$. By the previous exercises, we obtain

$$\left| \int_c^b e^{2\pi i f(x)} dx \right| \ll \frac{1}{(c-a)\delta}.$$

Breaking the original integral into two parts as:

$$\int_a^b e^{2\pi i f(x)} dx = \int_a^c e^{2\pi i f(x)} + \int_c^b e^{2\pi i f(x)},$$

and using the trivial estimate of $c - a$ for the first one, and the estimate of $O(1/(c-a)\delta)$ for the second one, we choose $c - a = 1/\sqrt{\delta}$ to derive the final estimate. If $f'(x) \geq 0$ throughout the interval, we break up the interval into two subintervals on which $f'(x)$ has constant sign. □

11.6.8 *Let $b - a \geq 1$. Let $f(x)$ be a real-valued function on $[a, b]$ with $f''(x) \geq \delta > 0$ on $[a, b]$. Show that*

$$\left| \sum_{a<n<b} e^{2\pi i f(n)} \right| \ll \frac{f'(b) - f'(a) + 1}{\sqrt{\delta}}.$$

Since $f''(x) \geq \delta > 0$ and $f'(x)$ is increasing, we may write the exponential sums as the (finite) sum,

$$\sum_m S_m,$$

where

$$S_m = \sum_{a<n<b, m-1/2<f'(n)<m+1/2} e^{2\pi i f(n)}.$$

We may write

$$S_m = \sum_{a_m<n<b_m} e^{2\pi i f(n)},$$

for certain integers a_m, b_m. Writing $F_m(x) = f(x) - mx$, and using the Euler-Maclaurin sum formula, we get that

$$S_m = \int_{a_m}^{b_m} e^{2\pi i F_m(x)} dx + \frac{1}{2}(e^{2\pi i F_m(a_m)} + e^{2\pi i F_m(b_m)}) +$$

$$\int_{a_m}^{b_m} B_1(x) 2\pi F_m'(x) e^{2\pi i F_m(x)} dx.$$

By the previous exercises, the first integral is at most $4/\sqrt{\delta}$ and the second integral is bounded since $|F_m'(x)| \leq 1/2$ in this range. Thus,

$$|S_m| \leq \frac{4}{\sqrt{\delta}} + 3.$$

There are at most $|f'(b) - f'(a) + 2|$ values of m for which S_m is non-zero and this completes the proof. □

11.6.9 *Show that the estimate in the previous exercise is still valid if* $f''(x) \leq -\delta < 0$.

This is clear by replacing $f(x)$ by $-f(x)$ in the previous exercise. □

11.6.10 *Show that the sequence* $\{\log n!\}_{n=1}^{\infty}$ *is u.d mod 1.*

By Stirling's formula and Exercise 11.1.14, it suffices to show that the sequence

$$(n + 1/2) \log n - n$$

is u.d. mod 1. Let $f(x) = (x + 1/2) \log x - x$. By Exercise 11.6.8,

$$\sum_{1 \leq N} e^{2\pi i m f(n)} \ll N^{1/2} \log N,$$

from which we deduce the u.d. mod 1 of the sequence $f(n)$. □

11.6.11 *Let* $\zeta(s)$ *denote the Riemann zeta function and assume the Riemann hypothesis. Let* $1/2 + i\gamma_1, 1/2 + i\gamma_2, ...$ *denote the zeros of* $\zeta(s)$ *with positive imaginary part, arranged so that* $\gamma_1 \leq \gamma_2 \leq \gamma_3 \cdots$. *Show that the sequence* $\{\gamma_n\}$ *is uniformly distributed mod 1.* The method of Exercise 8.4.8 can be adapted to show that the result still holds if n is replaced by $x > 1$ where we set $\Lambda_F(x) = \Lambda(n)$ if $x = n$ and zero otherwise. With $F = \zeta$ and $x = e^{2\pi m}$, the corresponding Weyl sums are all shown to be sufficiently small. □

11.6.12 *Let* A_n *be a sequence of sets of real numbers with* $\#A_n \to \infty$. *We will say that this sequence is* **set equidistributed** *mod 1 (s.e.d. for short) if for any* $[a, b] \subseteq [0, 1]$ *we have*

$$\lim_{n\to\infty} \frac{\#\{t \in A_n : a \leq (t) \leq b\}}{\#A_n} = b - a.$$

The usual notion of u.d. mod 1 is obtained as a special case of this by taking $A_n = \{x_1, ..., x_n\}$. *Show that the sequence of sets* A_n *is s.e.d. mod 1 if and* *only if for any continuous function $f : [0,1] \to \mathbb{C}$, we have*

$$\lim_{n\to\infty} \frac{1}{\#A_n} \sum_{t\in A_n} f(t) = \int_0^1 f(x)dx.$$

It suffices to consider real-valued functions f. The necessity is clear by approximating the continuous function f by step functions as in the solution of Exercise 11.1.3. For the converse, we again proceed as in Exercise 11.1.3 and approximate the characteristic function of an interval by continuous functions. □

11.6.13 *Show that the sequence of sets A_n is s.e.d mod 1 if and only if for every non-zero integer m, we have*

$$\lim_{n\to\infty} \frac{1}{\#A_n} \sum_{t\in A_n} e^{2\pi imt} = 0.$$

This is again immediate from the fact that any continuous function can be uniformly approximated by a finite trigonometric polynomial. The proof follows closely that given for the Weyl criterion. □

11.6.14 *Let A_n be the finite set of rational numbers with denominator n. Show the sequence A_n is set equidistributed mod 1.*

By the previous exercise, it suffices to check

$$\sum_{(t,n)=1} e^{2\pi imt/n} = o(\phi(n)),$$

for non-zero m. But the exponential sum is a Ramanujan sum $c_n(m)$ which equals $\mu(n/d)\phi(n)/\phi(n/d)$ where $d = (m,n)$. For m fixed, d is bounded since it must be a divisor of m. As n tends to infinity, $\phi(n/d)$ tends to infinity and the result is now clear. □

11.6.15 *A sequence of sets A_n with $A_n \subseteq [0,1]$ and $\#A_n \to \infty$ is said to have* **set asymptotic distribution function** *(s.a.d.f. for short) $g(x)$ if*

$$\lim_{n\to\infty} \frac{\#\{t \in A_n : 0 \le t \le x\}}{\#A_n} = g(x).$$

Show that the sequence has s.a.d.f. $g(x)$ if and only if for every continuous function f, we have

$$\lim_{n\to\infty} \frac{1}{\#A_n} \sum_{t\in A_n} f(t) = \int_0^1 f(x)dg(x).$$

This is again similar to the proof of Theorem 11.3.1. □

11.6.16 (Generalized Wiener-Schoenberg criterion) *Show that the sequence of sets* $\{A_n\}_{n=1}^{\infty}$ *with* $A_n \subseteq [0,1]$ *and* $\#A_n \to \infty$ *has a continuous s.a.d.f. if and only if for all* $m \in \mathbb{Z}$ *the limit*

$$a_m := \lim_{n\to\infty} \frac{1}{\#A_n} \sum_{t\in A_n} e^{2\pi imt}$$

exists and

$$\sum_{m=1}^{N} |a_m|^2 = o(N).$$

This too follows closely the proof of Theorem 11.3.3 where the arguments are replaced by the appropriate limits over sets. □

References

[A] R. Ayoub, *An introduction to the Analytic Theory of Numbers*, Mathematical Surveys, Number 10, American Mathematical Society, Providence, 1963.

[CG] B. Conrey and A. Ghosh, *On the Selberg class of Dirichlet series; small degrees*, Duke Math. Journal, 72(3) (1993), 673–693.

[D] H. Davenport, *Multiplicative Number Theory*, Second Edition, Graduate Texts in Mathematics, Vol. 74, Springer-Verlag, 1980.

[E] W. Ellison and F. Ellison, *Prime Numbers*, Wiley Interscience Publication, Hermann, Paris, France, 1985.

[EM] J. Esmonde and M. Ram Murty, *Problems in Algebraic Number Theory*, Graduate Texts in Mathematics, Vol. 190, Springer-Verlag, 1999.

[HR] H. Halberstam and H. E. Richert, *Sieve Methods*, Academic Press, 1974.

[HW] G. H. Hardy and E. M. Wright, *An introduction to the Theory of Numbers*, Fifth Edition, Clarendon Press, Oxford, 1985.

[K] N. Koblitz, *p-adic Numbers, p-adic Analysis and Zeta Functions*, Graduate Texts in Mathematics, Vol. 58, Springer-Verlag, 1977.

[KN] L. Kuipers and H. Niederreiter, *Uniform distribution of sequences*, Wiley, New York, 1974.

[Mo] H. L. Montgomery, *Ten lectures on the interface between analytic number theory and harmonic analysis*, CBMS Regional Conference Series in Mathematics, 84, 1994, American Mathematical Society, Providence, Rhode Island.

[M] M. Ram Murty, *Selberg conjectures and Artin L-functions*, Bulletin of the American Math. Society, 31(1) (1994), 1–14.

[MM] M. Ram Murty and V. Kumar Murty, *Non-vanishing of L-functions and applications*, Progress in Mathematics, Vol. 157, Birkhäuser-Verlag, 1997.

[MS] M. Ram Murty and N. Saradha, *On the sieve of Eratosthenes*, Canadian Journal of Math., 39(5) (1987), 1107-1122.

[S] A. Selberg, *Old and new conjectures and results about a class of Dirichlet series*, in Collected Papers, Vol. 2, 47–64, Springer-Verlag, 1991.

[Se] J.-P. Serre, Abelian ℓ-adic representations and elliptic curves, Benjamin, 1967.

[W] L. Washington, *Introduction to Cyclotomic Fields*, Graduate Texts in Mathematics, Vol. 83, Springer-Verlag, 1980.

Index

Graduate Texts in Mathematics

(*continued from page ii*)

191 LANG. Fundamentals of Differential Geometry.
192 HIRSCH/LACOMBE. Elements of Functional Analysis.
193 COHEN. Advanced Topics in Computational Number Theory.
194 ENGEL/NAGEL. One-Parameter Semigroups for Linear Evolution Equations.
195 NATHANSON. Elementary Methods in Number Theory.
196 OSBORNE. Basic Homological Algebra.
197 EISENBUD/HARRIS. The Geometry of Schemes.
198 ROBERT. A Course in p-adic Analysis.
199 HEDENMALM/KORENBLUM/ZHU. Theory of Bergman Spaces.
200 BAO/CHERN/SHEN. An Introduction to Riemann-Finsler Geometry.
201 HINDRY/SILVERMAN. Diophantine Geometry: An Introduction.
202 LEE. Introduction to Topological Manifolds.
203 SAGAN. The Symmetric Group: Representations, Combinatorial Algorithms, and Symmetric Functions.
204 ESCOFIER. Galois Theory.
205 FELIX/HALPERIN/THOMAS. Rational Homotopy Theory. 2nd ed.
206 MURTY. Problems in Analytic Number Theory. *Readings in Mathematics.*
207 GODSIL/ROYLE. Algebraic Graph Theory.
208 CHENEY. Analysis for Applied Mathematics.
209 ARVESON. A Short Course on Spectral Theory.
210 ROSEN. Number Theory in Function Fields.
211 LANG. Algebra. Revised 3rd ed.
212 MATOUek. Lectures on Discrete Geometry.
213 FRITZSCHE/GRAUERT. From Holomorphic Functions to Complex Manifolds.
214 JOST. Partial Differential Equations. 2nd ed.
215 GOLDSCHMIDT. Algebraic Functions and Projective Curves.
216 D. SERRE. Matrices: Theory and Applications.
217 MARKER. Model Theory: An Introduction.
218 LEE. Introduction to Smooth Manifolds.
219 MACLACHLAN/REID. The Arithmetic of Hyperbolic 3-Manifolds.
220 NESTRUEV. Smooth Manifolds and Observables.
221 GRÜnbaum. Convex Polytopes. 2nd ed.
222 HALL. Lie Groups, Lie Algebras, and Representations: An Elementary Introduction.
223 VRETBLAD. Fourier Analysis and Its Applications.
224 WALSCHAP. Metric Structures in Differential Geometry.
225 BUMP. Lie Groups.
226 ZHU. Spaces of Holomorphic Functions in the Unit Ball.
227 MILLER/STURMFELS. Combinatorial Commutative Algebra.
228 DIAMOND/SHURMAN. A First Course in Modular Forms.
229 EISENBUD. The Geometry of Syzygies.
230 STROOCK. An Introduction to Markov Processes.
231 BJÖrner/BRENTI. Combinatorics of Coxeter Groups.
232 EVEREST/WARD. An Introduction to Number Theory.
233 ALBIAC/KALTON. Topics in Banach Space Theory.
234 JORGENSON. Analysis and Probability.
235 SEPANSKI. Compact Lie Groups.
236 GARNETT. Bounded Analytic Functions.
237 MARTÍnez-AVENDAÑo/ROSENTHAL. An Introduction to Operators on the Hardy-Hilbert Space.
238 AIGNER, A Course in Enumeration.
239 COHEN, Number Theory, Vol. I.
240 COHEN, Number Theory, Vol. II.
241 SILVERMAN. The Arithmetic of Dynamical Systems.
242 GRILLET. Abstract Algebra. 2nd ed.
243 GEOGHEGAN. Topological Methods in Group Theory.
244 BONDY/MURTY. Graph Theory.
245 GILMAN/KRA/RODRIGUEZ. Complex Analysis.
246 KANIUTH. A Course in Commutative Banach Algebras.